Photoionization and Other Probes of Many-Electron Interactions

NATO ADVANCED STUDY INSTITUTES SERIES

A series of edited volumes comprising multifaceted studies of contemporary scientific issues by some of the best scientific minds in the world, assembled in cooperation with NATO Scientific Affairs Division.

Series B: Physics

RECENT VOLUMES IN THIS SERIES

The series is published by an international board of publishers in conjunction with NATO Scientific Affairs Division

A	Life Sciences	Plenum Publishing Corporation
B	Physics	New York and London
C	Mathematical and Physical Sciences	D. Reidel Publishing Company Dordrecht and Boston
D	Behavioral and Social Sciences	Sijthoff International Publishing Company Leiden
E	Applied Sciences	Noordhoff International Publishing Leiden

Photoionization and Other Probes of Many-Electron Interactions

Edited by

F. J. Wuilleumier
University of Paris-Sud
Orsay, France

PLENUM PRESS • NEW YORK AND LONDON
Published in cooperation with NATO Scientific Affairs Division

Library of Congress Cataloging in Publication Data

NATO Advanced Study Institute on Photoionization and Other Probes of Many-electron Interactions, Carry-le-Rouet, France, 1975.
Photoionization and other probes of many-electron interactions.

(NATO advanced study institutes series: Series B, Physics; v. 18)
Includes indexes.
1. Photoionization—Congresses. 2. Electrons—Scattering—Congresses. 3. Problem of many bodies—Congresses. I. Wuilleumier, F. J. II. Title. III. Series.
QC701.7.N37 1975   539.7'54   76-16512

ISBN-:978-1-4684-2801-8   e-ISBN-13: 978-1-4684-2799-8
DOI: 978-1-4684-2799-8

Proceedings of the NATO Advanced Study Institute on
Photoionization and Other Probes of Many-Electron Interactions
held at the Centre "Les Cigales" in Carry-le-Rouet, France,
August 31-September 13, 1975

© 1976 Plenum Press, New York
Softcover reprint of the hardcover 1st edition 1976
A Division of Plenum Publishing Corporation
227 West 17th Street, New York, N.Y. 10011

PREFACE

The Advanced Study Institute on "Photoionization and Other Probes of Many-Electron Interactions" was held at the Centre "Les Cigales" in Carry-le-Rouet (France), from August 31st till September 13th 1975. The Institute was sponsored by the Scientific Affairs Division of NATO. The "Centre National de la Recherche Scientifique" (France) gave also partial support to the French participants and the National Science Foundation (U.S.A.) to the American participants. A total of 18 lecturers, and 54 students selected among more than 120 applicants, attended the Institute.

Over the last few years, substantial progress has been made in the experimental study of photon- or electron interactions with atoms. In particular, the growing number of facilities created to use the synchrotron radiation makes now possible the realization of new types of experiments. The accumulation of new results showed clearly it was necessary to introduce electron correlations in the theoretical models in order to explain the existence and the probability of a large number of processes, in particular multiple processes. Thus large progress has also been made in the theoretical description of the excitation of the electronic systems and their interactions. It was the purpose of this Institute to bring together theoreticians and experimentalists in order to provide an opportunity to present in details the state of the art, in experiment as well as in theory, and to favor discussions on future experimental and theoretical studies.

The subject of the Institute involved photoionization and collision processes, neutral and ionic systems, outer and inner shell excitations in a wide range of energy exchanges. However the main theme was the study of ionization and excitation processes under photon and electron impact in the simplest atomic species, namely the neutral free atom, in the soft X-ray region and the same importance is given to this theme in this book. The theoretical part treats the single electron model and the more sophisticated models that take into account the correlation effects. The manifestations of these effects through the various experimental techniques (absorption - electron - ion spectroscopy) are presented. The comparison with ionization by heavy charged particles or by nuclear transitions is made. Then the influence of electron correlations on the decay processes following inner shell ionization, namely radiative and Auger transitions, is studied. Finally the cases where the atom is preionized in outer shell or is engaged in a solid or in a molecule is treated by comparison with the neutral free atom.

I would particularly like to thank Professor U. Fano and Dr M.O. Krause who gave me their encouragements and support from the very beginning, when the idea of holding such a Summer Course came into my mind in the middle of 1973. Their help was constant for the preparation of the scientific program of this Institute. Special thanks are also due to Professor M. Inokuti, who acted efficiently for the coordination of a number of lectures and to the other members of the Organizing Committee who brought their contribution to the preparation of the Institute : Dr P.Jaeglé, Professors J.P.Briand, W. Mehlhorn, T. Åberg, and Dr M.Van der Wiel. I would also like to thank the lecturers for their collaboration, not only for preparing their lectures and manuscripts, but also for their work during the whole two weeks at Carry-le-Rouet. They were present all the time and thus the students were able to have informal discussions with them very frequently.

The secretarial tasks of the Institute were extremely heavy. The Secretary of the Institute, Mrs Chantal Wuilleumier, must receive special thanks for her outstanding contribution to the practical organization and to the editing of this book. Mrs Marie Françoise Arcas also contributed partly to the preparation of the Institute and Miss Annie de Corte to the finishing of this volume while typing a number of manuscripts. To these coworkers I express here my sincere gratitude.

Several students contributed also to the success of this Summer Course by way of practical help : Drs P. Dhez, A. Johnson, M. Tavernier and J.P. Rozet. All of them are here sincerely thanked.

I would also like to thank Dr T. Kester, Head of the NATO Scientific Affairs Division for his helpful   assistance. I gratefully acknowledge the financial support of this organism, of the "Bureau de la Formation Permanente du Centre National de la Recherche Scientifique" (France) and of the National Science Foundation (United States). Their support made it possible to invite a panel of distinguished lecturers and to provide a substantial number of grants for participants.

Finally I wish to express my thanks to the housing staff of the Centre "Les Cigales" who contributed to make our stay very agreable and profitable.

François J. Wuilleumier
Chargé de Recherche au CNRS
University of Paris-Sud, Orsay (France)
Director of the Advanced Study Institute

February 1976

LIST OF LECTURERS

T. Åberg
> Laboratory of Physics, Helsinki University of
> Technology, 02150 Espoo 15, Finland.

M. Barat
> Laboratoire des Collisions Atomiques, Université
> Paris-Sud, Bât. 220, 91405 - Orsay, France.

J.P. Briand
> Institut du Radium, Université Paris VI, 11 rue
> Pierre et Marie Curie, 75231 Paris Cedex 05,
> France.

T.A. Carlson
> Oak Ridge National Laboratory, P.O.Box X,
> Oak Ridge, Tennessee 37830, U.S.A.

K. Codling
> The University of Reading, Physics Department,
> Whiteknights, Reading, U.K.

J.W. Cooper
> National Bureau of Standards, Washington, D.C.
> 20234, U.S.A.

K.T. Dolder
> Department of Atomic Physics, The University,
> Newcastle Upon Tyne, NE1 7RU, England.

S. Doniach
> Department of Applied Physics and Stanford Synchrotron Radiation Project, Stanford University, California

U. Fano
> Department of Physics, The University of Chicago, Chicago, Illinois 60637, U.S.A.

M.S. Freedman
> Chemistry Division, Argonne National Laboratory, 9700 South Cass Av., Argonne, Illinois 60439, USA

M. Inokuti
> Argonne National Laboratory, 9700 South Cass Av. Argonne, Illinois 60439, U.S.A.

P. Jaeglé
> Equipe de Recherche du CNRS n° 184 "Spectroscopie Atomique et Ionique", Université Paris-Sud, Bât. 350, 91405 - Orsay, France.

H.P. Kelly
> Department of Physics, University of Virginia, Charlottesville, Virginia 22901, U.S.A.

M.O. Krause
> Transuranium Research Laboratory, Oak Ridge National Laboratory, Oak Ridge, Tennessee 37830, U.S.A.

W. Mehlhorn
> Fakultät für Physik, Universität Freiburg, 78 Freiburg, Germany.

M. Van der Wiel
> F.O.M. Institute for Atomic and Molecular Physics, Amsterdam, The Netherlands.

G. Wendin
> Institute of Theoretical Physics, Fack, S-40220 Göteborg 5, Sweden.

R. Haensel, from the Institut für Experimental Physik der Universität Kiel, gave at the Institute a lecture entitled "Photo-absorption spectra of solids". Unfortunately, owing to some conditions independent of his willing, he was prevented to prepare his manuscript in time for publication in this volume.

Six participants were asked in advance to prepare seminars mainly concerned with their own work :

F. Combet Farnoux (Université Paris-Sud, Orsay, France)

J.P. Desclaux (Institut Laue-Langevin, Grenoble, France)

D. Dill (Boston University, U.S.A.)

J.A.R. Samson (Nebraska University, Lincoln, U.S.A.)

V. Schmidt (Freiburg Universität, Germany)

A. Starace (Nebraska University, Lincoln, U.S.A.)

Finally during the Institute, three other participants presented informal seminars that are not published in this volume space lacking : H. Hotop (Freiburg Universität, Germany), about his recent photoelectron spectroscopy experiments in Ba, C.M. Lee (Pittsburgh University, U.S.A.) about the R-matrix theory and H. Petersen (DESY, Hamburg, Germany) about the shape of the K absorption edge in Li.

CONTENTS

INTRODUCTION TO THE SCHOOL PROGRAM[*]

U. Fano

The University of Chicago

Chicago, Illinois 60637, U.S.A.

I. The Subject and Its Unity

Atomic theory developed initially through independent particle models, which account to a large extent for spectra and for inelastic collisions. Combination of these models with treatment of particle interactions by lowest order perturbation theory has extended the range of successful applications, particularly to include inelastic collisions by fast particles and the Auger effect. In the last 10-15 years, on the other hand, experimental progress has presented us with a rapidly expanding body of detailed evidence on phenomena in which particle interactions play a more elaborate role, exceeding the scope of ordinary perturbation treatments. Reviewing and sharing our present knowledge of these more complex phenomena — which are often characterized as "correlation effects" — is the objective of this School.

Foremost among these phenomena is — at least to my mind — the broad field of inelastic collisions at low energy, because it includes the processes of chemical transformation. Unraveling the mechanisms of these processes may be regarded as the central challenge of low energy physics. Accordingly, I look at the manifold processes to be discussed here as sources of insight on the largely unknown pathways through which many-electron systems evolve from one stable (or metastable) state to another one. Recall, in this connection, Bohr's emphasis on how the discrete nature of atomic spectra underlies molecular stability. We are concerned here with the pathways by which this stability is circumvented.

1

This point of view lends remarkable unity to our subject which might otherwise seem utterly diverse and disconnected. Indeed our subject involves radiative and collision processes, ionic and neutral systems, valence- and inner-shell and even nuclear processes, and energy exchanges ranging from fractions of 1 eV to many keV. As an aid to perceiving both the diversity and the unity, we may consider the many-branched reaction diagrams

$$h\nu + A \rightleftarrows \begin{matrix} h\nu + A^* \\ \uparrow\downarrow \\ A^{**} \\ \uparrow\downarrow \\ e + A^+ \end{matrix} \rightleftarrows e + A^{+*} \ , \quad h\nu + A^- \rightleftarrows \begin{matrix} \\ A^{-**} \\ \uparrow\downarrow \\ e + A \end{matrix} \rightleftarrows e + A^* \ , \quad (1)$$

and their isoelectronic analogs with positive charge. Diversity of the subject results from the multiplicity of alternative entrance and exit channels and of alternative excitation levels, its unity from the common role of the central complex, which is generally multiply excited as indicated by the double asterisk.

Studies based on independent particle models have nevertheless a role in our School, namely, that of indicating which phenomena can be accounted for in that way and — by implication — which phenomena imply a more complex pattern of interactions. This aspect will be covered by Cooper for photoionization and by other speakers for their respective fields. The theory of shake phenomena, to be discussed in Åberg's first lecture, may be regarded as an extension of independent particle models because it does not quite require explicit consideration of electron correlations.

II. Classes of Processes and Their Interconnections

A. Photoabsorption and its special role. The photoabsorption process, represented in the diagrams of Eq. (1) by the → sign in the left-hand channel, is characterized for our purposes by one experimental and one theoretical feature. Experimentally, the resolving power of optical monochromators enables us to form complexes with better energy definition than is afforded by other entrance channels to a reaction. Theoretically, the weakness of interaction of light with matter enables us to treat the initial step of photoabsorption by first order perturbation theory. Therefore the complex aspects of particle interactions, which are important to us, do not affect the entrance channel of photoreactions but only their central complex and the exit channels. The ensuing conceptual simplifications may be summarized by the statement that a photoprocess amounts to a "half scattering" process. For this reason, and also because we are mostly interested in double excitations that lie in the ionization continuum, photoionization has a prominent part in our School.

The photoprocesses of interest are thus concentrated in the spectral range variously known as far uv or soft x-ray range. Major technical developments in this field of spectroscopy, which have been decisive for its rapid growth, will be reviewed by Codling. The observation of photoabsorption alone fails to single out the exit channel of a photoprocess; hence it must be complemented by more detailed observations. Foremost among these are the data of electron spectroscopy to be reviewed by Krause.

B. Particle collisions. This field, whose general aspects will be discussed by Inokuti, is subdivided into fast and slow collisions. Collisions by fast charged particles have a basic feature in common with photoprocesses, namely, that the interaction which transmits energy to a target molecule can be treated by first order perturbation energy. Inokuti will describe the somewhat complementary relationship of fast particle collisions and of photoprocesses; the extensive evidence provided by fast collisions will then be reviewed by van der Wiel. The field of low energy electron-atom collisions — which forms for me the core of our subject — will be reviewed by Dolder. This subject presents special aspects when the collision takes place in a plasma and the target atom is highly ionized; these aspects will be covered by Jaegle together with other phenomena which are characteristic to atomic phenomena in plasmas.

Low energy collisions between atoms or ions also present very special aspects in that both collision partners usually carry a complement of electrons, whereby the collision involves the interpenetration of electron shells from the two partners. This field, to be outlined by Barat, is very rich in experimental and theoretical evidence; it would lie near the center of attention of this School were it not that time limitations have caused us to concentrate on single-atom processes.

C. Inner-shell phenomena. Processes involving an inner-shell vacancy generally resolve into a sequence of two separate processes, one leading to formation of the vacancy and one to its filling by transfer of an external electron. This separation holds insofar as the time interval between the two subprocesses, usually of the order of 10^{-14} sec , exceeds considerably the duration of each of them. The vacancy may result from ejection of an inner-shell electron by photoionization or by collision processes — mentioned above under A and B — which are of particular interest to us when they display effects of many-electron interactions. It may result alternatively from nuclear processes of internal conversion or electron capture; these will be surveyed by Freedman. His lecture will also deal with the ejection of atomic electrons which may result from the nuclear emission of charged particles, α or β rays.

Once a vacancy has been formed, the atom finds itself in an excited state of definite energy, as though it had been formed by photoabsorption. The excitation energy suffices to start additional processes which may involve complex electron interactions and lead to the emission of one or more Auger electrons and/or of fluorescence x rays. The study of these processes will be introduced by Åberg and developed by Mehlhorn for Auger electrons and by Briand for x-ray fluorescence.

D. Molecular and solid state effects. As noted above, the subject of this School is focussed on single-atom processes for the sake of attaining sufficient depth within a limited time. The phenomena of interest extend, however, to molecular and to larger aggregates of atoms. Naturally our current understanding tends to decrease with increasing complexity of the system. Lectures by Carlson on molecules and by Doniach and Haensel on solids have been included in our program for purposes of orientation.

E. Intrashell and intershell interactions. It has been considered as a thumb rule that many-electron interactions are confined to electrons of the same shell, in the sense that the interaction of electrons of different shells can generally be taken into account by first order perturbation treatment, as in the Auger effect, and has thus only a weak effect. More recently, computational and experimental evidence has emerged showing exceptions to this rule.[1] Many of the exceptions occurred in reaction channels of low fractional probability, but a more striking one has been provided by Brehm and Hoefler's discovery that the Ba ionization produced by 21.2 eV photons involves predominantly a two-electron transition.[2] The circumstances of this phenomenon, clarified further by a very recent experiment by Hotop and Mahr,[3] indicate that it results from tight coupling of 5p and 6s electrons; they will be reviewed in a seminar by Hotop.

III. Theoretical Descriptions. Scattering Matrix and Its Variants

The theoretical analysis of atomic phenomena often starts from the complete Schroedinger equation for the system of interest, or from the equation of a simplified model, and derives from it analytical or numerical predictions to be compared with experimental data. However, many of the diverse theoretical procedures that will be mentioned at this School may be fitted into a single frame which is constructed by proceeding in opposite direction. One begins by representing experimental data in terms of the parameters of some general theoretical expression, typically of a scattering matrix. Then one may eliminate from these parameters the influence of any element of the problem that are theoretically well understood and hence trivial for our purposes; the main elimination procedures will be surveyed in this section. The residual parameters generally form a smaller set than the initial ones and thus represent a distillation of the experimental evidence of particular interest. Calculation of this smaller set of parameters constitutes — from this point of view — a more immediate target of basic theory. Moreover comparison of theory and experiment at this level is less subject to distortion or to obfuscation by irrelevant circumstances.

A. Scattering matrix. The entrance and exit channels of a collision (or photo–) process can be characterized by the relative momenta, $\vec{k}_i$ and $\vec{k}_j$, of the reactants and by the additional quantum numbers (angular momenta, etc.) that identify their initial and final states, χ_i and χ_j. If we call $\vec{r}$ the relative position of the two reactants, whether in the initial or the final channel, the wave function of the system at large distances r has the general form

$$\lim_{r=\infty} \psi = \chi_i e^{i\vec{k}_i \cdot \vec{r}} - \Sigma_j \chi_j e^{i\vec{k}_j \cdot \vec{r}} S_{ji}, \qquad (2)$$

where S_{ji} indicates the probability amplitude that a collision starting in the channel i will end in the channel j. The aggregate of the probability amplitudes S_{ji} for all alternative pairs of entrance and exit channels (j, i) constitutes the scattering matrix of the system at the given energy. In the simplest example of elastic scattering of spinless particles (single channel, with $k_i = k_j = k$), Eq. (2) reduces to the familiar expression

$$\chi \{ e^{i\vec{k} \cdot \vec{r}} + (2ikr)^{-1} e^{ikr} \Sigma_\ell (2\ell+1)(e^{2i\delta_\ell} - 1)P_\ell(\cos\theta) \}, \qquad (3)$$

in terms of the phase shifts δ_ℓ and of the Legendre polynomials $P_\ell(\cos\theta)$ which depend on the scattering angle θ between $\vec{k}_i$ and $\vec{k}_j$.

B. Diagonalization of the scattering matrix. Even the simple Eq. (3) includes elements extraneous to the particle interactions, which actually determine only the phase shifts δ_ℓ. Elimination of such extraneous elements is achieved by taking advantage of the conservation of orbital momentum. It suffices to consider separately each standing wave with orbital quantum number ℓ, and specifically the amplitude ratio of its outgoing (i.e., scattered) component exp(ikr) and of its ingoing component exp(–ikr). The set of such scattering amplitudes, $\{\exp(2i\delta_\ell)\}$, constitutes the diagonal form of the matrix S_{ji}.

For a multichannel system, conservation of angular momentum still enables us to obtain a partial diagonalization of S_{ji}, which resolves this matrix into a set of square arrays corresponding to alternative angular momenta; the rows and columns of each array correspond to the alternative initial and final states of reactants and to alternative couplings of their spins. Complete diagonalization of S_{ji} is also possible, in which case the dynamical information of interest is represented by the eigenvalues and eigenvectors of this matrix. However, this form of the information is transparent only insofar as one can label the eigenvalues by additional constants — or at least approximate constants — of the collision besides the angular momentum; thus, e.g., the total spin label (singlet or triplet) is a good quantum number of e-H scattering.

C. Standing wave formulation. K matrix. Whereas the S matrix relates the amplitudes of outgoing waves to those of ingoing waves, equivalent information can be represented by relating the amplitudes of different standing waves. In the example of a single channel (elastic scattering) the radial standing wave function with orbital momentum ℓ behaves, for large distances r of the colliding particles, as

$$\sin(kr - \ell\pi/2 + \delta_\ell) = [\sin(kr - \ell\pi/2) + \cos(kr - \ell\pi/2)\,tg\delta_\ell]\cos\delta_\ell. \qquad (4)$$

The first term in the brackets represents the wave function that would hold in the absence of interaction between the reactants, while the cosine function represents a wave with a standard 90° phase shift with respect to the sine wave. The effect of the interaction is represented by the relative amplitude of the cosine and sine waves, $tg\delta_\ell$, together with the incidental normalization coefficient $\cos\delta_\ell$.

In the nontrivial case of many channels the expression on the right-hand side of Eq. (4) is replaced, to within normalization, by the set of wave functions

$$\chi_i\sin(k_i r - \ell_i\pi/2) + \Sigma_j\chi_j\cos(k_j r - \ell_j\pi/2)K_{ji}, \qquad (5)$$

with different i, where (χ_i, χ_j) indicate the internal states of the reactants in the respective channels. Each of these functions consists of the sine wave that would hold in one channel in the absence of interactions and of cosine waves in all channels j with amplitudes K_{ji}. This set of amplitudes constitutes a "reaction matrix" which contains the same information as the S matrix; the two matrices are related by $S = (1 + iK)/(1 - iK)$. The K matrix, being real and symmetric, is somewhat easier to calculate directly than the S matrix itself; it is also more directly related to the interactions.

D. Coulomb basis functions. Quantum Defect Theory. The S and the K matrix have been introduced here to describe how particle interactions modify the free motion of reactants that would prevail in their absence; hence they embody the effect of all interactions. One may, however, take advantage of the fact that the effects of some interactions, notably of the long range Coulomb force between charged reactants, have known analytical representations. To this end one replaces the free particle wave functions, e.g., in Eqs. (2), (4), and (5), by the corresponding Coulomb field wave functions. The S and K matrices thus redefined no longer represent the effect of all interactions, but only the effect of interactions diminished by the Coulomb forces that may prevail between the reactants in the several channels. They are thus more closely related to the many-particle interactions of interest to us and they are also often easier to calculate insofar as the residual interactions act only at short distances. Moreover these modified S and K matrices are smoother functions of energy, because their analytic structure does not include the occurrence of Rydberg series or of other details of spectra or collisions which result exclusively from the Coulomb field at large distances. The treatment of spectra or collisions in terms of such matrices is called Quantum Defect Theory and will be described in a seminar by Starace.[4]

E. Elimination of threshold singularities. R matrix. Non-analyticities occur in the energy dependence of the S and K matrices at any threshold energy, at which the reactants in any one channel barely separate with zero velocity at large distances. These singularities are extraneous to our interaction problem, since they relate to the normalization of continuum wave functions which depends on the eventual separation velocity, the velocity itself being in turn an irrational function of the energy. Singularities that are apparent in experimental data can be eliminated by appropriate reduction of the data, because their analytical character is determined by the properties of each channel wave function. The reduction is performed by changing the normalization of the radial wave function of each channel from the standard basis, which depends on the reactants' velocity and on their wave function's behavior for large distances r, to a basis which is characteristic of the behavior in the small-r limit and is energy independent in this limit. This renormalization induces a non-unitary transformation of the S and K matrices which depends on the presence and strength of long range forces. The transform of the K matrix, called the R matrix, depends on energy analytically through any threshold; residual singularities, e.g., those inherent in $\text{tg}\,\delta_\ell$ functions, can also be eliminated by considering the eigenphaseshifts themselves rather than their tangent functions.

This transformation, combined with the use of a Coulomb basis or of any other basis that may be appropriate, serves to make the R matrix independent of all long range interactions and thus to achieve conceptual and practical advantages. Conceptually it emphasizes the effect of many-particle interactions, which have a short range. Practically it limits the range of any

numerical integration over radial variables. Accordingly calculations of the
R matrix are playing an increasing role in our field of interest and will be discussed in seminars by Combet-Farnoux and by Lee.[5]

IV. Analysis and Calculation of Interaction Effects

The theory of many-electron interaction effects has achieved substantial
but still quite fragmentary results. There is, for example, no generally applicable scheme for describing and classifying the electron correlations which result from the interactions; the study of correlations is itself a useful intermediate step in the analysis of the observable effects of interactions.

We may distinguish at this stage the correlations of two (or more) electrons that are excited to the extent of moving mostly outside the radius of an
atom in its ground state, from the correlations that are confined within the
valence or inner shells. The study of the former was pioneered by Wannier's
analysis of the process of threshold ionization and has progressed mainly
through qualitative analysis rather than through accurate calculations. This
study will be presented by Fano.

A variety of interaction effects which are confined within filled shells
have been subjected to extensive calculations by methods of many-body theory.
These methods study directly the net electron density and currents which produce observable effects, instead of attempting to describe the motion of individual electrons. Two somewhat different streams of work have developed.
One, which utilizes the RPA method, has been followed by Amusia's group and
by Wendin; it will be surveyed here by Wendin. The other, called Many-Body
Perturbation Theory, has been followed by Kelly's and Poe's groups and will be
surveyed by Kelly.

Important results have been achieved, particularly during the 60's, by
the Close Coupling Method of calculation, whose implementation was spear-
headed by Burke[6] and by Smith.[7] This method treats one electron accurately;
on the other hand, practical considerations force it to allow only limited flexibility to the effects of interaction between this electron and the rest of the
atom, in the sense that the wave function of all other electrons is restricted
within a predesigned function space with a few (say, 3-6) dimensions only.
This restriction can be relaxed by splicing the method with other techniques.
An outline of the close coupling work will be presented in a seminar by
Combet-Farnoux.

Many-electron interactions can also be studied by Multi-Configuration
Hartree-Fock calculations, to an extent which is only limited by the choice

and size of the basis set. Their earlier applications, e.g., those of Weiss, were confined to bound states. This limitation has now been lifted by adaptations that permit the direct calculation of R matrices. Extensive and detailed applications of this approach have been forthcoming from Burke's group, a somewhat different one from Lee,[5] as will be indicated in Combet-Farnoux's and Lee's seminars. The evermore refined calculations of theoretical spectroscopy, dealing most often with the lower part of the discrete spectrum, belong conceptually to the same line of work. So do also the efforts[8] aimed at interpreting the recent highly accurate data on the fine and hyperfine splittings of Rydberg levels. However, this work has thus far remained disconnected from the main subject of this School and has accordingly been left out of its program.

Early attempts at accurate calculation of correlation effects, particularly of small atoms, utilized variational wave functions with explicit dependence on interelectron distances. Remarkable accuracy was thus achieved in some test studies, particularly in the well known calculations of two-electron systems by Hylleraas, Pekeris and Schwartz. Their extension to many-electron systems would, however, be extremely laborious. Furthermore the variational wave functions that yield very accurate values of one quantity — typically an energy or phase shift — need not prove accurate for other applications, such as the calculation of transition amplitudes.[9]

References

*Work supported by U. S. Energy Research and Development Administration, Contract No. COO-1674-112.

1. U. Fano, Comments At. Mol. Phys. $\underline{4}$, 119 (1973) and references therein.

2. B. Brehm and K. Hoefler, Int. J. Mass Spectry. $\underline{18}$, 000 (1975).

3. H. Hotop and D. Mahr, J. Phys. B $\underline{8}$, 000 (1975).

4. See also U. Fano, J. Opt. Soc. Amer., September 1975, and references therein.

5. P. G. Burke, Comments At. Mol. Phys. $\underline{4}$, 157 (1973); P. G. Burke and W. D. Robb, Adv. Atom. Mol. Phys. $\underline{11}$, 000 (1975) and references therein; C. M. Lee, Phys. Rev. A $\underline{10}$, 584 (1974).

6. P. G. Burke, Adv. Atom. Mol. Phys. $\underline{4}$, 173 (1968).

7. K. Smith, The Calculation of Atomic Collision Processes (J. Wiley, New York, 1971).

8. I. Lindgren, Comments At. Mol. Phys. $\underline{4}$, 163 (1973).

9. See, e.g., the discussion by S. A. Adelman, Phys. Rev. A $\underline{5}$, 508 (1972).

CORRELATIONS OF EXCITED ELECTRONS[*]

U. Fano

The University of Chicago

Chicago, Illinois 60637, U.S.A.

Inelastic collisions of an electron with an atom or molecule, at energies within several eV of an excitation or ionization threshold, lead to the temporary formation of a complex whose available energy is shared in comparable amounts between the incident electron and the one being excited. Analogous complexes are formed when two electrons are excited simultaneously by incident light or by an analogous external agent. The properties of such complexes, including the rate and modality of their prompt or delayed decomposition, depend on the correlation of the two excited electrons. Their study — experimental and theoretical, qualitative and quantitative — has yielded considerable understanding. Further development of this understanding and of its quantitative applications constitute, I think, a prerequisite for an adequate, general theory of low energy collision processes.

These lectures review the current status of the subject with particular emphasis on achieving a graphical description of the correlations. However, since our material has been covered from the same point of view in three earlier conference reports[1,2,3] the present version consists only of an outline and refers the reader to those reports for a fuller treatment of the various matters to be considered.

In particular, Ref. 2 has articulated the goals of the whole investigation through a series of questions which are presented at the end of its Sec. 1. Preliminary answers to those questions are then outlined in the concluding section of Ref. 2.

I. Successive Stages of Development

a) The widespread occurrence and the main properties of doubly excited complexes, mainly of autoionizing (or autodetaching) states of atoms and molecules (including negative and positive ions), were recognized in 1963–65 through a most rapid sequence of observations of resonances in photoabsorption and in electron collision spectra. The experimental observations were quickly complemented and extended by various types of calculation, which solved the Schroedinger equation numerically for two-electron systems, notably for numerous doubly excited states of He and H⁻. The results of these initial calculations were analogous to the experimental results in that they provided raw data on the energy and other properties of doubly excited states but no real interpretation of the remarkable systematics of such data. The structure and interrelationships of the spectra of doubly excited states of He are illustrated in Figs. 1 and 2.

As discussed in Ref. 1, a startling conclusion could be drawn from the combined evidence on doubly excited quasi-stationary states of neutral atoms and molecules, namely, that states with equal orbital and spin momenta and equal parity may form different Rydberg series converging smoothly to the same limit but with quite different probabilities of excitation and decay. Analogous indications were obtained for negative ions. It was inferred that the classification of each of these series should include the value of one or more additional, novel quantum numbers in terms of which one should formulate selection rules on the rates of excitation and decay. Because the doubly-excited states are not stable, such quantum numbers should characterize the value of approximate, rather than exact, constants of the motion. Identification of these quasi-constants was viewed as a major step towards the analysis of electron correlations.

b) Turning now to theoretical analysis, we note that while the wave functions of successive states of a Rydberg series extend over successively larger ranges of radial distances from the center of an atom, this variation appears to preserve the characteristic quasi-constants of each series. This remark suggested that the main characteristics of each series are rather independent of the radial extension of each state. Hence they might be identified by solving the Schroedinger equation for He or H⁻ while keeping constant a radial coordinate that represents an average radial distance of two electrons from the nucleus. The coordinate $R = [r_1^2 + r_2^2]^{\frac{1}{2}}$ served for this purpose in exploratory calculations by Macek.[1,4] He obtained approximate wave functions for doubly excited states of He, having the form $\Phi_\mu(R; \Omega) F_{\mu n}(R)$, where μ is a series index, Ω a set of two-electron coordinates independent of R, Φ_μ is calculated numerically for a number of fixed values of R, and the $F_{\mu n}(R)$ are eigenfunctions of a radial equation corresponding to the alternative levels

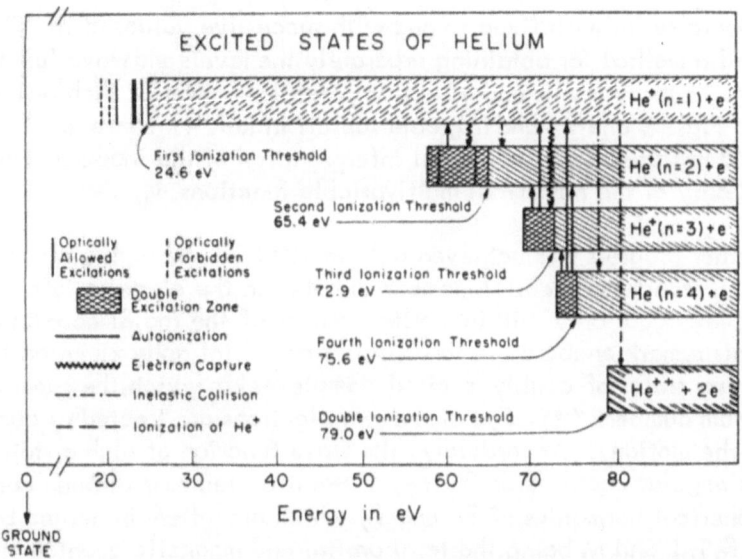

Fig. 1. Diagram showing several channels in the spectrum of helium
and their interconnection by autoionization processes. Also
shown are other interchannel transition processes which are
implicitly related to autoionization. (From Ref. 1.)

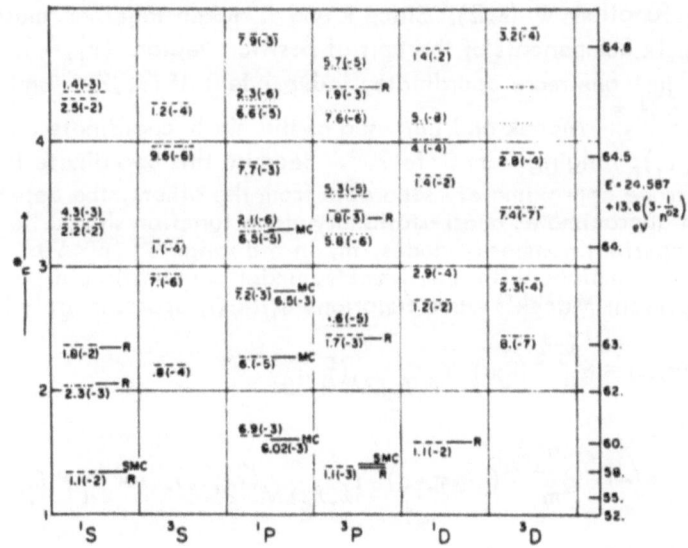

Fig. 2. Diagram of He** levels below 65 eV. Note non-linear energy
scale. ———— Experiment (R - Rudd [21], SMC [22] MC [17]),
- - - w = 1, -·--·- w = 2, ····· w = 3. Close coupling
calculations: Burke et al. [9], [23], [24]. (Results of
other calculations do not differ significantly.). Numerical
entries $\gamma(E)$ represent reduced width $\Gamma n*^3/(27.21 \text{ eV}) = \gamma \times 10^E$
(Courtesy A.R.P. Rau). (From Ref. 1; the numbers in brackets
refer to references cited therein.)

$E_{\mu n}$ of a Rydberg series with fixed μ and with successive values of n. This work provided a method for obtaining separately the levels and wave functions of each series. Energy levels and effective potential curves of such series are illustrated in Figs. 3 and 4 and in greater detail in Ref. 4. However the results did not provide any physical interpretation of the index μ owing to the complexity of the numerical multivariable functions Φ_μ.

c) Further progress was achieved only in 1973-74, through Lin's remark that substantial transfers of orbital momentum between the electrons of a doubly excited complex occur only within limited ranges of the radial coordinate R.[2,5,6] This remark enabled Lin to construct an initial approximation to the quasi-stationary states of doubly excited complexes, in which the separate orbital quantum numbers (ℓ_1, ℓ_2) of the two electrons are treated as quasi-constants of the motion. Accordingly, the wave function of such a state depends on the angular coordinates $(\hat{r}_1, \hat{r}_2)$ of the two electrons through combinations of spherical harmonics of $\hat{r}_1$ and $\hat{r}_2$ which are often indicated by $Y_{\ell_1 \ell_2 LM}(\hat{r}_1, \hat{r}_2)$, L and M being the total orbital and magnetic quantum numbers of the electron pair.

Consider now that each of the unit vector symbols $\hat{r}_1$ and $\hat{r}_2$ represents two independent coordinate parameters, which are included in the set Ω of Macek's wavefunction $\Phi_\mu(R;\Omega)$. Since R and Ω, taken together, must be equivalent to the six components of the pair of position vectors $(\vec{r}_1, \vec{r}_2)$, the set Ω must include just one more coordinate, independent of $(\hat{r}_1, \hat{r}_2)$ and of $R = [r_1^2 + r_2^2]^{\frac{1}{2}}$. Macek and Lin used as the sixth coordinate an angle $\alpha = \arctan(r_2/r_1)$, ranging from $0°$ to $90°$. Because this coordinate is understood by Lin to be approximately separable from the others, the dependence on α of each approximate, quasi-stationary wave function should be characterized by a constant number of nodes, m, in the range $0° < \alpha < 90°$. These considerations, combined with antisymmetry under permutation of particles, led Lin to represent Macek's wave functions $\Phi_\mu(R;\Omega)$ approximately by

$$\Phi_{\ell_1 \ell_2 m}(R;\Omega) = g_m^{\ell_1 \ell_2}(R;\alpha) \; Y_{\ell_1 \ell_2 LM}(\hat{r}_1, \hat{r}_2)$$

$$+ (-1)^S g_m^{\ell_1 \ell_2}(R; \tfrac{1}{2}\pi - \alpha) \; Y_{\ell_1 \ell_2 LM}(\hat{r}_2, \hat{r}_1), \text{ for } \ell_1 \neq \ell_2, \quad (1a)$$

$$\Phi_{\ell \ell m}(R;\Omega) = [1 + (-1)^{S+L+m}] g_m^{\ell \ell}(R;\alpha) \; Y_{\ell \ell LM}(\hat{r}_1, \hat{r}_2), \text{ for } \ell_1 = \ell_2 = \ell, \quad (1b)$$

where S is the spin quantum number of the electron pair. Note that this formulation characterizes the dependence on the five coordinates $(\hat{r}_1, \hat{r}_2, \alpha)$ by the complete set of five quantum numbers $(\ell_1, \ell_2, L, M, m)$. The angular

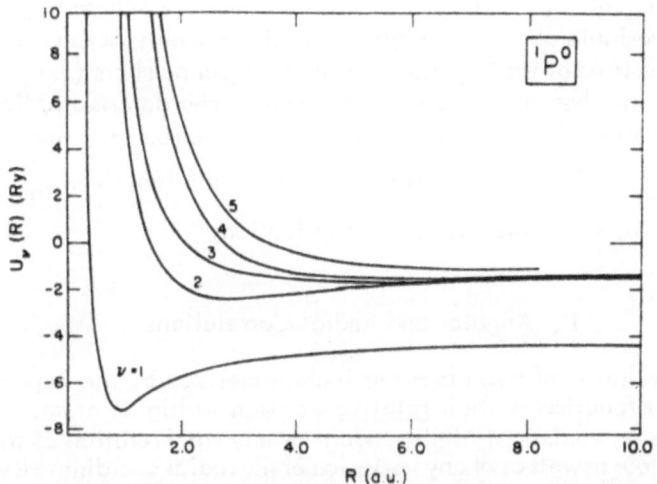

Fig. 3. Effective potentials $U_\nu(R)$ for $^1P^O$ He levels. Data from
     Macek (Courtesy K.-T. Lu). (From Ref. 1.)

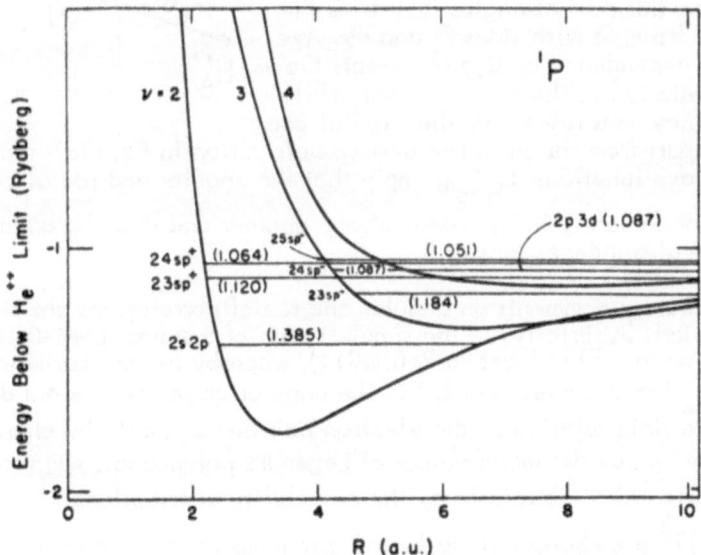

Fig. 4. Detail of Fig. 3 showing level positions below He$^+$ (n = 2).
     (From Ref. 1.)

functions $Y_{\ell_1\ell_2 LM}$ are known analytically; the functions $g_m^{\ell_1\ell_2}(R;\alpha)$ remain to be calculated by solving—at constant R—an appropriate Schroedinger equation in the single variable α. Accordingly, a quasi-stationary autoionizing state is completely classified by the angular and spin quantum numbers $(\ell_1\ell_2 LM, SM_s)$, by the quantum number m — which will be seen to characterize radial correlations — and by Macek's radial excitation quantum number n. The angular factor $Y_{\ell_1\ell_2 LM}$ of its wave function is standard, the functions $g_m^{\ell_1\ell_2}(R;\alpha)$ and $F_{\mu n}(R)$ have to be calculated for each application.

II. Angular and Radial Correlations

The correlation of two electrons is characterized by the dependence of their joint wave function on their _relative_ position within an atom. It may thus be regarded as approximately independent of any rigid rotation of the pair about the nucleus as well as of any variation of the radial coordinate R which merely changes all interparticle distances by a common factor. There remains a dependence on the angle θ_{12} between the directions $\hat{r}_1$ and $\hat{r}_2$ of the two electrons, and on the angle α which identifies the relative radial distance of the two electrons; this pair of angles identifies the _shape_ of the triangle with sides $\vec{r}_1$ and $\vec{r}_2$. We say that the dependence on θ_{12} represents the angular correlation of the two electrons, while the dependence on α represents their radial correlation. Apart from the effect of antisymmetrization in Eq. (1a), Lin's approximate wave functions $\Phi_{\ell_1\ell_2 m}$ imply that the angular and radial correlations are approximately independent of one another and that the angular correlations are also independent of R.

Introductory comments on angular and radial correlations are presented in Sec. 2 of Ref. 3. Briefly, in the simplest case of S states (L=M=0, $\ell_1=\ell_2=\ell$) the Y function in (1b) reduces to $P_\ell(\cos\theta_{12})$, whereby the probability distribution of θ_{12} has ℓ nodes. For $L \neq 0$, the angular correlation is not disentangled from the rigid rotation of the electron pair and can only by characterized indirectly, e.g., by the mean values of Legendre polynomials $\langle P_k(\cos\theta_{12})\rangle$. With regard to radial correlations, the probability distributions $[g_m^{\ell_1\ell_2}(R;\alpha)]^2$ are characterized by the arrangement of their m nodes and antinodes in the plane $(R,\alpha) \equiv (r_1, r_2)$; simple examples are shown in Figs. 5, 6, and 7.

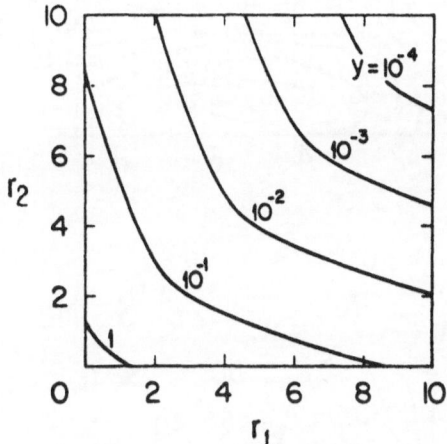

Fig. 5. Contour line plot of the approximate Eckart wavefunction for the He ground state, $y(r_1,r_2) = \exp(-\beta r_1 - \gamma r_2) + \exp(-\gamma r_1 - \beta r_2)$ with $\beta = 2$ and $\gamma = 1.34$ (Courtesy C. E. Theodosiou). (From Ref. 3.)

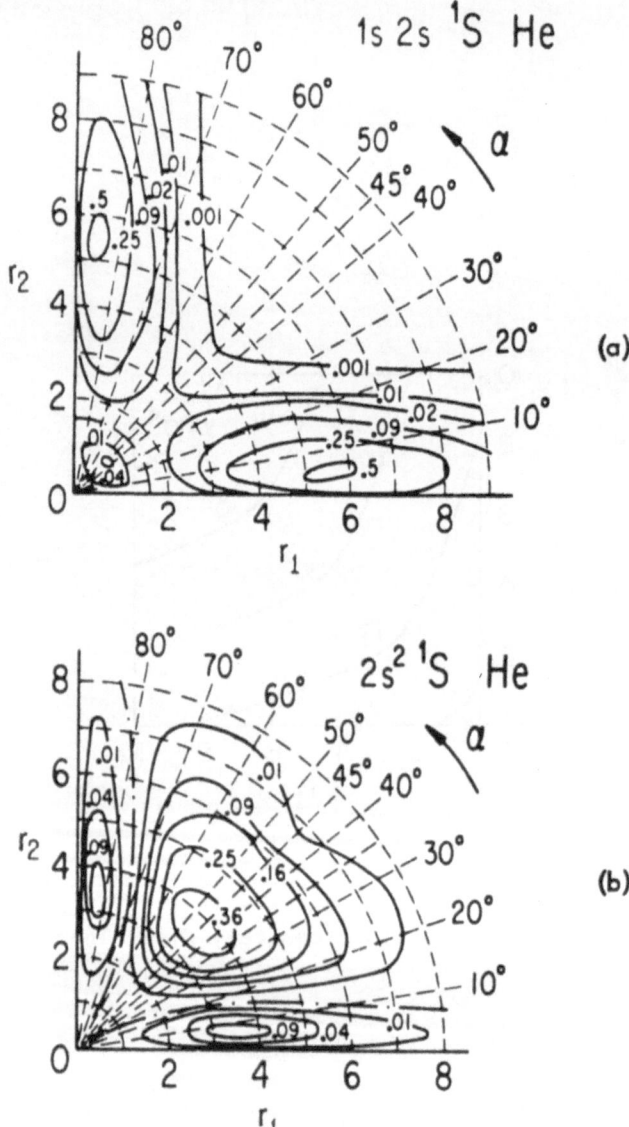

Fig. 6. Contour plots of approximate probability distribution $|\Psi(\vec{r}_1, \vec{r}_2)|$ for He. a) 1s2s ^{1}S, b) 2s^2 ^{1}S. Angular correlations represented by admixture of p^2, d^2, etc., are disregarded.) —— equidensity contours; –·–·– nodal lines. (From Ref. 2.)

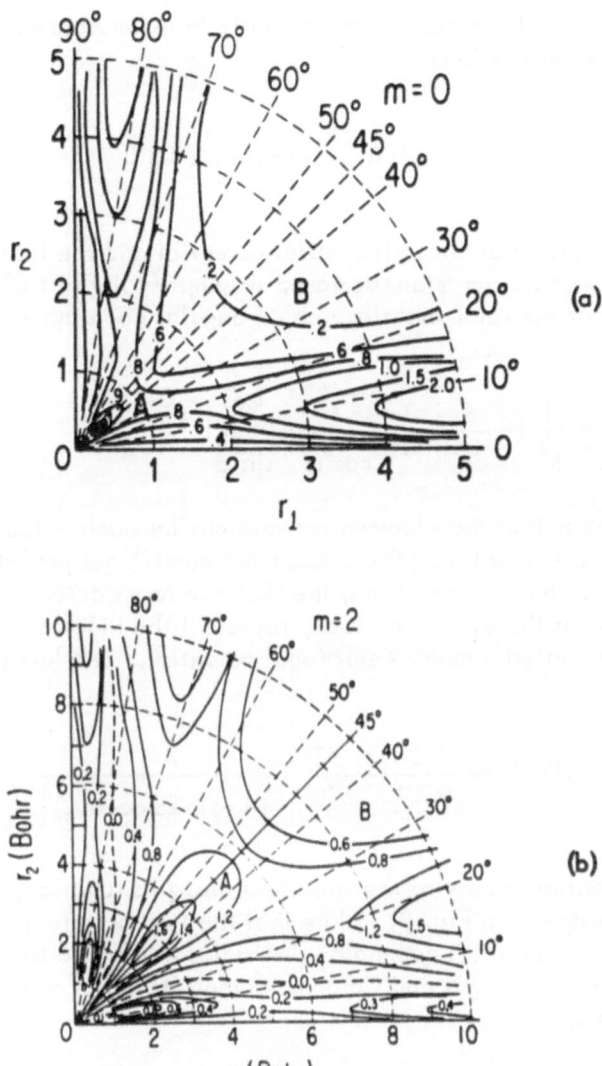

Fig. 7. Contour plots of $\left|g_m(R;\alpha)\right|^2$ for helium ^1S channels;
a) m = 0, b) m = 2. (Plots of Fig. 6 represent $\left|g_m\right|^2$
multiplied by appropriate radial functions.) (From
Ref. 2.)

III. Analytical Formulations

The Schroedinger equation for a pair of electrons in the field of a fixed nucleus with charge Ze is expressed conveniently, in hyperspherical coordinates and in atomic units, in the form

$$[\frac{d^2}{dR^2} + 2E - U](R^{5/2} \sin\alpha \cos\alpha \; \Psi) = 0 \; , \tag{2}$$

where U plays the role of an effective radial potential and the factors combined with the wave function Ψ are designed to simplify the rest of the equation. In fact U is an operator that depends on angular coordinates,

$$U = \frac{1}{R^2} [-\frac{d^2}{d\alpha^2} - \frac{1}{4} + \frac{\vec{\ell}_1^2}{\cos^2\alpha} + \frac{\vec{\ell}_2^2}{\sin^2\alpha}] - \frac{C(\alpha, \theta_{12})}{R} \; , \tag{3}$$

whose structure determines the electron correlations for each value of R. The last two terms in the brackets of (3) represent the centrifugal potentials due to the orbital motion of the two electrons; the first two terms derive from the relative radial motion of the electrons, i.e., represent the kinetic energy of this motion and may be called a mock-centrifugal potential. The last term of (3), with

$$- C(\alpha, \theta_{12}) = - \frac{2Z}{\cos\alpha} - \frac{2Z}{\sin\alpha} + \frac{2}{(1 - \sin 2\alpha \cos\theta_{12})^{\frac{1}{2}}} \; , \tag{4}$$

represents the potential energy of the two electrons; this function, which is very important, is mapped in Fig. 8. Note that the two main terms of the potential operator U in (3) depend on our radius R in proportion to R^{-2} and to R^{-1}, just as the centrifugal and Coulomb potentials do for a single electron in a hydrogenic atom.

Following Macek[4] we reduce Eq. (2) to a system of coupled radial equations by setting

$$R^{5/2} \sin\alpha \cos\alpha \; \Psi = \sum_\mu \Phi_\mu (R;\Omega) F_\mu (R) \; . \tag{5}$$

This expansion reduces (2) to the form

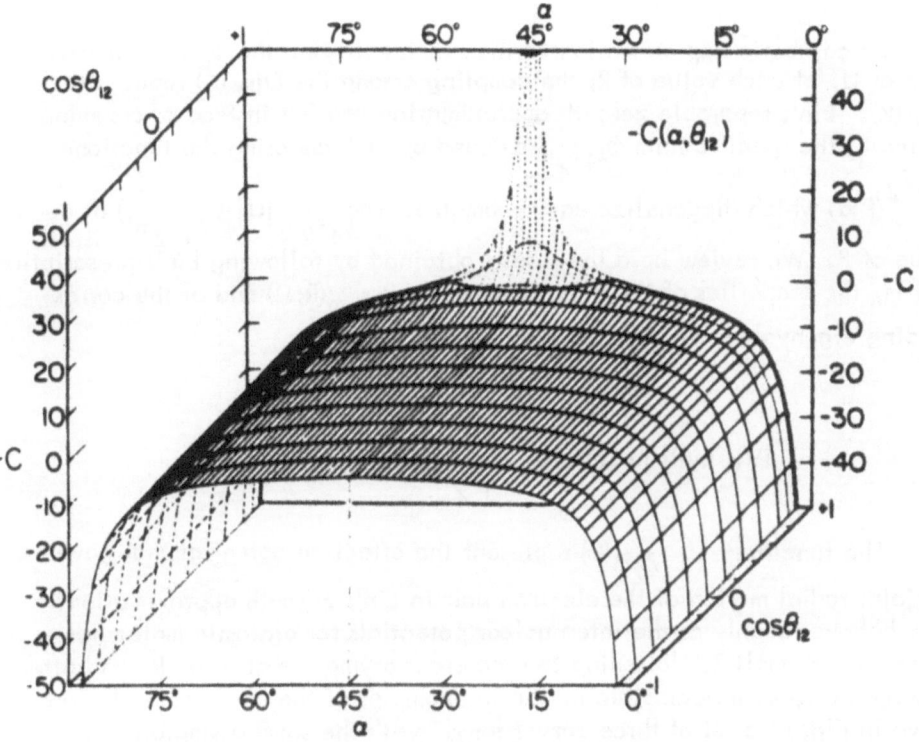

Fig. 8. Relief plot of $-C(a,\theta_{12})$ with $Z = 1$. The ordinates
represent a potential surface in Rydberg units at
$R = 1$ bohr (Courtesy of C. E. Theodosiou). (From Ref. 2.)

$$\left[\frac{d^2}{dR^2} + 2E \right] F_\mu(R) - \Sigma_{\mu'} \left[U_{\mu\mu'}(R) - W_{\mu\mu'}(R) \right] F_{\mu'}(R) = 0 \qquad (6)$$

with the coupling potential matrices

$$U_{\mu\mu'} = (\Phi_\mu | U | \Phi_{\mu'}) \ , \quad W_{\mu\mu'} = 2(\Phi_\mu | \frac{\partial}{\partial R} \Phi_{\mu'}) \frac{d}{dR} + (\Phi_\mu | \frac{\partial^2}{\partial R^2} \Phi_{\mu'}). \quad (7)$$

Macek's adiabatic approximation consists of identifying the Φ_μ as eigenfunctions of U, at each value of R; the coupling among the Eqs. (6) reduces then to $W_{\mu\mu'}(R)$. Lin's separable zero-th approximation consists instead of assuming for the Φ_μ the specific form $\Phi_{\ell_1 \ell_2 m}$ defined by (1), choosing the functions $g_m^{\ell_1 \ell_2}(R;\alpha)$ which diagonalize each submatrix $(\Phi_{\ell_1 \ell_2 m} | U | \Phi_{\ell_1 \ell_2 m'})$ at each value of R. We review here the results obtained by following Lin's prescription, that is, the properties of the eigenfunctions $\Phi_{\ell_1 \ell_2 m}(R;\Omega)$ and of the corresponding eigenvalues $U_{\ell_1 \ell_2 m}(R)$.

IV. Lin's Functions $U_{\ell_1 \ell_2 m}(R)$ and $g_m^{\ell_1 \ell_2}(R;\alpha)$

The functions $U_{\ell_1 \ell_2 m}(R)$ represent the effective potential that governs the joint radial motion of the electron pair in Lin's zero-th approximation. They behave roughly as the internuclear potentials for diatomic molecules, rising toward small R, flattening to a constant asymptote at large R, and often forming a well at intermediate R. Some examples of these curves for He are shown in Fig. 9; a set of three curves for H⁻ with the same asymptote is shown here in Fig. 10.

The asymptote at large R represents an energy eigenvalue of a single electron that remains attached to the nucleus when the other one escapes with zero residual energy. As R approaches this limit one sees readily[2] that

$$U_{\ell_1 \ell_2 m}(R) \xrightarrow[R \to \infty]{} - \frac{Z^2}{n^2} - \frac{2(Z-1)}{R} \qquad (8)$$

in Rydberg units and to within terms of order R^{-2}. The principal quantum number n of the electron that remains bound is found to be $n = \frac{1}{2}(\ell_1 + \ell_2 + m + \frac{3}{2} \pm \frac{1}{2})$ where the + or − sign is selected so that n be an integer.

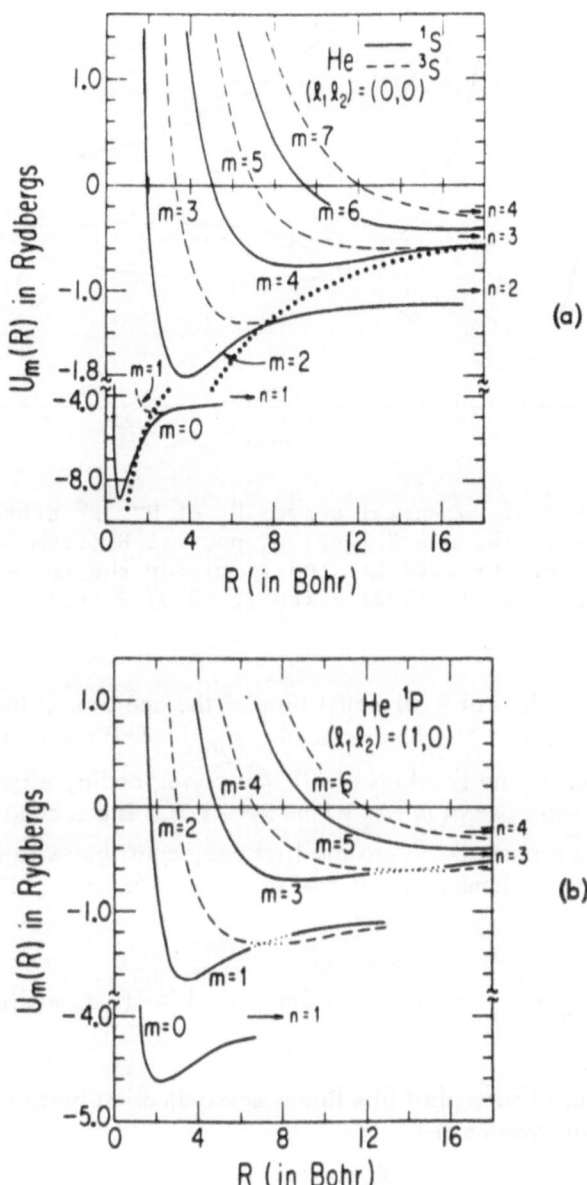

Fig. 9. Potential curves $U_m(R)$ for S and ^{1}P channels of He.
a) ——— s^2 ^{1}S; ---- s^2 ^{3}S; .... Eq. (9); b) ——— sp ^{1}P+;
---- sp ^{1}P- (From Ref. 2.)

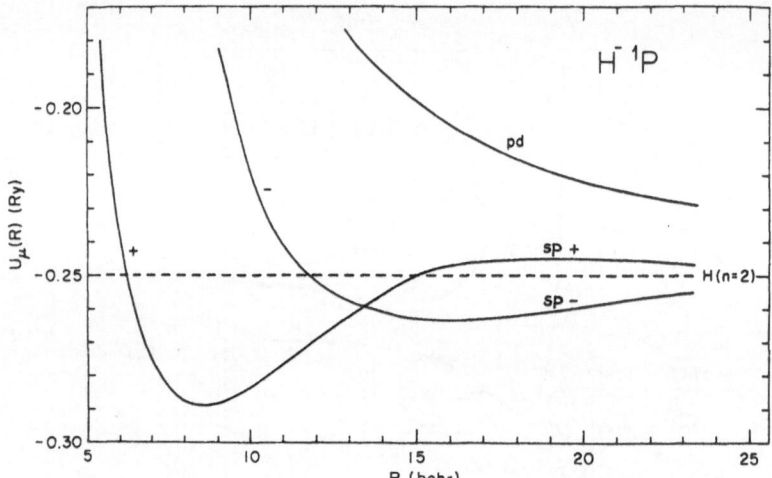

Fig. 10. The set of potential curves U_μ of H^- $^1P^o$ states that con-
         verge to the n = 2 level of neutral H (from Ref. 7).
         Lin's approximate labels $(\ell_1\ell_2 m)$ for the three curves are
         $(101) \equiv (sp+)$, $(102) \equiv (sp-)$, $(210) \equiv (pd)$.

For small values of R, the first term of the operator U in (3) dominates over the second term. The functions $\Phi_{\ell_1\ell_2 m}$ for R=0 are thus eigenvectors of the expression in the brackets of (3); the corresponding eigenvalues have the analytical expression $(\ell_1+\ell_2+2m+2)^2-1/4$. The second term of (3), when treated as a perturbation to the first one, contributes to (3) only through its diagonal matrix elements evaluated at R=0. Thus one obtains the simple approximation

$$R^2 U_{\ell_1\ell_2 m}(R) \xrightarrow[R \sim 0]{} (\ell_1+\ell_2+2m+2)^2 - 1/4 - (\ell_1\ell_2 m|C|\ell_1\ell_2 m)_{R=0}\, R. \quad (9)$$

The plots in Fig. 11 show that this linear approximation holds up to surprisingly large values of R.

The wave functions $g_m^{\ell_1\ell_2}(R;\alpha)$ have also characteristic behavior in the two limits of large and small R. This aspect of our problem has been discussed and illustrated in Ref. 2 and further in Sec. 3 of Ref. 3. At large R, the fact that either one of the two electrons remains attached to the nucleus is repre-sented by the probability $|g_m^{\ell_1\ell_2}(R;\alpha)|^2$ being concentrated in either of the extreme ranges of α, near 0° or near 90°. These are the ranges of α where the potential coefficient $-C(\alpha;\theta_{12})$ is large and negative. At small R, on the contrary, where the kinetic energy term of U predominates and the energy $U_{\ell_1\ell_2 m}(R)$ is positive, the wave function $g_m^{\ell_1\ell_2}$ oscillates throughout the range of α indicating that the two particles indeed form one complex.

The transition from the small-R to the large-R behavior occurs in a rather narrow range of R, in which the energy $U_{\ell_1\ell_2 m}(R)$ drops to negative values lower than the saddle point of the potential function $-C(\alpha,\theta_{12})/R$, that is, when

$$U_{\ell_1\ell_2 m}(R) \sim -\sqrt{2}(4Z-1)/R , \tag{10}$$

as discussed in Ref. 2. The locus of such transition points is indicated by a dotted line in Figs. 9 and 11. Passing through this transition represents the most critical point in the evolution of a doubly excited complex from an integrated to a quasi-dissociated form. The transition is evidenced particularly by the rather localized "bumps" in the full drawn curves in Fig. 11, which have the character of "avoided crossings."

The discussion of Ref. 2 centers on the remark that these avoided crossings represent presumably the locus of electron configurations at which the state of the complex is most likely to perform a transition from one channel to another, e.g., to become doubly excited or, conversely, to autoionize. This remark is regarded as the main result of the studies reported in these lectures. A subsidiary, but also important, result emerges from the remark that the eigenvalue plots of $R^2 U_{\ell_1\ell_2 m}(R)$ which are represented by dashed lines in Fig. 11 are far smoother than the others in the transition region, which implies a lesser probability of transitions between channels. This behavior, associated with particular combinations of the quantum numbers, is shown to depend on the occurrence of a node of $g_m^{\ell_1\ell_2}(R;\alpha)$ in the region of the saddle point of $-C(\alpha;\theta_{12})$ and to account for the selection rules mentioned in Sec. 1a of this article. The presence of the nodes has the effect of smoothing out the evolution of the wave function as R varies.

We return now to the energy functions $U_{\ell_1\ell_2 m}(R)$, and specifically to the relationship between their values and the number of nodes, m, of the corresponding eigenfunctions $g_m^{\ell_1\ell_2}(R;\alpha)$. According to general experience with wave mechanics, one would anticipate U to increase with increasing m, as the mean kinetic energy of the motion along α increases with the number of nodes. This expectation is indeed verified in general, but with notable exceptions

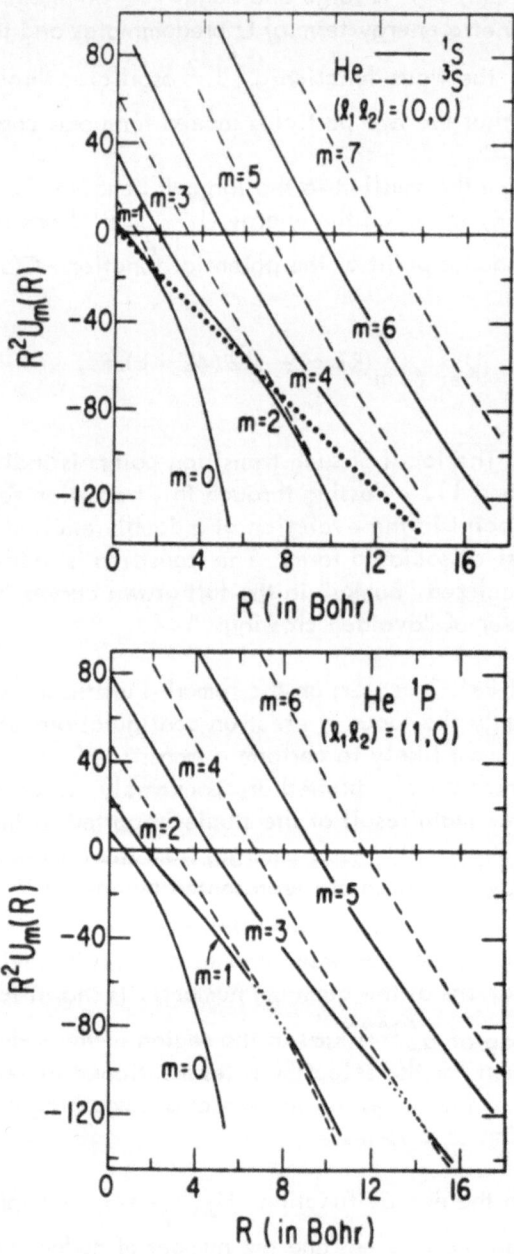

Fig. 11. Plots of $R^2 U_m(R)$; data and notation as in Fig. 9. (From Ref. 2.)

manifested by curve crossings in the lower portions of Figs. 9 and 11. The exceptions occur for pairs of curves that approach the same limit (8) at large R and whose wave functions differ by the occurrence of a single node in the region of vanishingly low values of $\alpha \sim 45°$ and large R. Here the difference in average kinetic energy of the two states is negligible, while the combination of angular and radial correlations may raise the average potential energy for the state with lower m, depending on the values of the relevant quantum numbers (see Sec. 3 of Ref. 3 and Refs. 4 and 5). An important consequence of this phenomenon has been pointed out recently by Lin[7] for the example of H^- shown in Fig. 10. In this example, the deeper well of the energy function $U_{101}(R)$ — indicated by sp+ in the figure — is narrowed by the rise of this function at large R, which reflects greater than average electron repulsion. The well is in fact so narrow that it cannot hold any bound state. Such a state would in fact have been autodetaching, with the fair rate of formation and decay that pertains to the deeper-welled (or "+") state of each pair. On the other hand, the shallower well of the function $U_{102}(R)$ — labeled "sp-" — extends to very large R and gives rise, in priciple, to an infinity of very weakly bound autodetaching levels, two of which have been calculated by entirely different approaches with equivalent results. These states, however, should be very difficult to observe, because they are extremely narrow, weakly bound, and with low rate of formation.

V. Macek's Energy Functions $U_\mu(R)$

Lin's approximate wave functions correspond to full quasi-separability of the variables. They are accordingly classified by as many quantum numbers as there are independent variables and they permit a most detailed and simple analysis of the electron correlations. The correlations become more involved as one proceeds to take into account interaction effects disregarded in Lin's approximation, but the resulting complications can still be surveyed insofar as these effects are concentrated in limited ranges of the variables. These ranges are, of course, those where two (or more) of Lin's energy functions $U_{\ell_1\ell_2m}(R)$ with different values of the quantum numbers are degenerate or nearly degenerate, because the corresponding eigenfunctions are then more susceptible to perturbation by previously neglected interactions. The occurrence and the effects of degeneracies are surveyed in Refs. 2, 3, and 4 and are illustrated particularly by Fig. 12. The main phenomenon disregarded by Lin is the transfer of orbital momentum between the electrons; its influence upon the wave functions can be described as a mixing of angular and radial correlations. Its influence on the energy functions is to modify Lin's $U_{\ell_1\ell_2m}(R)$ by a phenomenon of spectral repulsion which spreads their values farther apart where they are closest and, in particular, eliminates the crossings of Lin's functions with different (ℓ_1, ℓ_2) but equal (L, S) quantum numbers.

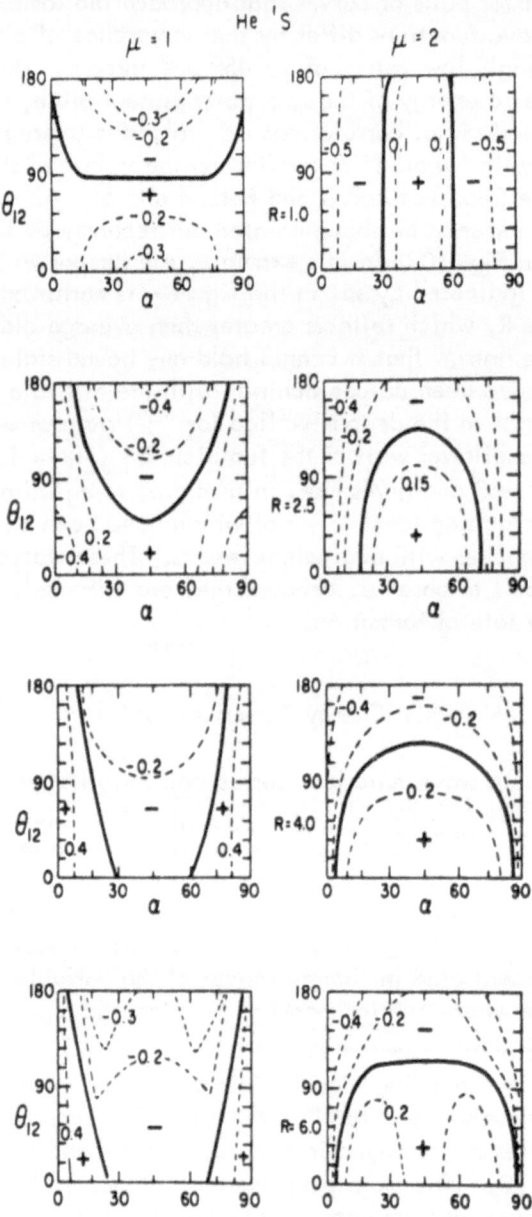

Fig. 12. Contour line plots of the joint radial and angular
correlation functions Φ_1 and Φ_2 of ^{1}S states of He for
different values of R (Courtesy C. D. Lin).   (From
Ref. 3.)

The resulting functions are those called $U_\mu(R)$ by Macek.[4] Their associated wave functions $\Phi_\mu(R;\Omega)F_{\mu n}(R)$ still do not provide an optimum representation of the two-electron states that are identified experimentally, as
shown by the following consideration. These functions correspond to states
for which the matrix $U_{\mu\mu'}$ in Eq. (6) is diagonal and which would accordingly
be stationary if the off-diagonal matrix elements $W_{\mu\mu'}(R)$ defined in Eq. (7)
were negligible. In fact, evaluation of sample values of $W_{\mu\mu'}(R)$ using the
calculated wave functions $\Phi_\mu(R;\Omega)$ shows that the radial equations (6) with
different μ indices are more strongly coupled than is indicated by the observed
rates of formation and decay of doubly excited states. Thus it appears that a
better procedure should exist for identifying and constructing wave functions
$\Phi_\mu(R;\Omega)$ and energy functions $U_\mu(R)$ that constitute a better approximation to
those of the actual channels. The search for such a procedure is a current
goal.

VI. Conclusions

The studies described in these lectures have enabled us to outline a ten-
tative description of the formation and decay of doubly excited complexes,
which is presented in Sec. 5 of Ref. 2. According to that picture the complex
may be regarded as originating in a particular channel μ of the system, with
given total angular momenta (L,S) of orbit and spin; this channel is that of
lowest energy when the atom is initially in its ground state. Transitions between two channels occur, with generally low probability, at values of the
radial coordinate R where their energy curves, $U_\mu(R)$ and $U_{\mu'}(R)$ show an
avoided crossing. That is, they occur in the transition region between their
small-R and large-R trends. Presumably, the transitions occur normally be-
tween channels for which $|U_\mu(R) - U_{\mu'}(R)|$ is smallest, consistently with
selection rules. The rules are that: a) L and S are normally constant, since
they change only under the influence of spin-orbit coupling, b) the pair of
separate orbital numbers of two electrons (ℓ_1, ℓ_2) changes only in limited
ranges of R which are normally different from those where transitions occur,
c) transitions between channels with the same (L, S, ℓ_1, ℓ_2) but without and
with a node near $a \sim 45°$ — i.e., transitions with a change of the $\pm$ quantum
number — are also unlikely. For further details, see Ref. 2.

The probability of channel transitions appears to increase, however,
with increasing values of the radial correlation quantum number m. A different
type of state with very strong radial correlation appears to occur, which is
represented by a coherent superposition of channel wave functions with different values of $m \gtrsim 5$-10. These states, whose wave functions have strong maxima near $a = 45°$, are believed to provide the pathway for reaching simultaneous
very high excitation or detachment of both electrons. They have been studied

in fair detail by Wannier and others,[8] and the main theoretical predictions have been verified experimentally, but the physico-mathematical mechanism by which they may arise from lower channel excitation has not yet been described in any detail.

References

*Work supported by U.S. Energy Research and Development Administration, Contract No. COO-1674-112.

1.   U. Fano, Atomic Physics 1 (Plenum Press, New York, 1969) p. 209.

2.   U. Fano and C. D. Lin, Atomic Physics 4 (Plenum Press, New York, 1975) p. 47.

3.   U. Fano, "Analysis of Electron Correlations", invited talk for the IX ICPEAC, Seattle, Washington, July 1975 (to be published).

4.   J. Macek, J. Phys. B 2, 831 (1968).

5.   C. D. Lin, Phys. Rev. A 10, 1986 (1974) and 12, 493 (1975).

6.   U. Fano and C. D. Lin, Physics of Electronic and Atomic Collisions ed. by B. C. Čobić and M. V. Kurepa (Institute of Physics, Beograd, Yugoslavia, 1973) p. 229.

7.   C. D. Lin, Phys. Rev. Lett. (in press).

8.   G. Wannier, Phys. Rev. 90, 817 (1953); A.R.P. Rau, Phys. Rev. A 4, 207 (1971); R. Peterkop, J. Phys. B 4, 513 (1971), and U. Fano, J. Phys. B 7, L401 (1974).

THE SINGLE ELECTRON MODEL IN PHOTOIONIZATION

John W. Cooper

National Bureau of Standards

Washington, D.C. 20234

1. INTRODUCTION

The single electron model serves three purposes: (a) it provides a "zeroth order" approximation for the direct calculation of cross sections for photoionization and for other atomic processes, (b) it provides physical insight into the mechanisms that control these processes and (c) it forms the basis for the development of theoretical methods dealing with many electron correlations. All of the above aspects will be treated in the following discussion, but first the appropriate formulas for photoionization will be developed in order to show the relationship between the model and many electron treatments.

2. THE PHOTOIONIZATION CROSS SECTION

The theory of atomic photoionization, as well as the basic theory of emission and absorption in atomic systems may be based on first order time dependent perturbation theory, the results of which can be expressed by Fermi's "golden rule" (1)

$$\tau = 2\pi \left| < \Psi_o^* \, H' \, \Psi_f > \right|^2 \rho \, (E) \tag{1}$$

which gives a prescription for calculating the probability per unit time, τ, for a transition taking place between two states of a quantum mechanical system (0 and f) due to a weak time dependent interaction. In almost all experimental situations the light intensity is low enough for this approximation to be valid far beyond the accuracy to which calculations are performed. However, correct application of the theory requires the wave functions Ψ_o and Ψ_f as well as the interaction operator H' to be specified precisely.

31

Atomic photoionization takes place from an initially bound atomic state (usually the ground state of the system) into some state of the continuum which represents an atomic ion core plus one (or possibly more) free electrons. In the derivation of the "golden rule" (Eq. 1), Ψ_o and Ψ_f represent these initial and final states. They are defined as exact eigenstates of an atomic Hamiltonian H_a which satisfy the equations:*

$$H_a \; \Psi_o \; = \; E_o \; \Psi_o \tag{2a}$$

$$H_a \; \Psi_f \; = \; E_f \; \Psi_f \tag{2b}$$

Ψ_o and Ψ_f are thus two of a complete set of eigenstates Ψ_j which determine the time development of the atomic system $\Phi(t)$ via the relation

$$\Phi(t) \; = \; \sum_j C_j(t) \quad \Psi_j \; e^{-iE_j t} \tag{3}$$

The normalization conditions on Ψ_o and Ψ_f are

$$\int \Psi_o^* \; \Psi_o \, d\tau = \; N \tag{4a}$$

and

$$\int \Psi_f^* \; \Psi_{f'} \, d\tau = \; N \delta \; \cdot \; (\; E_f - E_f' \;) \tag{4b}$$

where (4a) and (4b) represent integration over the configuration space of all N electrons in the atom and E_f and E_f' refer to continuum states of different energy.

The perturbation operator for the interaction depends upon photon intensity, polarization and direction. Specifically it is

$$^{(2)} H' = \; \underline{A} \quad \alpha \, e^{-i\underline{k}\omega \cdot \underline{r}_j} \; \underline{\varepsilon} \cdot \nabla_j \tag{5}$$

where $|\underline{A}|$ represents the intensity of a beam of photons with direction $\hat{\underline{k}}_\omega$ and polarization direction specified by $\underline{\varepsilon}$; $\underline{r}_j$ and ∇_j are coordinate and gradient operators with respect to the jth electron and α is the fine structure constant.

Using Eq. 5 in Eq. 1 the probability for a transition between two definite initial and final states is

$$\tau_{of} = |\underline{A}|^2 \; 2 \, \pi \alpha^2 \left| < \Psi_o^* \; \sum_j e^{-i\underline{k}\omega \cdot \underline{r}_j} \; \underline{\varepsilon} \cdot \nabla_j \; \Psi_f > \right|^2 \tag{6}$$

Since the number of photons/unit area-unit time is a beam of photons is

$$P = \; \underline{k}_\omega/2 \, \pi \cdot \left| \underline{A} \right|^2 \tag{7}$$

the cross section for photoionization is

$$\sigma \; (\text{or} \; \frac{d\sigma}{d\Omega} \;) = \frac{\tau_{of}}{P} \; = \; \frac{4 \, \pi^2 \alpha^2}{k_\omega} \left| < \Psi_o^* \; \sum_j e^{-i\underline{k}\omega \cdot \underline{r}_j} \; \underline{\varepsilon} \cdot \nabla_j \; \Psi_f > \right|^2 \tag{8}$$

---

*The development may be either relativistic or non-relativistic corresponding to Dirac or Schrodinger Hamiltonians. The treatment here will be non-relativistic and will ignore all effects of electron spin. All formulas will be in atomic units.

Equation 8 (compare Ref. 3, Eqs. 69.2 and 69.5) is the basic starting point for the study of photoionization of atomic systems. The steps in its derivation have been sketched in order to bring out the following points.

(a) The density of states factor ρ_E is set equal to 1. The density of final states is correctly accounted for by the normalization/unit energy range implied by Eq. 4b.

(b) The same formula is used for "total" σ and differential cross sections $\frac{d\sigma}{d\Omega}$. This is because the process under consideration is specified by the final state Ψ_f. The final state wave function may alternatively represent an ion core plus an electron moving in a specified direction or an ion core plus an electron with definite quantum numbers ℓ and m (we ignore spin here). The total cross section may thus be represented alternatively as the incoherent sum of partial cross sections $\sigma_{\ell m}$ given by Eq. 8 or as the integral over $\frac{d\sigma}{d\Omega}$. Note that in evaluating differential cross sections the sums over ℓ and m appear <u>within</u> the matrix element since an electron moving in a specific direction may be represented by a partial wave expansion.

(c) The matrix element for photon absorption depends explicitly on photon direction and polarization direction relative to the electron coordinates.

(d) The states Ψ_o and Ψ_f are exact eigenstates of the same Hamiltonian H_a.

Conservation of energy requires that the photon energy k_ω/α be related to the energy difference between initial and final states via the Einstein relation

$$E_f - E_o = k_\omega/\alpha \tag{9}$$

Provided $|\underline{k}_\omega \cdot \underline{r}_j| \ll 1$ the factor $e^{i\underline{k}_\omega \cdot \underline{r}}$ in the matrix element may be replaced by 1 (dipole approximation). If

$$< [\sum_j \underline{r}_j \cdot H_a] >_{of} = (E_f - E_o) < \sum_j r_j >_{of} \tag{10}$$

within the dipole approximation the matrix element of Eq. 8 may be written in the alternative forms

$$< \Psi_o^* \sum_j \underline{\varepsilon} \cdot \nabla_j \Psi_f > = (E_f - E_o) < \Psi_o^* \sum_j \underline{\varepsilon} \cdot \underline{r}_j \Psi_f > \tag{11}$$

    dipole velocity   dipole length

Note that within the dipole approximation the matrix element no longer depends upon the direction of the photon beam. The two forms of Eq. 11 are often used in the same calculation as a check as to whether or not the approximations that have been made are valid. However, the agreement of cross sections obtained using the two forms does not imply that the calculation is realistic. In particular <u>any</u> calculation made using a central field Hamiltonian separable in the electron coordinates will satisfy Eq. 10 and thus length and velocity cross sections will agree if no further approximations are made.

The range of validity of the dipole approximation may be illustrated by comparing total cross sections for the one case where calculations may be carried out exactly, the hydrogen atom, as shown in the following table.

Incident Energy (keV)	Exact (10^{-24}cm^2) (Ref. 4)	Dipole (10^{-24}cm^2) (Ref. 5)
1.	11.5	11.4
10.	4.56×10^{-3}	4.55×10^{-3}
100.	1.66×10^{-6}	1.55×10^{-6}
1000.	1.96×10^{-9}	0.51×10^{-9}

3. THE SINGLE ELECTRON MODEL

The single electron model may be considered as a special case of the general treatment given above. The only difference is that the initial and final states which appear in the photo-ionization matrix elements of Eqs. 8 and 11 are taken to be exact eigenstates of a single electron Hamiltonian

$$H(1) \; \Phi_o = \varepsilon_o \; \Phi_o \tag{2a'}$$

$$H(1) \; \Phi_f = \varepsilon_f \; \Phi_f \tag{2b'}$$

All of the above formulas remain the same with the replacement of Ψ_o, $_f$ by ϕ_o, $_f$, E_o and E_f by ε_o and ε_f, N in Eq. 4a by 1 and the deletion of the sum over electron coordinates within the matrix element. The success or failure of the single electron model thus depends upon how realistic the single electron Hamiltonian H(1) is in describing the initial and final states of the photoionization process. This description will be incomplete due to the nature of the model which explicitly excludes <u>all</u> two electron processes such as core excitation and autoionization. However, the model does include the effects of the other electrons in an N electron system implicitly in the definition of H(1).

4. APPLICATION OF THE SINGLE ELECTRON MODEL TO SIMPLE SYSTEMS: PHOTOIONIZATION OF Na ATOMS

The formalism sketched above will now be applied to some simple systems to illustrate the methods employed. Consider first photoionization of the valence electron of a sodium atom in its ground state $1s^2 2s^2 2p^6 3s \; ^1S_o$. The Hamiltonian for the sodium atom (non-relativistic, spin independent) is

$$H_a = \sum_j \nabla_j^2 / 2 + \frac{Z}{r} - \sum_{i \neq j} \frac{1}{r_{ij}} \quad (Z=11) \tag{12}$$

and in the single electron approximation this is replaced by

$$H(1) = \nabla_1^2 / 2 + Z_{eff} / r \tag{13}$$

This, of course, is the key approximation and will be dealt with in detail in the next section. The ground state wave function of the 3s electron can be written as

$$\Phi_{3s}(\underline{r}) = R_{3s}(r)\ Y_{00}(\theta,\Phi)$$

where R_{3s} satisfies the radial equation

$$(\frac{d^2}{dr^2} + 2Z_{eff}/r + \ell(\ell+1)/r^2 + 2\varepsilon_{3s})R_{3s}(r) = 0 \quad (\ell=0) \quad (14)$$

with the normalization condition $\int_0^\infty R_{3s}^2 dr = 1$. Note that Eq. 14 defines both the ground state wave function and its energy. The ionization potential is simply $I = -\varepsilon_{3s}$ and the final state electron energy is given by Eq. 9 as

$$\varepsilon(=k^2/2) = \varepsilon_f = \varepsilon_{3s} + k_\omega/a \quad (15)$$

where k_ω/a is the photon energy. For $\varepsilon>0$ there will be an infinite number of final states which are eigenfunctions of Eq. 13 of the form

$$\Phi_{\varepsilon\ell m}(\underline{r}) = R_{\varepsilon\ell}(r)\ Y_{\ell m}(\theta,\Phi) \quad (16)$$

where the radial functions $R_{\varepsilon\ell}$ satisfy the equation

$$(\frac{d^2}{dr^2} + \frac{2Z_{eff}}{r} - \frac{\ell(\ell+1)}{r^2} + 2\varepsilon)R_{\varepsilon\ell}(r) = 0 \quad (17)$$

The normalization condition (Eq. 4b) determines the asymptotic form of each radial function (see Ref. 3, pp 22,23)

$$R_{\varepsilon\ell}(r) = \sqrt{\frac{2}{\pi k}}\ \cos(kr + \frac{Ln(2kr)}{r} + \delta_c + \delta_z) \quad (18)$$
$$r \rightarrow \infty$$

Note that asymptotically the functions $R_{\varepsilon\ell}$ are represented by Coulomb wave functions since at large distances $Z_{eff} = 1$, i.e., an outgoing electron in atomic photoionization is described as moving in the field of the ionic core. δ_c and δ_z are phase shifts due to the Coulomb field and the effective central potentials, respectively.

An electron moving in the direction $\hat{\underline{k}}$ can be represented by a wave function of the form

$$\Phi_{\hat{\underline{k}}} = \sum_\ell i^\ell(2\ell+1)R_{\varepsilon\ell}(r)\ Y_{\ell 0}(\theta,\Phi)e^{i(\delta_c+\delta_z)} \quad (19)$$

where θ and ϕ are relative to the $\hat{\underline{k}}$ axis and $R_{\varepsilon\ell}$, δ_c and δ_z are defined by Eq. 18. Using the above wave functions in the dipole length matrix element, the differential cross section for the valence electron is

$$\frac{d\sigma_{3s}}{d\Omega} = \pi\alpha(\varepsilon-\varepsilon_{3s})\left|<R_{3s}\ r\ R_{\varepsilon 1}>\right|^2 \cos^2\theta \quad (20)$$

where $\theta = \underline{E}\cdot\hat{\underline{k}}$ is the angle between the polarization direction and the direction of electron ejection. The total cross section will be

$$\sigma_{3s} = \frac{4\pi^2\alpha}{3}(\varepsilon-\varepsilon_{3s})\left|<R_{3s}\ r\ R_{\varepsilon 1}>\right|^2 \quad (21)$$

Note the following:

(a) The angular distribution is expressed relative to the

direction of photon polarization. Since the initial state is $^1S_0$ (no orientation) this is the only specified direction.

(b) The same result (but not the same formula) would be obtained using the dipole velocity form of the matrix element since H(1) (Eq. 13) satisfies Eq. 10 exactly.

(c) The total cross section (Eq. 21) can be obtained in two ways, either by integrating Eq. 20 over all angles or by using the general form for final states (Eq. 16) and performing the sums over ℓ and m over the squared matrix element. Both procedures give the same result.

Consider now an atom with a 3p valence electron (e.g., Al or an excited state of Na). The same procedures may be used but now the central field description of the valence electron is degenerate, i.e

$$\Phi_{3p}(r) = R_{3p}(r) Y_{1m}(\theta,\Phi) \quad (m = -1,0,1) \qquad (22)$$

where R_{3p} and ε_{3p} are obtained from Eq. 14 as before. Provided

the atom has no preferred orientation, the individual cross sections for these initial states must be averaged.

The dipole length matrix element for 3p ionization will be

$$< R_{3p}(r) Y_{1m}(\theta,\Phi) \underline{\varepsilon} \cdot \underline{r} \sum_\ell i^\ell (2\ell+1) R_{\varepsilon\ell}(r) Y_{\ell o}(\theta,\Phi) e^{i(\delta_c+\delta_z)} >$$

Note that now the dipole operator $\underline{\varepsilon} \cdot \underline{r}$ will result in two non-vanishing partial waves when the matrix element is evaluated, i.e. the final state will be of the form

$$\Phi_\varepsilon = R_{\varepsilon o} e^{i\delta_c} - 5 Y_{20}(\theta,\Phi) e^{i\delta_z} R_{\varepsilon 2} \qquad (23)$$

where $\delta_{0,2} = \delta_c + \delta_z$ for $\ell = 0$ and 2, respectively. Since these two terms appear within the matrix element the differential cross section will depend on the radial matrix elements $\langle R_{31} r R_{\varepsilon 0} \rangle = \sigma_0$, $\langle R_{31} r R_{\varepsilon 2} \rangle = \sigma_2$ and also on the phase shift difference $\delta = \delta_2 - \delta_0$. The differential and total cross sections averaged over the initial states m = -1,0,1 will be in this case

$$\frac{d\sigma_{3p}}{d\Omega} = \frac{\sigma_{3p}}{4\pi} (1 - \beta P_2(\cos\theta)) \qquad (24)$$

$$\sigma_{3p} = 4\pi^2\alpha/9 (\varepsilon + \varepsilon_{3p}) (\sigma_0^2 + 2\sigma_2^2) \qquad (25)$$

where $\qquad \beta = 2 (1 - 2 \sigma_0/\sigma_2 \cos\delta) \qquad (26)$

Note that the differential cross section depends upon the phases δ_0 and δ_2 but that the total cross section does not.

The same procedures can be used to obtain cross sections for the inner subshell electrons 1s, 2s and 2p. The only differences are (a) the radial equation for the bound state wave function (Eq. 14) must be solved for a 1s, 2s or 2p radial wave function (and binding energy) and (b) the cross sections must be multiplied by the number of electrons in the subshell. The final wave functions will be the same as in the above examples.

The same methods can, of course, be applied to initial state electrons with $\ell \neq 0,1$. The general formulas corresponding to

Eqs. 24-26 are $d\sigma_{n\ell}/d\Omega = \sigma_{n\ell}/4\pi\,(1+\beta\,P_2(\cos\theta\,))$ (24')

$$\sigma_{n\ell} = 4\pi^2\alpha/(2\ell+1)\,(\varepsilon+\varepsilon_{n\ell}\,)\left[\ell\sigma_{\ell-1}^2+(\ell+1)\,\sigma_{\ell+1}^2\right]$$ (25')

$$\beta = \frac{\ell(\ell-1)\,\sigma_{\ell-1}^2+(\ell+1)(\ell+2)\sigma_{\ell+1}^2 - 6\,\ell(\ell+1)\,\sigma_{\ell+1}\,\sigma_{\ell-1}\cos(\delta_{\ell+1}-\delta_{\ell-1})}{(2\ell+1)\left[\ell\sigma_{\ell-1}^2 + (\ell+1)\,\sigma_{\ell+1}^2\right]}$$ (26')

for an initial state electron with quantum numbers nl. The use of a central field Hamiltonian (Eq. 13) is not necessary for Eqs. 24'-26) to be valid. The only requirement is the separability of angular and radial coordinates (Eq. 16 and 22). Eq. 24' also applies to the case of unpolarized light with the simple replacement of β by $\frac{\beta}{2}$ if the angle θ is interpreted as that between incident photon and final electron directions. Finally, the angular dependence of the differential cross section given by Eq. 24' will be the same independent of the use of a single electron model (see Ref. 6 for further discussion).

5. DEFINITION OF THE SINGLE PARTICLE HAMILTONIAN HARTREE-FOCK METHODS

Single electron wave function of the form $R_{n\ell}(1)Y_{\ell m}(\theta,\phi)$ for electrons in the ground states of atoms may be obtained by a solution of the Hartree-Fock equations for the entire atom and provide a realistic description of both the charge distributions and of the binding energies of electrons in individual subshells. It is thus quite reasonable to base the definition of the one electron Hamiltonian to be used in single electron model calculations on a Hartree-Fock solution of the atomic ground state.

The radial equation obtained by the Hartree-Fock procedure (see Ref. 7 for details) are similar to Eq. 14. For example, the 3s radial equation of a sodium atom can be written as

$$\left(\frac{d^2}{dr^2} + \frac{2\,Z_{3s}\,R_{n\ell}(r)}{r} + 2\,\varepsilon_{3s}\right)R_{3s}(r) = X_{3s}(R_{3s}(r),R_{n\ell}(r)\,)$$ (27)

Similar equations may be written for the 1s, 2s and 2p radial wave functions with the effective central potential $Z_{3s}(R_{n\ell}(r))$ and exchange terms $X_{3s}(R_{3s},R_{n\ell})$ different in each equation. In the Hartree-Fock procedure all these equations must be solved simultaneously since the equation for each radial orbital depends on the other radial orbitals $R_{n\ell}(r)$. In Eq. (27), Z_{3s} is given approximately by the relation

$$Z_{3s}(r_1)/r_1 \cong \frac{Z}{r_1} - \int\frac{\rho(r_2)}{r_{12}}\,dr_2$$ (28)

where $\rho(r_2)$ is the r dependent electron density which can be obtained from the radial wave functions.

An approximate solution of the Hartree-Fock equations can be obtained by replacing the exchange terms in each equation by an

average exchange potential $\sim \rho(r)^{1/3}$ and using the approximate
form of the effective central potential given by Eq. (28). All
radial wave functions then obey the equation

$$(\frac{d^2}{dr^2} + \frac{2 Z (\rho(r))}{r} - \frac{\ell(\ell +1)}{r^2} + 2 \varepsilon_{n\ell}) \ R_{n\ell}(r) = 0 \qquad (29)$$

The radial equations for each orbital must still be solved self
consistently since $Z(\rho(r)$ depends upon all other orbitals $R_{n\ell}$.

The Hartree-Fock-Slater procedure outlined above (see Ref.
8 for further details) provides a central potential for each atom
which can then be used to define the single electron Hamiltonian
of Eq. 13. This method of defining a single electron potential
and its relativistic extension (see Ref. 9) have been widely used
in calculation of photoionization (4,5,9-12) over the last decade.

There is an alternative method of defining a single electron
Hamiltonian based on the Hartree-Fock procedure which accounts
for exchange effects explicitly. To illustrate the method consider a calculation of the transition rate of the resonance
transition 3s-3p of sodium via the Hartree-Fock method. In order
to perform this type of calculation Hartree-Fock solutions for

the ground state $1s^2 2s^2 2p^6 3s$ and for the excited state $1s^2 2s^2 2sp^6 3p$
would be performed. Using the wave functions and total energies
obtained from these calculations the transition rate could be
calculated directly using Eq. 6. Note that such a calculation is
actually a many electron calculation of the transition rate.
Although the single electron matrix element $\langle \phi_{3s} \varepsilon \cdot r \phi_{3p} \rangle$ will
account for the bulk of the transition rate, the full calculation
will involve the 1s, 2s and 2p orbitals which will differ slightly
in the two Hartree-Fock calculations. If the effects of core
relaxation are ignored, i.e., if the same core orbitals are used
in both calculations, the Hartree-Fock procedure provides a
method for performing calculations with the one electron model
which can be applied to photoionization. The method involves
performing a full Hartree-Fock calculation for the ground state
but applying the method to only a single orbital of the final
state. For a 3s-εp transition the continuum state radial equation
using this procedure will be exactly the same as that of the 3p
radial Hartree-Fock equation, i.e.,

$$(\frac{d^2}{dr^2} + \frac{2 Z_{3p}(r)}{r} - \frac{2}{r^2} + 2 \varepsilon) R_{\varepsilon 1}(r) = X (R_{n\ell}, R_{\varepsilon 1}) \qquad (30)$$

The above procedure in principle defines a single electron
Hamiltonian. However, Eq. 10 is no longer satisfied since the
exchange terms will be different in initial and final states and
consequently length and velocity forms of the matrix elements
(Eq. 14) will no longer agree (13).

6.   EMPIRICAL METHODS

The above discussion has concentrated on methods of obtaining realistic one electron wave functions and binding energies

for photoionization calculations based on ground state Hartree-Fock calculations. Other techniques have been used, including the following:

(a) A central potential obtained from the charge distribution of a ground state Hartree-Fock solution will not give the correct binding energy for an electron in the ground state or excited states of the same symmetry. Such a central potential may be modified by including extra terms which may then be adjusted to produce the same binding energies that are obtained from spectral data. (See Ref. 14).

(b) Rather than carrying out a detailed numerical calculation, the basic cross section formulas are often evaluated using approximate initial and final state wave functions obtained from various sources. Thus bound states may be approximated by hydrogenic wave functions with a value of the nuclear charge Z adjusted to account for the screening effect of other atomic electrons. Continuum states can be represented by the same approximation (Coulomb waves) or even by free wave solutions (Z=0). Such calculations are extremely useful for establishing limiting forms of cross sections at high energy (see Ref. 9), in establishing systematic trends or in exploring new phenomena.

7. RELATIONSHIPS WITH MANY BODY APPROACHES

The one electron model for photoionization has been defined by Eqs. (2a') and (2b') and by using only single electron wave functions in the cross section formula, Eq. 8. The one electron aspect of the model stems from the fact that only one electron wave functions are used in the matrix elements. However, the most successful methods of obtaining these one electron wave functions rely upon Hartree-Fock methods which require a full, but approximate solution of the N electron problem for each atom. Thus to a certain extent many electron effects (i.e., screening and exchange effects of the other electrons) are "built into" the one electron model.

The one electron model serves as a starting point for many body procedures such as many body perturbation theory or the random phase approximation. All of these procedures require in principle a complete set of eigenstates of a single particle Hamiltonian (Eqs. 2a' and 2b') as a starting point. The single electron cross section as defined by Eq. 8 thus becomes a "zeroth order" estimate which is systematically corrected. The corrections in any specific calculation will then depend upon which "zeroth order" estimate is used as a starting point (see Ref. 15, sec. 3.4 for further discussion).

8. DISTRIBUTION OF OSCILLATOR STRENGTHS

Perturbation theory was used in Sec. 2 to obtain the transition probability for specific transitions. An alternative application considers the response of an atomic system in its ground state to a weak perturbation $E_o \cos \omega t$. An alternative

application of perturbation theory (see Ref. 16) yields the
frequency (ω) dependent polarizibility

$$\alpha(\omega) = \sum_s |<\Psi_0^* \sum_j Z_j \Psi_s>|^2 \left(\frac{1}{\omega_s - \omega} - \frac{1}{\omega_s + \omega} \right) \tag{31}$$

where as before $H_a \Psi_{0,s} = E_{0,s} \Psi_{0,s}$ and $\omega_s = E_s - E_0$ and the sum over
s included all possible final states of energy E_ε.

Eq. 31 may be written as

$$\alpha(\omega) = \sum_s \frac{|<\Psi_0^* \sum_j Z_j \Psi_s>|^2 \; 2\omega_s}{\omega_s^2 - \omega^2} \tag{32}$$

which is the same formula one would obtain for the polarizibility
(dipole moment/unit field strength) of a classical harmonic
oscillator of strength

$$f_s = 2\omega_s |<\Psi_0^* \sum_j Z_j \Psi_s>|^2 \tag{33}$$

Note that $\alpha(\omega)$ represents the response due to all such oscil-
lators in Eq. 31. Since the summation of Eq. 31 includes an
integration over continuum states, the proper definition for
continuum states will be

$$\frac{df_s}{d\omega_s} = 2\omega_s |<\Psi_0^* \sum_j Z_j \Psi_s>|^2 = \frac{1}{2\pi} \frac{df_s}{dE_s} \tag{34}$$

where now the states Ψ_s must be normalized per unit energy (or
frequency) range via Eq. 4b.

Comparing Eq. 34 with the expression for the total photoionization cross section in dipole length approximation, i.e.,
$\sigma = 4\pi^2 \alpha\omega_s |<\Psi_0 \sum_j Z_j \Psi_s>|^2$ the total cross section for photoionization may be defined in terms of the differential oscillator
strength/unit energy range (df/dE) as

$$\sigma = \pi \alpha \, df/dE \tag{35}$$

The concept of oscillator strength provides a link between
the transition rates (or oscillator strengths) for discrete
transitions and the total photoionization cross section. Note
that its definition (eq. 34) implies (a) a nonrelativistic treatment and (b) the validity of the dipole approximation.

There are a number of sum rules which relate energy moments
of the oscillator strength distribution to quantities that may be
calculated from the atomic ground state; e.g. the Thomas–Kuhn
rule

$$\sum_s f_s + \int_I^\infty df/dE = N \tag{36}$$

and the dipole moment sum rule

$$\sum f_s / \omega_s + \int_I^\infty \frac{1}{\omega_s} df/dE = 2 |<\Psi_0 (\sum_j Z_j^2) \Psi_0>|^2 \tag{37}$$

Similar rules exist for $<\omega_s^q f_s>$ with $q = -2, 1$ & 2. These sum
rules are exact since their derivation depends only on the
validity of Eq. 10 which holds for a nonrelativistic, spin
independent atomic Hamiltonian (e.g., Eq. 12). A similar set of
sum rules holds in the single electron model. (See Ref. 17, Sec.

2.5 and 4.2.)

9. PHOTOIONIZATION AT HIGH ENERGIES

The single electron model has proven to be remarkably successful in predicting total absorption due to photoionization at high energies over broad spectral ranges. Calculations of this type have been performed for all elements of the periodic system for photon energies greater than 1keV (see Refs. 9 and 15). The methods used include relativistic effects both in the definition of the central potential used and in the explicit calculation of the initial and final state wave functions. Furthermore, in calculations of this type the dipole approximation (Eq. 11) is not used.

An example of the agreement that is typical of this type of calculation and the absorption cross sections as measured by a number of experimental groups at various energies is shown in Fig. 1. Note that the agreement between theory and experiment is extremely good (typically better than 10%) over the entire spectral range even though the theory is based on atomic calculation and the measurements were made in solid materials. The major discrepancies are near the thresholds for M and L absorption.

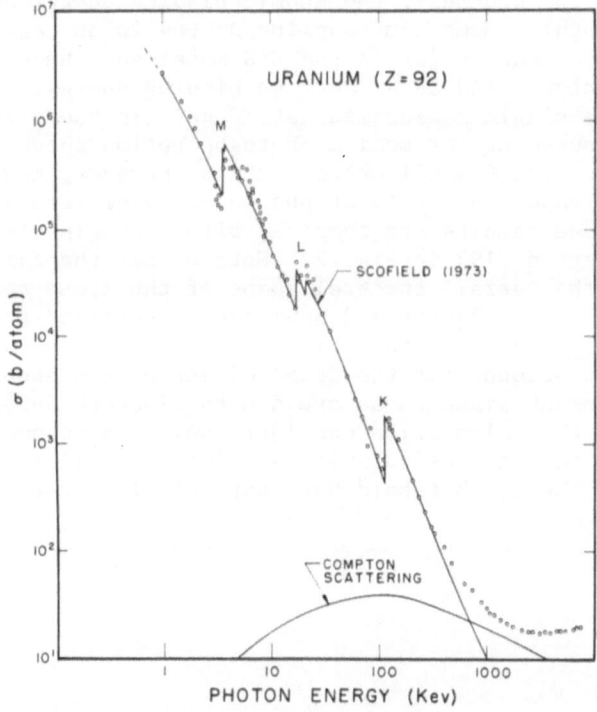

Fig. 1. Photon Absorption Cross Sections for Uranium. The experimental points are from 18 different sources.

10. INTERMEDIATE AND LOW ENERGIES: SODIUM ABSORPTION

The sodium atom provides a useful comparison between experiment and theory in the intermediate and low energy ranges. For purposes of orientation, the binding energies (eV) of the various subshells as obtained from experiment (ESCA and spectroscopic data) and via the nonrelativistic Hartree-Fock-Slater method (ref. 8) are listed below

Subshell	$1s^2$	$2s^2$	$2p^6$	$3s$
H.F.S.	1062.	64.3	36.3	5.16
Solid	1072.	53.	31.	--
Atom	?	?(71.)	$38.0(^3P_2)$	5.14
			$38.1(^3P_1)$	
			$38.2(^1P_0)$	
			$38.5(^1P_1)$	

Note the following:

(a) The single electron binding energy for the 3s electron predicted by the HFS one electron model is in very good agreement with the known ionization potential.

(b) There are large differences in binding energies in the atom and the solid.

(c) For the $2p^6$ subshell, the atomic binding energy actually depends on the angular momentum coupling of the $2p^5 3s$ core. This splitting is not accounted for in the HFS model and the calculated energy must be interpreted as an average binding energy.

Except for a single measurement at 10 keV (in the solid) there are no measurements of sodium photoabsorption above the K ionization limit. The L shell cross section, however, has been measured in both vapor and solid at photon energies from 30 to 250 eV (18). These results are compared with a single electron HFS model calculation (19) in Fig. 2. Notice that the calculation predicts the overall spectral shape of the cross section (the measurements are relative and have been normalized arbitrarily in the figure) for both atom and solid. The calculation, of course, cannot account for the detailed structures observed in the vapor measurement since these are due to discrete autoionizing transitions, and the calculation considers only continuous absorption. Note also that the calculation predicts a small increase in absorption at the L_I threshold but that this is not observed.

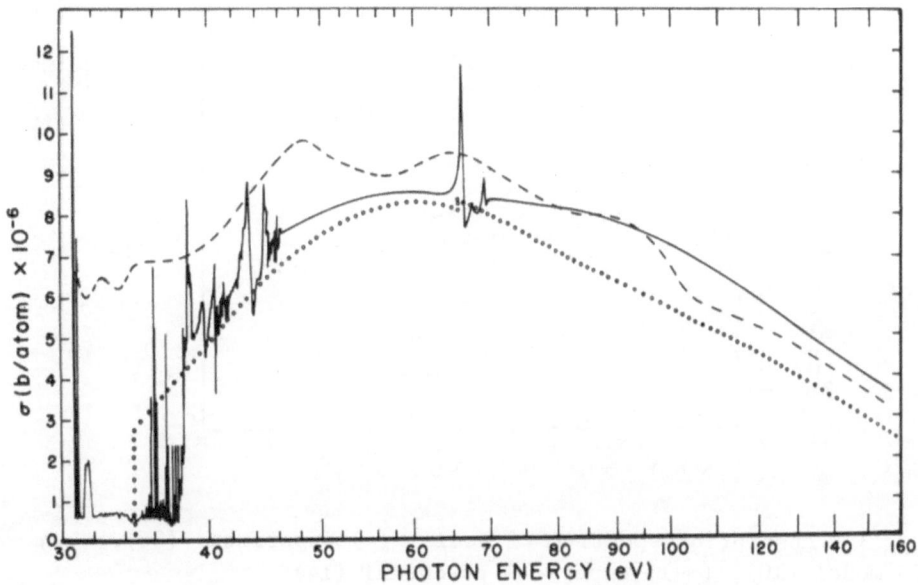

Fig. 2. Absorption cross sections for sodium in vapor (-) and solid (....) phases (Wolff et al. 1972), compared with atomic calculation (OOO) McGuire, 1970).

The cross section for photoionization of the valence electron in sodium has been measured and has been calculated using a number of one-electron models. Since the bulk of the oscillator strength distribution for 3s excitation (95-99%) is in the resonance transition 3s-3p the cross section for photoabsorption is extremely small and difficult to measure. One would expect the single electron approximation to provide a good estimate of the cross section.

The results of two single electron calculations are compared with experiment in Fig. 3. Both calculations correctly account for the minimum in the cross section immediately above threshold but underestimate the cross section at higher energies. Recent many body calculations have partially accounted for this discrepancy (20).

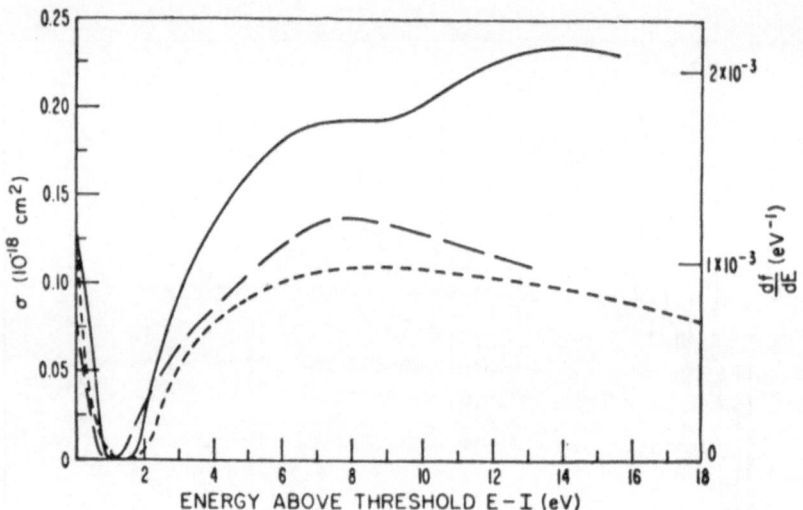

Fig. 3. Absorption spectrum of sodium (–) Experiment (26), (– –)
HFS model (10), (--), emperical potential (14).

The examples cited above illustrate the usefulness of the
one electron model in obtaining estimates of total cross sections. At high energies the model provides cross sections that
seem to be adequate for most purposes. At lower energies and
particularly at energies near thresholds the model provides an
initial estimate of the cross section. Comparison with experimental data can then be made to determine where more sophisticated calculations may be needed. The same remarks hold true for
differential cross sections.

11. PROPERTIES OF RADIAL WAVE FUNCTIONS

Within the framework of the single particle model the
photoionization cross section (and distribution of oscillator
strength) depends upon the radial wave functions $R_{n\ell}$ and $R_{\varepsilon\ell}$
defined by Eq. 17. If a central potential is used for model
calculations these in turn depend upon the central potential
$Z(r)/r$ so that essentially all of the physical information is
contained within the potential. A study of the general properties of the potential and of the wave functions derived from it
can thus shed some light on the atomic dynamics of the processes
involved.

Consider the general form for the central potential

given by Eq. 17, i.e., $\frac{Z(r)}{r} - \ell(\ell+1)/r^2$. In the HFS method
outlined in Sec. 5, $Z(r)$ is determined by solving for the lowest

N bound states of a given atom. Note that since there are only N electrons in the ground state, the procedure for determining $Z(r)$ also determines the fractional occupancy of the various subshells. From this standpoint the central potential can be viewed as a potential whose strength increases with increasing Z in such a way as to bind electrons in the atomic ground state in the correct order.

For a given central potential the solutions of Eq. (13) that are bound states for given ℓ are not the only bound state solutions that exist. For example, for sodium there will be radial solutions corresponding to R_{1s}, R_{2s} and R_{3s} which, of course, correspond to bound states, but also excited state solutions corresponding to R_{ns} with n>3. Most of the charge distribution for these excited discrete states will lie "outside the atom," i.e., in the region $r<r_0$ where $Z(r) \stackrel{\sim}{=} 1$. For $r>r_0$ the radial wave function may be represented exactly by

$$R_{n\ell} = \cos \pi \sigma_{n\ell} F_{n\ell} (r) + \sin \pi \sigma_{n\ell} G_{n\ell} (r) \qquad (38)$$
$$r > r_0$$

where $F_{n\ell}$ and $G_{n\ell}$ are regular and irregular Coulomb wave functions, i.e., solutions of Eq. 13 with $Z(r) = 1$. For $r<r_0$ the solutions of Eq. 17 with different values of n are almost the same apart from normalization. Consequently the factor $\sigma_{n\ell}$ (the quantum defect) is almost constant for large values of n.

For continuum states of low energy the same considerations apply. The continuum solutions of Eq. 17 for $r>r_0$ (cf. Eq. 18) will be

$$R_{\varepsilon\ell} = \cos \delta_{\varepsilon\ell} F_{\varepsilon\ell} (r) + \sin \delta_{\varepsilon\ell} G_{\varepsilon\ell} (r) \qquad (39)$$
$$r > r_0$$

where $\delta_{\varepsilon\ell}$ are energy dependent phase shifts and $F_{\varepsilon\ell}$ and $G_{\varepsilon\ell}$ are free Coulomb waves of energy ε.

The above considerations are important since they make it possible to do approximate calculations of discrete oscillator strengths and photoionization cross sections in a very simple way. The quantum defects for a particular Rydberg series are determined by the binding energies of these states via the Rydberg formula

$$\varepsilon_{n\ell} = - \frac{1}{2} \frac{1}{(n - \sigma_{n\ell})^2} \qquad (40)$$

To the extent that the quantum defects are slowly varying functions of n and that most of the radial wave function lies outside r_0 approximate values for oscillator strengths of discrete transitions can be determined solely from a knowledge of the quantum defects $\sigma_{n\ell}$. The method can be extended to the continuum via extrapolation using the relation implied by Eqs. 38 and 39, i.e.,

$$\pi \sigma_{n\ell} = \delta_{n\ell} \qquad (41)$$

(See Ref. 21 for further details.)

The free wave solutions of Eq. 17 are, of course, oscillating functions. With increasing energy ε the nodes of any free wave solution $R_{\varepsilon\ell}$ will move to smaller values of r. This is what causes photoionization cross sections to decrease with increasing energy at high energies, since each matrix element $< R_{n\ell} \; r \; R_{\varepsilon\ell} >$ represent an integral over a function that is independent of energy ($rR_{n\ell}$) and a rapidly oscillating function. Since the bound state radial orbitals also have nodes (except for $n = \ell$), at certain energies the matrix element may vanish due to complete cancellation of positive and negative portions of the integrand. This is the reason for the minimum in the sodium photoionization cross section shown in Fig. 3.

Near the origin all radial wave functions $R_{n\ell}$ and $R_{\varepsilon\ell}$ will have approximately the same radial dependence (provided ε is small) since the central potential will be large compared to $\varepsilon_{n\ell}$ or ε. This approximate independence can be made explicit by writing $C_{\varepsilon\ell} \; \bar{R}_{\varepsilon\ell} = R_{\varepsilon\ell}$ where $\bar{R}_{\varepsilon\ell}$ is a constant which must be evaluated by the rule for normalization of continuum wave functions, Eq. 18. Rapid variation of $C_{\varepsilon\ell}$ as a function of energy will then lead to rapid variations of the matrix element and hence the cross section. Note, however, that such variations depend upon the normalization procedure which is carried out at $r > r_0$ even though most of the matrix element integrand lies at radial distances $r < r_0$. An example of such rapid variations is given in a calculation of the 3d-εf transitions from the M shell of xenon calculated in a HFS potential (10) as shown in the following table.

(Rydbergs)	$\delta_{\varepsilon f}$	C_{3f}	$< R_{3d} \; r \; \bar{R}_{\varepsilon f} >$
0.	.14	53.	$.15 \times 10^{-3}$
0.6	2.2	1530.	.145 "
1.0	2.8	1310.	.142 "
2.0	3.3	908.	.135 "
4.0	3.5	818.	.123 "

Note that the large increase in C_{3f} in the region immediately above threshold is accompanied by a rapid change in phase shift.

For xenon the HFS potential is not strong enough to bind a 4f electron within the region $r < r_0$ due to a small potential barrier in the region $r \cong r_0$ (22). The rapid change given in the table above is due to the penetration of this weak barrier (see Ref. 17, Sec. 4.4 for further discussion).

Actually, the single electron model gives a poor approximation for photoionization cross sections in cases such as that shown above. Rapid variations in the cross sections for np^6 and nd^{10} subshells in the rare gases and neighboring elements predicted by single electron model calculations have been observed experimentally, but the changes with energy are invariably less than those predicted by the calculations. Thus, while the single electron model can be used to predict where such variations may

occur, good agreement with experiment is obtained only by resort-
ing to a many body calculation.

Finally, some of the physical ideas that arise from a study
of the single particle model provide a framework for further
theoretical developments. For example, the representation of a
single electron wave function "outside the atom" by a quantum
defect or phase shift (Eqs. 38 and 39) may be extended to a
number of such single electron channels and the interactions
between them studied. The multi-channel quantum defect theory
(23,24) developed in this way provides a convenient representation of oscillator strengths and parameters for autoionizing
states and photoionization cross sections in the spectral region
near two or more thresholds for photoionization.

As mentioned previously the single particle model provides
a complete set of states which serves as a starting point for the
systematic improvement of calculations via many body techniques.
These procedures ordinarily require radial wave functions over
the entire range of r. An alternative procedure, R matrix
theory (25), takes advantage of the fact that all single electron
radial wave functions may be represented for $r > r_0$ by Eqs. 38
and 39. Hence, a complete set of single particle states need
only be obtained numerically in the region $r < r_0$ using this
procedure.

REFERENCES

The references listed here consist of published papers, review articles and books. Larger references that contain material pertinent to the subjects treated here are marked by an asterisk and those sections that are particularly relevant are indicated in parentheses.

1. E. Fermi, Rev. Mod Phy. $\underline{4}$, 87 (1932).
2*. W. Heitler, "The Quantum Theory of Radiation," Oxford, 1954, third ed. (Secs. 17 and 28).
3*. H. A. Bethe and E. E. Salpeter, "Quantum Mechanics of One and Two Electron Systems," Academic Press, New York, 1957 (Secs. 59, 61, 69, 72).
4. J. H. Scofield, Lawrence Livermore Lab. Rep. UCRL51326 (1973).
5. W. J. Veigele, Atomic Data $\underline{5}$, 51 (1973).
6. J.W. Cooper and S. T. Manson, Phys. Rev. $\underline{177}$, 157 (1969).
7*. D. R. Hartree, "The Calculation of Atomic Structures," Wiley & Sons, New York (1957) (Chaps. 3 and Sec. 6.6).
8. F. Herman and S. Skillman, "Atomic Structure Calculations," Prentice Hall, Englewood Cliffs, New Jersey (1963).
9*. R. H. Pratt, A. Ron and H. K. Tseng, Rev. Mod. Phys. $\underline{45}$, 273 (1973) (Sec. 2.2, 3,6).
10. S. T. Manson and J. W. Cooper, Phys. Rev. $\underline{165}$, 126 (1968).
11. E. J. McGuire, Phys. Rev. $\underline{175}$, 20 (1968).
12. F. Combet Farnoux, J. Phys. (Paris) $\underline{30}$, 521 (1969).
13. D. J. Kennedy and S. T. Manson, Phys. Rev. $\underline{A5}$, 217 (1972).
14. A.E. Boyd, Planetary Space Sci. $\underline{12}$, 769 (1964).
15*. J. W. Cooper, Atomic Inner Shell Processes, Vol. 1, Academic Press, New York, 1975. (Sec. 3.2, 3.3, 3.4, 3.6)
16. A. Dalgarno and W. D. Davison, "Advances in Atomic and Molecular Physics," Vol. 2, Academic Press, New York, 1966.
17*. U. Fano and J. W. Cooper, Rev. Mod. Phys. $\underline{40}$, 441 (1968) (Sec. 2.1, 2.2, 2.4, 2.5, 4.2, 4.5, 4.8, 4.9, 5.1, 5.2, 5.3).
18. H. W. Wolff, K. Radler, B. Sonntag and R. Haensel, Z. Phys. $\underline{257}$, 353 (1972).
19. E. J. McGuire, Sandia Lab. Rep. No. SC-RR-721 (1970).
20. T. N. Chang, Physical Review, to be published.
21. M. J. Seaton, Mon. Not. Roy. Astro. Soc. $\underline{118}$, 504 (1958). A. Burgess and M. J. Seaton, Mon. Not. Roy. Astro. Soc. $\underline{120}$, 121 (1960).
22. J. W. Cooper, Phys. Rev. $\underline{128}$, 681 (1962).
23. M. J. Seaton, Proc. Phys. Soc. $\underline{88}$, 801 (1966).
24. K. T. Lu and U. Fano, Phys. Rev. $\underline{A2}$, 81 (1970).
25. P. G. Burke and W. D. Robb, "Advances in Atomic and Molecular Physics," to be published, 1975.
26. R. D. Hudson and V. L. Carter, J. Opt. Soc. Am. $\underline{57}$, 651, 1471 (1967).

SHAKE THEORY OF MULTIPLE PHOTOEXCITATION PROCESSES

T. Åberg*

Institut du Radium and Université Pierre et Marie Curie

11, rue Pierre et Marie Curie, 75231 Paris Cedex 05, France

1. INTRODUCTION

In this lecture we consider photoionization processes in which the ionization of an electron is accompanied by direct excitation or ionization of additional electrons. The indirect processes of autoionization like the Auger effect which become possible when the photons have enough energy to produce singly-ionized inner-shell hole states above the double ionization threshold are discussed in the second lecture.

The direct multiple photoexcitation processes such as shake-off and -up [1,2,3] are manifestations of electron-electron correlation effects since they cannot be described by the one-electron frozen-core model of the target. This follows from the fact that the photon-electron interaction is described by a one-electron operator. However, multiple photoexcitation is not in general a weak process. In fact, at certain photon energies and in particular cases the probability of multiple processes can even exceed the probability of single ionization [4,5]. Total relative probabilities of the order of 0.20 are not unusual for the excitation and ionization of outer electrons accompanying inner-shell ionization [1,2]. Hence it is important to distinguish between total and single ionization cross sections [6].

Theoretical work on multiple photoexcitation processes involving the ionization of at least one electron has been briefly discussed in Ref. 1 which covers most of the literature up to about 1972. Here we would like to mention subsequent works on this subject. The photoionization of He and H^- has been treated above and close to the $n = 2$ threshold using very accurate ground state and close-coupling final state wave functions [5,7-9]. These calculations have been repeated using Hartree-Fock continuum state wave functions [10,11]. Many-Body-Perturbation Theory has been applied to double photoionization of Ne and He [12,13] and to excitation accompanying ionization of Fe [14].

Total relativistic shake probabilities have been calculated for noble gases [15]. The single-electron cross section has been interpreted in terms of shake probabilities [16] and the influence of initial and final state configuration interaction on shake-up probabilities has been examined [17,18].

The energy and angular distributions of the ejected electrons in the double ionization process of He have been treated by employing Coulomb wave and Green functions in the momentum representation [19]. The electron propagator method has also been applied to excitation accompanying ionization in molecules [20-23]. Shake-up satellites in transition metals have also been analyzed theoretically [24,25].

2. ADIABATIC AND SUDDEN LIMITS IN THE PHOTOIONIZATION PROCESS

As a consequence of the photon-electron interaction there is a dissociation of target atoms into various ionic states and free electrons whose linear momenta $\bar{k}_i$ are measured with a photoelectron spectrometer. The corresponding kinetic energies $\epsilon_i = \frac{1}{2}k_i^2$ fulfil the Einstein relation

$$\epsilon_i = \omega - I_i, \tag{1}$$

where ω is the photon energy and where I_i represents the various possible ionization energies. Hence, before the interaction the electrons are described by a Hamiltonian $H(N)$ and at the time of the detection of a photoelectron by the Hamiltonian $H(N-1)+K(1)$, where $K(1)$ is the kinetic energy operator.

If ω is slightly larger than one of the single-ionization energies I_j the corresponding photoelectron leaves the atom slowly. According to the adiabatic theorem [26] the N-1 electrons will readjust themselves to the Hamiltonian $H(N-1)$ by passing into the eigenstate φ_j which corresponds to the eigenenergy

$$E_j(N-1) = I_j + E_0(N). \tag{2}$$

The photoelectron spectrum will exhibit one peak corresponding to this energy.

If ω is much larger than I_j, then the corresponding photoelectron leaves the ion very quickly. Consequently there is almost a discontinuous change from $H(N)$ to $H(N-1)+K(1)$. Since the N-electron ground state ψ_0 is not an eigenstate of $H(N-1)+K(1)$ the photoelectron spectrum will correspond approximately to the resolution of ψ_0 into these eigenstates[27]. The probability of observing a photoelectron which propagates with the momentum $\bar{k}$ after having left the ion in state φ_n is

$$P_n(\bar{k}) = \frac{\left| \int g_n(X) v_{\bar{k}}^*(X) dX \right|^2}{\iint v_{\bar{k}}^*(X) \gamma(X|X') v_{\bar{k}}(X) dX\, dX'} \qquad (3)$$

according to the sudden approximation [27,28]. The probability $P_n(\bar{k})$ is called the generalized shake probability in the following with the understanding that the probability of no shaking which corresponds to a normal ground state or inner-shell hole configuration of the ion is included. The possible k values are given by Eqs. (1) and (2). The free-electron wave function is denoted by $v_{\bar{k}}(X)$, where $X = (r, \zeta)$ is the position-spin co'ordinate. In Eq. (3)

$$g_n(X) = \sqrt{N} \int \varphi_n^*(X_1 \ldots X_{N-1}) \psi_0(X_1 \ldots X_{N-1} X) dX_1 \ldots dX_{N-1} \quad (4)$$

is the generalized overlap amplitude, and

$$\gamma(X|X') = N \int \psi_0^*(X_1 \ldots X_{N-1} X) \psi_0(X_1 \ldots X_{N-1} X') dX_1 \ldots dX_{N-1} \quad (5)$$

the reduced first-order density matrix of the ground state of the N electron system.

From the considerations given above it follows that in going from the adiabatic to the sudden limit the photoelectron spectrum starts to exhibit new peaks and continua in addition to the adiabatic peak which corresponds to single ionization. The intensity distribution will change as a function of ω but the positions of the individual peaks and continua will remain unchanged. In the sudden limit the relative intensities are given approximately by Eq. (3).

This interpretation of the photoelectron spectrum is based on the approximation that the N-1 electron system is left in stationary states. Note also that the energy conservation requirements (1) and (2) represent additional constraints on the sudden approximation result (3).

Relativistic effects have not been considered and the influence of the incident photon momentum $\bar{k}_i$ has been neglected. If $\bar{k}_i$ is taken into account $-\bar{k}$ in Eq. (3) has to be replaced by the momentum transfer vector $\bar{q} = \bar{k}_i - \bar{k}$. Hence the photoionization probability is determined in the limit of large q $\cong$ k by the absolute value of the generalized overlap amplitude in the momentum space at $\bar{q}$. To the extent the Compton electron can be represented by a plane wave the same quantity determines the Compton scattering cross section but evaluated at $\bar{q} = \bar{k}_i - \bar{k}_f - \bar{k}$ since the momentum $\bar{k}_f$ of the scattered photon has to be taken into account. However, note that because ω is usually much larger than any binding energies in a Compton scattering experiment many excited states are involved in addition to the ionic ground state. Consequently, the total cross section is essentially determined by the reduced first-order density matrix (5) in the momentum space [29] . Within the framework of the impulse approximation the (e, 2e) experiment is also described by the momentum representation of the generalized overlap amplitude. In this case the momentum transfer vector is $\bar{q} = \bar{k}_0 - \bar{k}_1 - \bar{k}_2$, where $\bar{k}_0$ is the momentum of the incident electron and $\bar{k}_i$ (i = 1, 2) the momenta of the ejected electrons [30] .

3. RELATIONSHIP BETWEEN THE GENERALIZED SHAKE PROBABILITY
AND THE DIFFERENTIAL PHOTOIONIZATION CROSS SECTION

For an incoming photon beam which is linearly polarized in the direction $\bar{e}$ the differential photoionization cross section is given by

$$\frac{d\sigma}{d\Omega} = \frac{4\pi^2 \alpha \, a_0^2}{\omega} \left| \langle \psi | \sum_{j=1}^{N} \, (\bar{p}_j \cdot \bar{e}) \, e^{i\bar{k} \cdot \bar{r}_j} \, | \psi_0 \rangle \right|^2 , \quad (6)$$

where the final state wave function ψ , normalized per unit energy, represents at least one outgoing electron in the direction of $\bar{k}$. Asymptotically, as k becomes large, ψ approaches an antisymmetric product of a plane wave $v_{\bar{k}}^-$ and an ionic state wave function φ_n

$$\psi \longrightarrow \sqrt{\frac{1}{N}} \sum_{j=1}^{N} \, (-1)^j \, v_{\bar{k}}^- (X_j) \, \varphi_n (X_1 \ldots X_{j-1} X_{j+1} \ldots X_N).(7)$$

By substituting this expression in Eq. (6) it can be shown that the leading term of $d\sigma/d\Omega$ in powers of k is

$$\frac{d\sigma}{d\Omega} \longrightarrow \frac{4\pi^2 \alpha \, a_0^2 (\hat{k}\cdot\bar{e})}{\omega} \left| \int g_n (X) \, v_{\bar{q}} (X) \, dX \right|^2 \qquad (8)$$

which is in accordance with the results of Sec. 2 [27,22] .

Since large k values ($\omega \cong \frac{1}{2}k^2$) correspond to small r values the behaviour of $g_n (X)$ close to the nucleus determines the asymptotic energy dependence of $d\sigma/d\Omega$. If the lowest-order radial part of $g_n (X)$ behaves as $Nr^\ell (1+ar)$ for small r, then

$$\frac{d\sigma}{d\Omega} = C(\hat{k})\omega^{-\ell - 7/2}, \qquad (9)$$

where the energy independent coefficient $C(\hat{k})$ is proportional to $N^2 a^2$. This result follows from Eq. (8) using the spherical wave expansions of $g_n (X)$ and $v_{\bar{q}} (X)$ and properties of the spherical Bessel functions [31] . In the one-electron frozen core model $g_n (X)$ reduces to the spin orbital u_λ which is vacated in the photoionization process. The central field potential gives $a = - Z/(\ell +1)$, where ℓ is the angular momentum quantum number of u_λ . Consequently, different central-field approximations yield high-energy cross sections which differ only by the normalization factor N^2. The normalization screening theory [32] in which a reliable one-electron high-energy ($\omega > 10$ KeV) cross sections is obtained from the hydrogenic cross section by scaling is essentially based on this fact.

Note that it is not always possible to conclude that the asymptotic cross section (9) of the subshell ($n\ell_0$) is associated with $\ell = \ell_0$ [19] . For example, the $(n\ell_0)^x \, {}^1S \longrightarrow (n\ell_0)^{x-2} \varepsilon_1\ell_1 \varepsilon_2\ell_2 {}^1P$ double ionization cross section $\left(x = 2(2\ell_0 + 1)\right)$ behaves at least in principle as $\omega^{-7/2}$ in the high-energy limit since there is a configuration mixing between the $(ns)(n\ell_0)^x$ and $(ns)^2 (n\ell_0)^{x-2}\varepsilon_1\ell_1$ configurations of the residual ion. On the other hand it is expected that the $(n\ell_0)^x \, {}^1S \longrightarrow (n\ell_0)^{x-1}\varepsilon\ell \, {}^1P$ single-ionization cross section behaves as $\omega^{-\ell_0 -7/2}$ since there is no configuration mixing between the $(ns)(n\ell_0)^x$ and $(ns)^2 (n\ell_0)^{x-1}$ configurations due to parity and symmetry reasons.

In the case of He and He-like ions the generalized shake probability (3) approaches asymptotically a $\bar{k}$ independent expression [1] which is related to $N^2 a^2$ in Eq. (9). This expression can be cast in an alternative form [33] by using a cusp condition at the nucleus [1]. Both forms have been used to demonstrate the sensitivity of the multiple excitation cross sections $\sigma(\varepsilon's, \varepsilon p)$ and $\sigma(n s, \varepsilon p)$ to radial correlation effects in the initial state and to test the accuracy of various initial state wave functions [34,35] .

The convergence of $\sigma(\varepsilon's, \varepsilon p)$ towards its high-energy limit seems to be slower than that of $\sigma(n s, \varepsilon p)$. For example, in He at an ω value which is about ten times the threshold, $\sigma(2s, 2p)/\sigma_t$ is only about 15% higher than the asymptotic value [10] whereas $\sigma(\varepsilon's, \varepsilon p)/\sigma_t$ is still about twice as large as the limit [1] .

4. SHAKE PROBABILITY FOR DEEP INNER-SHELL IONIZATION AND INTERPRETATION OF KOOPMANS' THEOREM

In Sec. 3 we considered the differential photoionization cross section and showed how it is related to the generalized shake probability. Some simplifications are possible if it assumed that the electron is ejected from a deep inner subshell $\lambda = (n_0 \ell_0)$ of a closed shell system [27] .

We assume that the correlation between an electron in the subshell and the rest of the electrons can be neglected. Consequently, the initial state wave function ψ_0 is given by

$$\psi_0 = \sqrt{\frac{1}{N}} \sum_{j=1}^{N} (-1)^j u_\lambda(X_j) \, X(X_1 \ldots X_{j-1} X_{j+1} \ldots X_N) \quad (10)$$

where u_λ is one of the spin orbitals λ and where X is the corresponding antisymmetrized $^2\ell_0$ wave function of the rest of the electrons. Note that X may include correlation among the electrons outside λ. In fact, if ψ_0 is expanded in a complete set of Slater determinants all determinants which contain the set $\{u_\lambda\}$ can be included in the approximate wave function (10).

In the total cross section the final state wave function is a $^1P^0$ wave function which is a linear combination of wave functions (7), where $v_{\vec{k}}^-$ is replaced by a central field one-electron continuum wave function $v_{\varepsilon m_\ell m_s}(X_j)$. The coefficients are the appropriate Clebsch-Gordon coefficients. It is possible that several functions φ_n belong to a given excited configuration and a given set of the quantum numbers ($LSM_L M_S$). They are distinguished in the following by the superscript Γ. Note that a φ_n^Γ is in principle an " exact " wave function of the Hamiltonian H(N-1). The wave function φ_0 corresponds to the normal configuration $(n_0 \ell_0)^{-1}$.

If it is assumed that $\langle u_\lambda | \varphi_n^\Gamma \rangle = 0$ in addition to $\langle u_\lambda | \chi \rangle = 0$, $\langle u_\lambda | u_\lambda \rangle = \langle \chi | \chi \rangle = 1$, then in the high-energy limit

$$\sigma_n^\Gamma (\omega) = \frac{8 \pi^2 \alpha \, a_0^2}{3} \, \omega \times D_1 (\varepsilon \ell, n_0 \ell_0)^2 \left| \langle \varphi_n^\Gamma | \chi \rangle \right|^2, \quad (11)$$

where $x = \ell_0 + 1$ or ℓ_0 $(\ell_0 > 0)$ depending on whether $\ell = \ell_0 + 1$ or $\ell_0 - 1$. The dipole matrix element is obtained from

$$D_j (\varepsilon \ell, n_0 \ell_0) = \int_0^\infty R_{\varepsilon \ell} (r) R_{n_0 \ell_0} (r) r^{j+2} dr, \quad (12)$$

where $R_{\varepsilon \ell}$ is the appropriate final state radial wave function corresponding to φ_n^Γ and $\varepsilon = \omega - I_n^\Gamma$. The monopole selection rules require that φ_n^Γ belongs to $L = \ell_0$, $S = 1/2$. The amplitude $\langle \varphi_n^\Gamma | \chi \rangle$ squared which is independent of the quantum numbers M_L and M_S represents the shake (-up or -off) probability. This probability can be easily obtained by the sudden approximation if the passive electrons are assumed to be frozen in χ during the sudden creation of a hole in the subshell λ. However, the requirement of a sudden event is probably unnecessary stringent if φ_n^Γ refers to a bound state or to a continuum state with a slowly outgoing electron, since the general cross section reduces to the expression (11) if $\langle v | \chi \rangle = 0$ in addition to $\langle u_\lambda | \varphi_n^\Gamma \rangle = 0$ [27]. If these orthogonality constraints are not exactly fulfilled the cross section corresponding to the φ_n^Γ with $L \neq \ell_0$ does not necessarily vanish and certain exchange terms appear also in the cross section for $L = \ell_0$ [27, 17].

If both φ_n^Γ and χ are described by the independent electron model i.e. by Slater determinants that correspond to given initial and final configurations then the probability reduces to the conventional shake-probability which can be expressed in terms of monopole matrix elements $D_0 (n_2 \ell, n_1 \ell)$ [27]. With regard to applications, excitation and ionization following K shell ionization have been treated both within the Hartree-Fock (Δ E_{SCF}) model [1, 15] and within the model which includes initial and final state correlation in the ionic shake-up states φ_n^Γ [27, 28].

From the behaviour of the monopole and dipole matrix elements as a function of energy it is clear that the most important cross sections σ_n^Γ correspond to single ionization and to ionization accompanied by shake-up and shake-off of slow electrons [1]. If ω is large, then $D_1 (\varepsilon \ell, n_0 \ell_0)$

should be approximately independent of n and Γ. Consequently, the summation over n and Γ yields

$$\sigma_t = \sigma_0 + \sum_{n=1} \sum_{\Gamma} \sigma_n^{\Gamma} = \frac{8\pi^2 \alpha\, a_0^2}{3}\, \omega \, \times \, D_1 (\varepsilon\ell, n_0 \ell_0)^2 \quad (12)$$

According to this result the one-electron cross section which is given by the frozen core model actually corresponds to the total cross section σ_t and not to the single ionization cross section σ_0 [16] .

The results of this section also offer an analogous interpretation of approximate inner-shell ionization energies which are based on the frozen core model like the eigenenergies of the Hartree-Fock equation (Koopmans' theorem). The frozen-core ionization energy

$$I_\lambda = \langle X \,|\, H(N-1) \,|\, X \rangle - \langle \psi_0 |\, H(N) \,|\, \psi_0 \rangle \quad (13)$$

of an electron in the subshell λ is an approximation of I_0 corresponding to φ_0. The definition

$$I_n^{\Gamma} = \langle \varphi_n^{\Gamma} \,|\, H(N-1) \,|\, \varphi_n^{\Gamma} \rangle - \langle \psi_0 |\, H(N) \,|\, \psi_0 \rangle \quad (14)$$

and the substitution of the expansion

$$X = \langle \varphi_0 \,|\, X \rangle \, \varphi_0 + \sum_{n,\Gamma} \langle \varphi_n^{\Gamma} |\, X \rangle \, \varphi_n^{\Gamma} \quad (15)$$

in Eq. (13) yield

$$I_\lambda = I_0 + \sum_{n,\Gamma} |\langle \varphi_n^{\Gamma} |\, X \rangle|^2 \, (I_n^{\Gamma} - I_0) \quad (16)$$

which shows that I_λ is the weighted average of the actual ionization energies [36] . The weights are the shake-up and -off probabilities (including the probability $|\langle \varphi_0 | X \rangle|^2$ of no shaking). Hence it is more appropriate to compare I_λ with the mean energy weighted by the intensity distribution of the whole spectrum from λ than with the energy of the adiabatic peak which corresponds to the single ionization energy I_0.

In the case, where I_λ is given by the unrestricted Hartree-Fock eigen-energy ($\mathcal{X}$ is a single determinant), the lowest order perturbation contributions to $I_\lambda - I_o$ have been determined [28]. A perturbation analysis of the unrestricted form of the sum in Eq. (16) shows that it gives rise to a similar lowest-order relaxation correction which corresponds to the conventional shake picture. However, there are two additional terms in the lowest-order perturbation expansion which are not fully accounted for by Eq. (16). These terms represent the pair correlation which vanishes upon the ionization of a λ electron and the change of the pair correlations of the rest of the electrons. Hence the shake approach discussed in this section is limited to cases where the relaxation is more important than the changes of the pair correlations among the electrons.

5. SUMMARY

The photoelectron spectrum has been analyzed in terms of the adiabatic and sudden limit. To the extent the photoionization and the subsequent decay processes are completely separable the spectrum reflects the statistical distribution of various ionic states which are created by the photon-electron interaction. In the sudden limit this distribution is determined by the generalized overlap amplitude between the various ionic states and the ground state. This probability is related to the high-energy limit of the photoionization cross section which behaves as $\omega^{-\ell-1/2}$ if the lowest order radial part of the corresponding overlap amplitude goes to zero with r as r^ℓ. There is also a close relationship between the generalized shake probability and the cross section of Compton scattering and (e, 2e) reactions.

At sufficiently high energies the cross section of deep inner-shell photoionization can be approximated by a product of a $\bar{k}$ independent shake (or no-shake) probability and a single-ionization cross section. The conventional one-electron monopole shake probability is a special case of the $\bar{k}$ independent shake probability. The factorization of the photoionization cross section leads to a physical interpretation of the frozen-core single-ionization cross sections and ionization energies in terms of the shake probabilities.

ACKNOWLEDGEMENTS

I am indebted to Dr. M. Inokuti for his enlightening comments on the shake concept and to Dr. R.L. Martin for communication of his encouraging results in advance of publication. I would also like to thank

Professor J.P. Briand and his colleagues for their kind hospitality at
Institut du Radium, where the writing of my lectures was accomplished.

REFERENCES

*Permanent address : Laboratory of Physics, Helsinki University of
Technology, 02150 Espoo 15, Finland.

1. T. Åberg, in Inner-Shell Ionization Phenomena and Future Applications,
   eds. R.W. Fink, S.T. Manson, J.M. Palms, and P.V. Rao, U.S.
   Atomic Energy Comm. Rep. Nr. CONF-720404, 1509 (1973).
2. M.O. Krause, in Inner-Shell Ionization Phenomena and Future Applications, eds. R.W. Fink, S.T. Manson, J.M. Palms, and P.V. Rao,
   U.S. Atomic Energy Comm. Rep. Nr. CONF-720404, 1586 (1973).
3. T.A. Carlson, in The Physics of Electronic and Atomic Collisions,
   eds. B.C. Cobić and M.V. Kurepa, (Institute of Physics, Beograd,
   1973), 205.
4. T. Åberg, in X-ray Spectra and Electronic Structure of Matter eds.
   A. Faessler and G. Wiech (München, 1973), 1.
5. H.A. Hyman, V.L. Jacobs and P.G. Burke, J. Phys. B $\underline{5}$, 2282 (1972).
6. F. Wuilleumier and M.O. Krause, Phys. Rev. A $\underline{10}$, 242 (1974).
7. V. Jacobs, Phys. Rev. A $\underline{3}$, 289 (1971).
8. V. Jacobs and P.G. Burke, J. Phys. B $\underline{5}$ L67 (1972).
9. V. Jacobs and P.G. Burke, J. Phys. B $\underline{5}$ 2272 (1972).
10. K.L. Bell, A.E. Kingston and I.R. Taylor , J. Phys. B $\underline{6}$, 1237 (1973).
11. K.L. Bell, A.E. Kingston and I.R. Taylor, J. Phys. B $\underline{6}$, 2271 (1973).
12. T.N. Chang and R.P. Poe, Phys. Rev. A $\underline{12}$, 1432 (1975).
13. B. Talukdar and M. Chatterji, Phys. Rev. A $\underline{11}$, 2214 (1975).
14. H.P. Kelly, Phys. Rev. A $\underline{6}$, 1048 (1972).
15. T.A. Carlson and J.W. Nestor, Jr, Phys. Rev. A $\underline{8}$, 2887 (1973).
16. C.S. Fadley, in Electron Spectroscopy, eds. R. Caudano and J. Verbist
    (Elsevier, Amsterdam 1974), 895 ; C.S. Fadley, Chem. Phys. Lett.
    $\underline{25}$, 225 (1974).
17. R.L. Martin and D.A. Shirley, Lawrence Berkeley Laboratory, Report
    No LBL-3468, 1974 and to be published.
18. R.L. Martin, B.E. Mills, and D.A. Shirley, Lawrence Berkeley Labo-
    ratory, Report No LBL-3469, 1975 and to be published.
19. M.Y. Amusia, E.G. Drukarev, V.G. Gorhkov and M.P. Kazachkov,
    J. Phys. B $\underline{8}$, 1248 (1975).
20. G.D. Purvis and Y. Öhrn, J. Chem. Phys. $\underline{60}$, 4063 (1974).
21. L.S. Cederbaum, Molec. Phys. $\underline{28}$, 479 (1974).
22. G.D Purvis and Y. Öhrn, J. Chem. Phys. $\underline{62}$, 2045 (1975).
23. L.S. Cederbaum, J. Chem. Phys. $\underline{62}$, 2160 (1975).
24. T. Robert and G. Offergeld, Chem. Phys. Lett. $\underline{29}$, 606 (1974).

25. S. Larsson, Chem. Phys. Lett. $\underline{32}$, 401 (1975).

26. A. Messiah, Quantum Mechanics Vol.II , (North-Holland, 1972) 740.

27. T. Åberg, Ann. Acad. Sci. Fenn. AVI, $\underline{308}$, 1 (1969).

28. B.T. Pickup and O. Goscinski, Molec. Phys. $\underline{26}$, 1013 (1973).

29. e.g. P. Eisenberger and P.M. Platzman, Phys. Rev. A $\underline{2}$, 414 (1970).

30. S.T. Hood, I.E. McCarthy, P.J.O. Teubner and E. Weigold, Phys. Rev. A $\underline{8}$, 2494 (1973) ; S.T. Hood, I.E. McCarthy, P.J.O. Teubner and E. Weigold, Phys. Rev. A $\underline{9}$, 260 (1974).

31. A.R.P. Rau and U. Fano, Phys. Rev. $\underline{162}$, 68 (1967).

32. R.H. Pratt, A. Ron and H.K. Tseng, Rev. Mod. Phys. $\underline{45}$, 272 (1973).

33. K. Kabir and E.E. Salpeter, Phys. Rev. $\underline{108}$, 1256 (1957) ; E.E. Salpeter and H.H. Zaidi, Phys. Rev. $\underline{125}$, 248 (1962).

34. T. Åberg, Phys. Rev. A $\underline{2}$, 1726 (1970).

35. J.J. Wendoloski and G.A. Petersson, J. Chem. Phys. $\underline{62}$, 1016 (1975).

36. This sum rule has been discussed in the literature in different contexts : P. Nozières, Theory of Interacting Fermi Systems (Benjamin, 1964) p 350 ; B.I. Lundqvist, Phys. Kond. Materie 9, 236 (1969) ; D.C. Langreth, Phys. Rev. B $\underline{1}$, 471 (1970) ; R. Manne and T. Åberg, Chem. Phys. Lett. $\underline{7}$, 282 (1970) ; H.W. Meldner and J.D. Perez, Phys. Rev. A $\underline{4}$, 1388 (1971).

THE RANDOM PHASE APPROXIMATION WITH EXCHANGE

Göran Wendin

Institute of Theoretical Physics

Fack, S-402 20 GÖTEBORG 5, Sweden

1. INTRODUCTION

1.1. General Philosophy

A suitable subtitle for these lectures would be "Many-electron theory of photoionization in atoms" in order to indicate that the random phase approximation with exchange (RPAE) is only one out of several possible approximations and also that we shall focus attention on continuum photoionization processes rather than on transitions to bound states. A continuum photoionization cross section studied over a large range of energies gives information about the dynamical behaviour of the atomic electrons and theoretical calculations have to account for the whole range between very slow and very fast photoelectrons. The purpose of these lectures is to discuss the applicability of the RPAE in relation to the type of atom, the type of transition and the photoelectron energy considered. We shall also see examples of how the RPAE can be improved by inclusion of double excitations accounting for relaxation of the frozen core, shake off and Auger transitions.

The main part of these lectures will make use of a simple diagrammatic representation of the many-electron correlations in the photoionization process, in particular those corresponding to the RPAE. However, one can easily formulate the RPAE within the framework of ordinary time-dependent atomic theory. In section 1.2 I shall briefly discuss the <u>linearized time-dependent Hartree-Fock equations</u> which are in fact identical to the RPAE equations. The advantage with the former approach is that the linearized TDHF has a very direct and physical interpretation in terms of mean field theory. The name random phase approximation (RPA) was originally introduced by Bohm and Pines [1] in their original discussions of the properties of a gas of electrons. However, this name can be misleading since it tends to obscure the fact that the RPA deals with the <u>average response</u> to an applied perturbation, i.e. the <u>mean field</u>. Furthermore it must be pointed out from the start and often repeated that it is wrong to think that the RPAE is a theory for an electron plasma and as such very suspect when it comes to treating atomic systems. The RPAE simply represents a certain level of approximation of the time-dependent Schrödinger equation and the physics of the approximation depends on which type of physical system one applies it to. In an atomic case, for instance, one uses atomic basis wave functions and the RPAE is then an approximation within ordinary atomic theory. Only when the RPAE is explicitly applied to a gas of free electrons with plane-wave basis functions does one get the plasma theory, leading to collective plasma oscillations. This is very important to remember later when we actually talk about collective resonances in atoms, namely that this atomic collective behaviour really is an <u>atomic</u> property, having nothing to do with plasma models for atomic systems.

The present lectures will be limited in scope in the sense that we discuss the physics of the RPAE and extensions of the RPAE with explicit application only to atomic Ba and La. For a broader review of methods, results and comparison with experiments I refer to refs. [2,3,4,5] and references therein.

1.2. Linearized Time-Dependent Hartree Approximation (Linearized TDH)

In order to make the physics as clear and simple as possible we shall start from the Hartree equations instead of the Hartree-Fock equations, only considering a rough outline of the theory (for details, see [6,7]. By straightforward generalization of the stationary Hartree equations one arrives at the time-dependent Hartree equations

$$\left[T + V_H(t) \right] u_i(rt) = i \frac{\partial}{\partial t} u_i(rt) \tag{1.1}$$

$$V_H(t) = \sum_{j \neq i}^{N} \int \frac{|u_i(r't)|^2}{|r-r'|} d^3r' \tag{1.2}$$

Here T denotes the kinetic energy plus the bare nuclear potential for one electron and V_H is the average electrostatic potential from the charge cloud of the <u>N-1 other electrons</u>, i.e. the Hartree potential. In the absence of any time-dependent external driving field the time-dependence of $V_H(t)$ solely originates from time-dependent changes in the electronic density $|u_i(r't)|^2$. Therefore, in this model the time evolution of a given orbital i only depends on the mean instantaneous Coulomb potential of all the other electrons.

We want to study the genuinely time-dependent solutions where the electronic charge density $|u_i(rt)|^2$ deviates from the stationary Hartree value. A standard method is then to displace the constituents of the system to non-equilibrium positions and then let the system develop in time. To study the eigenfrequency spectrum of the TDH eqs.(1.1) and (1.2) we therefore construct small deformations of the one-electron Hartree orbitals via hybridization, mixing in small amounts of excited one-electron orbitals (cf. hybridization and directed bonds in chemistry). This is contained in the following general statement

$$\text{Deformation of the wave} \atop \text{function, charge density} \Bigg\} \Longleftrightarrow \Bigg\{ \text{Creation of electron-hole pairs} \atop \text{in the corresponding orbital(s)}$$

In order to study how perturbation of the charge density will develop in time we therefore construct truly time-dependent hybrids

$$U_i(rt) = \left[U_i(r) + e^{-i\omega t} \sum_m x_{mi}\, U_m(r) + e^{+i\omega t} \sum_m y^*_{mi}\, U_m(r) \right] e^{-i\xi_i t} \quad (1.3)$$

Eq.(1.3) describes how the i-th Hartree orbital $u_i(r)$ oscillates with frequency ω due to admixture of excited states m. Introducing eq.(1.3) into the TDH eqs.(1.1) and (1.2), keeping terms linear in x_m and y_m (small amplitudes!) one can derive the linearized TDH equations

$$(\varepsilon_n - \varepsilon_i - \omega) x_{ni} + \sum_m \langle jn|v|mi \rangle x_{mi} + \sum_m \langle nm|v|ij \rangle y_{mi} = 0$$

$$(\varepsilon_n - \varepsilon_i + \omega) y^*_{ni} + \sum_m \langle mi|v|jn \rangle y^*_{mi} + \sum_m \langle ij|v|nm \rangle x^*_{mi} = 0$$

$$(1.4)$$

These equations are usually written on a compact matrix form

$$\begin{pmatrix} A & B \\ B^* & A^* \end{pmatrix} \begin{pmatrix} X \\ Y \end{pmatrix} = \omega \begin{pmatrix} X \\ -Y \end{pmatrix}$$

$$(X)_{mi} = x_{mi} \; ; \; (Y)_{mi} = y_{mi}$$

$$A_{ni,\,mj} = \delta_{mn}\,\delta_{ij}\,(\varepsilon_n - \varepsilon_i) + \langle jn|v|mi \rangle$$
$$B_{ni,\,mj} = \langle nm|v|ij \rangle$$

The coupling matrix elements in (1.5b) have the diagrammatic form

$$\langle jn|v|mi \rangle = \underset{j\;\;m}{\overset{i\;\;n}{\diagup\!\!\!\diagdown}} \quad ; \quad \langle nm|v|ij \rangle = \overset{i\;\;n}{\bigvee} \quad (1.6)$$

so that the interaction part of the matrix (1.5a) takes the form

$$(1.7)$$

This demonstrates that the linearized TDH ($\cong$ RPAE) involves a very special type of electron-electron interaction which is basically dipole-dipole interaction between charge density displacements. When one calculates the response of the system to an external field, e.g. photons, in principle one has to invert the matrix and consequently one could think of the RPAE as an infinite perturbation expansion in the matrix elements in (1.6). This means hooking together the matrix elements in (1.7) in all possible ways, generating the well known RPAE diagrams consisting of strings of "bubbles", the so called "bubble diagrams" (see fig.2).

In a completely consistent scheme one should actually start from the <u>time-dependent Hartree-Fock</u> equations. This introduces into eqs. (1.6) and (1.7) another type of matrix element describing electron-hole scattering (see e.g. eq.2.5b). Since this matrix element represents the <u>exchange</u> of the bubble matrix element one finds that there is complete equivalence between the linearized time-dependent Hartree-Fock and the RPAE. Equations (1.4)-(1.7) can also be derived via linearization of the Heisenberg equation of motion for the density matrix. There is a substantional amount of work along these lines going on at present and I refer to refs. [8,9,10,11,12] for further reading.

2. GENERAL DISCUSSION OF PHOTOIONIZATION

2.1. Single excitations

Below are shown two equivalent pictures of the fundamental one-electron photoionization process in a frozen core approximation:

(i) <u>"Ordinary" description in real space</u>: The photon kicks one electron into an excited orbital n, leaving a hole in orbital i. The electron-hole pair has an excitation energy $\omega_n = \varepsilon_n - \varepsilon_i$ and carries a dipole moment

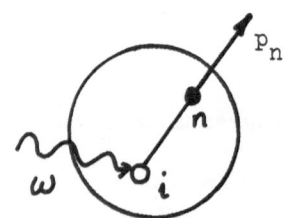

$$p_n = \langle i|z|n \rangle_r = \int P_i(r) r P_n(r) dr \qquad (2.1)$$

Due to energy conservation the photoelectron must have kinetic energy such that $\omega_n = \omega$. This value of ω_n we somewhat sloppily denote by ω.

(ii) <u>The corresponding Goldstone diagram</u> (see Kelly's lectures):

Energy conservation

$$\delta(\omega_n - \omega)$$
$$\Rightarrow \omega_n = \omega \qquad \Rightarrow$$

In this order, the excitation amplitude is nothing but the ordinary dipole matrix element

$$D(\omega) = p_\omega = \int P_i(r)rP_n(r)dr \Big|_{\omega_n \neq \omega} \tag{2.2}$$

From this one gets the photoionization cross section (for details
see ref. [13] and the lecture by J.W. Cooper)

$$\sigma(\omega) \sim \omega \, |D(\omega)|^2 \tag{2.3}$$

The actual choice of zeroth-order one-electron wave functions
is largely a matter of philosophy, being a question of whether one
wants to include many-electron effects through

(i) frequency dependent corrections to the unperturbed excitation amplitude that <u>explicitly</u> reveal the dynamics of the system
(e.g. via resonant energy denominators). One may then have to go to
infinite order in the perturbation expansion.

(ii) improved wave functions that <u>implicitly</u> describe some of
the dynamical effects. One can then often manage with low order perturbation theory.

The photoionization process can conveniently be described in
terms of a dipole that grows in time as illustrated in fig.1. The
originally created dipole will polarize the electronic density of
the atom, inducing additional dipole moments in the medium (i.e. displacements of the other orbitals; cf. linearized TDH(F)). The induced
dipoles will act back, modifying the potential seen by the original
excitation and the whole process should be considered in a self-consistent manner. In particular, the RPAE will consider the size of
the HF ground state orbitals to be unchanged (frozen core approximation) and will only consider the change in the mean field due to
displacements of the ground state HF orbitals with respect to the
nucleus. The SCF-character can be accomplished in two ways:

(i) Infinite order perturbation theory. This gives dynamical
corrections in explicit form.

(ii) SCF wave functions. Much of the dynamics is now implicit.
Small or moderate corrections.

Let us now explicitly discuss possible one-electron basis wave
functions for closed shell atoms. The ground state then is singlet S
and, considering only internal Coulomb interaction, dipole excita-

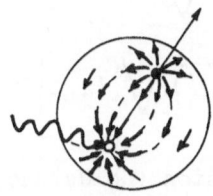

Fig.1. Classical picture of the dipole field associated with
photoionization of an atomic electron. The big circle symbolizes
the atomic ground state radius, i.e. the picture is in real space.

tion will lead to a singlet P final state. A convenient single particle basis can be chosen in the following way:

    1. Ground state: One-electron HF orbitals.

    2. Excited states: Hartree-Fock type of potential built from
the frozen HF ground state with a hole in the initial state orbital.

$\Longrightarrow$ V^{N-1} ionic potential

However, this does not suffice to define the potential, one has to
give further specifications. There are typically two cases, namely
V^{N-1} HF$_{av}$ [14,15,16,17] and V^{N-1} HF ^{S}L [2,4,15,18,19,20,21,22,23]

    (i) The V^{N-1} HF$_{av}$ potential: One takes a spherical average of
the potential (or equivalently, averages over all possible m_l-values
of the hole. If one also averages over m_s-values one gets the configuration average potential, the difference being not very important).
The excited orbitals in this potential are then restricted HF$_{av}$ orbitals and the basis is called φ^N by Amusia and coworkers [2,15] (N,
because the excitation does not change the total number of electrons).
The potential is central and non-local and the excited state wave
functions are characterized by the quantum numbers nlm_lm_s, e.g.$4dm_lm_s$.

    Physically, this basis implies that the excitation of an elec-
tron-hole pair does not induce any response in the rest of the system. Therefore, all effects of dynamical interaction (intra- and interchannel, shake up/off, Auger processes etc.) are transferred to
the perturbation expansion and have to be accounted for explicitly.
In particular, one has to study the mean field response by perturbation expansion resulting in the RPAE bubble diagrams in fig.2 (the
exchange diagrams have not been drawn but are assumed to be included).

    (ii) The V^{N-1} HF ^{S}L potential: One no longer considers the elec-
tron and the hole separately. Instead one mixes the m_l and m_s values
of the whole configuration to eigenstates of total L and S. Taking
the $4d\text{-}nf$ ^{1}P excitation as an example one then considers the $4d^9nf$
configuration and the mixing gives

Fig.2. The RPAE "bubble" expansion (Goldstone time-ordered diagrams,
time axis vertical; inclusion of exchange understood). TDAE stands for
Tamm-Dancoff Approximation with Exchange [7] and involves interactions
induced by the external perturbation. The correlation part involves
pair fluctuations in the unexcited sea of electrons (ground state
fluctuations) and coupling to externally excited electron-hole pairs.

$$4d^9 nf \left\{ m_l \; m_s \right\} \quad \Longrightarrow \quad 4d^9 nf \; {}^{S}L \; ({}^3P, {}^3D, {}^1P)$$

The physical picture is the following: As before the photon creates an electron-hole pair but now we demand that the remaining 4d-electrons should displace themselves self-consistently in response to the excitation so that the overall symmetry be ^1P (the dipole polarizability is the important thing; cf. also the variational method of the TDHF). It turns out that this extremely important effect can be included without much effort, simply by changing the numerical coefficient in front of the exchange potential in the one-electron HF-equation which determined the φ^N basis. Considering the expectation value of the exchange term one gets [17,21]

$$
\begin{array}{lll}
V^{N-1} \; HF_{av}: & -3/7 \; R^1(4dnf;nf4d) & \text{(a)} \\
V^{N-1} \; HF \; {}^1P: & +11/7 \quad -\text{''}- & \text{(b)}
\end{array}
\qquad (2.4)
$$

The V^{N-1} HF ^1P (and generally ^{S}L) basis is called $\varphi^{N(LS)}$ by Amusia and coworkers [2,15] . Since the two potentials in eq.(2.4) only differ by a numerical coefficient both the average and the LS-potential must be central potentials. My personal picture of the dipole excitation is shown in the figure below, suggesting an analogy with the

 problem of pulling a body out of an "elastic" liquid. The excitation breaks the original spherical symmetry and the medium rearranges itself self-consistently in accordance with the new dipole symmetry. This SCF-rearrangement is now taken into account in the $\varphi^{N(LS)}$ wave function for excited states and as a consequence there is no longer any direct 4d-nf intrachannel interaction. Expressed in terms of diagrams, the wave function includes the effects of the bubble and exchange diagrams

$$
\text{(a)} \qquad \qquad ; \qquad \text{(b)} \qquad \qquad \qquad (2.5)
$$

The infinite series of diagrams built on (2.5) will propagate strictly forwards (or backwards) and is called the Tamm-Dancoff Approximation with Exchange (TDAE) as shown in fig.2. It follows that the 4d-nf HF ^1P basis automatically sums the infinite TDAE series of 4d-nf transitions and left in the RPAE are only the so called (ground state) correlation diagrams which essentially describe the pair correlations in the HF ground state in the absence of any external perturbation.

2.2. Beyond the RPAE: Double Excitation, Relaxation

In fig.3 is shown one connection between the electronic rearrangement around the core-hole and the photoelectron and elementary
diagrammatic processes. Here we have omitted the dipole-dipole correlations of the RPAE-type discussed in section 2.1 and rather consider
the electron and the hole to be so far apart that they act as separate charges with negligible effects of dipole-dipole interaction.
Fig.3a then describes how the core-hole attracts surrounding electrons and how the photoelectron repels them. This means that the electronic density will be distorted around the hole and the electron.
Recalling that the displacement of charge can be expanded in terms
of electron-hole excitations it follows that the rearrangement processes can be described in terms of the core-hole and the photoelectron creating secondary electron-hole pairs as shown in fig.3b,c.
Combining these processes with the single electron excitation induced
by the photon we get the double excitation/ionization diagrams shown
in fig.3d,e (there are also additional processes [39]).

Usually, however, the most important process is that the electronic medium quietly relaxes around the core-hole and the photoelectron, transferring the relaxation energy to the photoelectron. The
end product (before possible Auger decay later on) then is a single
electron-hole pair, meaning that the double excitation processes in
fig.3d,e have to be virtual (non-energy conserving). The corresponding digrams for single ionization with relaxation are shown in
fig.7 (see also e.g. [40,41]).

It should be noted that the close connection between relaxation
and multiple excitation/ionization leads to very important relations
(sum rules) between the strength and position of the relaxed core-
level and the intensity of the multiple excitation spectrum (see e.g.
[5,36,37] and the lectures by T. Åberg and M.O. Krause).

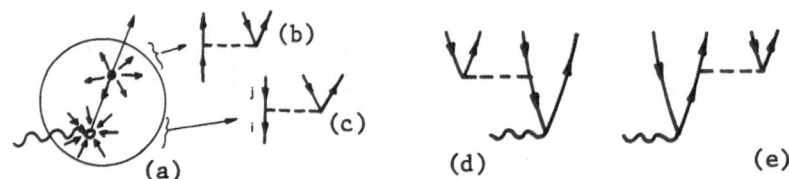

Fig.3. Pictures of the double excitation mechanism. (a) Physical
picture of rearrangement in response to an electron-hole excitation.
(b,c) Corresponding elementary diagrammatic processes. In (c) i=j describes shake up/off due to the positive charge of the core-hole;
i≠j describes Auger types of processes. (d) Photoionization with
shake up, Auger. (e) Photoionization with inelastic scattering of
the photoelectron.

2.3. When is the RPAE Important [4] ?

(i) <u>Deep inner shells</u>: <u>The RPAE is not important</u>. The photoelectron leaves the region of the core-hole very fast, having very small
overlap with the initial state. The dipole moment is then very small
and so is also the polarizability of the shell. One might perhaps say
that (cf. fig.1) the dipole of the excitation is much larger than the
size of the core-hole region so that the main character is that of
two separate charges. The important thing is then not the dipole
polarizability and the RPAE but the <u>relaxation of the frozen core
including double excitations</u>. This corresponds to the well known fact
that a Herman-Skillman or even a hydrogenic model gives good account
of the shape of the photoionization cross section for deep inner
shells but that the ionization threshold in the frozen HF approximation can be wrong by tens of electron volts.

(ii) <u>$1 \rightarrow 1-1$ transitions</u>: <u>The RPAE is not important</u>. The argument is essentially the same as in (i). Phrased differently, the
principle quantum number n has to change and the spacial overlap must
be quite small. The polarizability is thus small for this type of
transition.

(iii) <u>$nl \rightarrow n'(l+1)$, $n < n'$: The RPAE is sometimes of considerable
importance</u>, especially in the ionization threshold region. Ex.: Xe
$3d - \varepsilon f$, Ba $3d-4f$, εf: The final state electron goes through a continuum resonance near threshold (Xe) or ends up in a localized bound
state (Ba) that carries a very large part of the total transition
strength of the channel. The overlap then becomes rather large and
the dipole-dipole interaction becomes quite important. Note however
that relaxation effects are <u>also</u> very important. Another ex.: Ar 3s-
np, n discrete or continuum (the autoionizing resonances of window
type in Ar): The RPAE is not important for the isolated 3s-np channel.
However, via interchannel (intershell) coupling to the $3p - \varepsilon d$ transitions (see below, point (iv)) the RPAE becomes very important.

(iv) <u>$nl \rightarrow n'(l+1)$, $n \lesssim n'$: The RPAE is of fundamental importance</u>.
<u>Sometimes very strong collective effects, even giant dipole resonan-
ces</u>. The condition $n=n'$ allows the photoelectron to become localized,
resonantly or permanently, in the region of the core-hole, resulting
in enormous overlap and dipole moment. One may say that there are
lowlying excited states that make the nl-shell extremely polarizable.
The effect of this large dipole polarizability has to be included in
a self-consistent manner via infinite order perturbation theory (φ^N
basis) or via improved one-electron wave functions ($\varphi^{N(LS)}$ basis)
plus reasonably low order perturbation expansion.

The maximum polarizability, and thus the maximum importance of
the RPAE, can be expected when the dominating excited levels (discrete or continuum) become strongly localized in the region of the core-
hole. This happens when the n(l+1) shell is about to start filling
and therefore the RPAE is especially important for the rare gases and
the elements that follow these, e.g. alkalis, alkaline earths and
transition (rare-earth) elements. A very interesting sequence is

Xe, Cs, Ba, La, Ce,.... In Ba we should consider the following transitions

$$6s \longrightarrow 6p, np, \quad p$$
Ba: $\qquad 5p \longrightarrow 5d, nd, \quad d$
$$4d \longrightarrow 4f, nf, \quad f$$

It should be noted that the importance of the RPAE has to be related to the velocity of the photoelectron. For <u>very high photoelectron energy</u> the electron leaves the core region very rapidly and the
situation is similar to that of deep inner shells, i.e. negligible
dipole-dipole interaction, collective effects without importance, the
φ^N and $\varphi^{N(LS)}$ basis' yielding the same result. In this region of
photoelectron energy (XPS, ESCA) the important processes are core
relaxation and multiple excitation/ionization.

For <u>low photoelectron energy</u>, however, the electron becomes resonantly localized in the core-hole region. The modification of the
excitation due to the dipole polarizability of the nl-shell then is
enormous and has to be calculated self-consistently one way or another. Collective effects are very strong and in the case of the 4d-
and 5p-shells in e.g. Ba one could even talk about a giant dipole
resonance (cf. nuclear physics [6,7]) in the photoionization cross
section. In such a case the difference between the φ^N and the $\varphi^{N(LS)}$
basis' is very large indeed!

3. 4d-PHOTOIONIZATION OF ATOMIC BARIUM AND LANTHANUM

3.1. Splitting of the $4d^9 4f$ Configuration

In this chapter we shall discuss the problem much in the same
spirit as Dehmer and Starace [24,25,26] . Using the φ^N basis the
4f-orbital becomes localized in the inner part of a double well potential in the same region of space as the 4d-electrons, the radial
overlap being $\langle 4d | 4f \rangle_r \cong 0.8$. The overlap with higher f-orbitals in
the outer part of the well is very small so that the fundamental
transition becomes

$$4d^{10} \, {}^1S_0 \longrightarrow 4d^9 4f \, {}^S L_1 \, ({}^3P, \, {}^3D, \, {}^1P) \qquad\qquad (3.1)$$

In the absence of spin-orbit interaction only the ${}^1P$ level can be
reached and this transition carries almost all of the oscillator
strength in the 4d-f channel. However, this one-electron spectrum refers to the ground state potential seen by a 4d-electron. The rest
of the system really does not know that there has been an excitation
so that the Coulomb energy has not been minimized for the excited
final state configuration. This can be done to some exent by first
order degenerate perturbation theory, calculating the multiplet splitting of the $4d^9 4f$ average configuration. We then have the picture
shown in fig.4 [27]

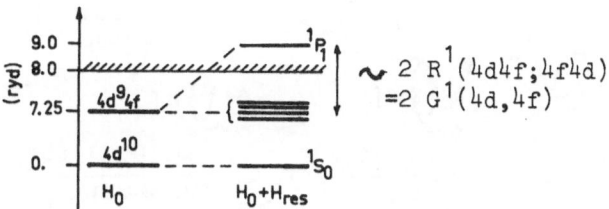

Fig.4. Multiplet splitting of the $4d^9 4f$ configuration in Ba.
$$H_{res} = \sum_{i<j} 1/r_{ij} - \sum_{i} V(r_i)$$

The $4d^9 4f$ ^1P level is a superposition of electron-hole pairs of the φ^N basis. The very large shift implies very strong mixing of these pairs (i.e. the m_l and m_s values) resulting in a <u>strongly collective level</u>, <u>all the electrons in the 4d-shell taking part in the excitation</u>. Note that in the present case the ^1P level is pushed up into the continuum, becoming a <u>continuum resonance of "4f-character"</u>. We also note, that the interaction that causes the shift is the TDAE matrix element (bubble matrix element) $2R^1(4d4f;4f4d)$. Therefore, solving the TDAE for the 4d-4f transitions accomplishes the diagonalization discussed in fig.4. If instead one applies the RPAE (fig.2) one retrieves the same picture, but in addition one finds that the ground state correlations will decrease the multiplet splitting by effectively reducing the magnitude of the Slater integrals.

3.2. The Reaction Matrix RPAE

A convenient way to understand the structure of the RPAE is to begin with the TDAE, omitting the ground state correlations in fig.2 (so called time reversed diagrams). One thus only keeps the direct intra- and interchannel interaction and the result should come out identical with the reaction matrix method [13,28,29].
Let us specifically study $4d-\varepsilon f$ ionization, neglecting interaction with other channel. The intrachannel interaction $V_{nn'}$ then is

$$V_{nn'} = \quad (a) \quad + \quad (b) \qquad (3.2)$$
$$K=1 \qquad K=2,4$$

$$V_{nn'} = \frac{9}{5} R^1(4dnf;n'f4d) - \frac{8}{35} R^2(4dnf;4dn'f) - \qquad (3.3)$$
$$- \frac{2}{21} R^4(4dnf;4dn'f)$$

In the following we shall only draw the dipole-dipole (bubble) part of eq.(3.2). The photoionization amplitude in the TDAE then becomes (for the details of the present discussion, choice of V^{N-1} potential, treatment of the perturbation expansion etc., see refs. [4,16,17])

$$D(\omega) = \text{[diagrams]} + \cdots \qquad (3.4)$$

$$D(\omega) = P_\omega - \underset{n}{S}\frac{P_n V_{n\omega}}{\omega_n - \omega - i\delta} + \underset{nn'}{S}\frac{P_n V_{nn'} V_{n'\omega}}{(\omega_n - \omega - i\delta)(\omega_{n'} - \omega - i\delta)} - \cdots \qquad (3.5)$$

We introduce the notation (cf. Kelly's lectures)

$$\frac{1}{\omega_n - \omega - i\delta} = \frac{PP}{\omega_n - \omega} + i\pi\,\delta(\omega_n - \omega) \qquad (3.6)$$

where PP denotes the principal part value, and take the imaginary part of each bubble in eq.(3.4). The first term is real but after that we get a geometrical series

$$\text{[diagrams]} \qquad (3.7)$$

or explicitly

$$-i\pi P_\omega V_{\omega\omega}\left[1 + (-i\pi V_{\omega\omega}) + (-i\pi V_{\omega\omega})^2 + \cdots\right] =$$
$$= i\pi P_\omega V_{\omega\omega}/(1 + i\pi V_{\omega\omega}) \qquad (3.8)$$

The omitted real parts are now reintroduced through the infinite series of principle parts

$$V_{\omega\omega}(\omega) \longrightarrow \text{[diagrams]} + \cdots \qquad (3.9)$$

$$P_\omega(\omega) \longrightarrow \text{[diagrams]} + \cdots \qquad (3.10)$$

This generates all the terms in the decomposition of eqs.(3.4) and (3.5). Eqs.(3.9) and (3.10) can be written as <u>integral equations</u>

$$V_{\omega\omega}(\omega) = V_{\omega\omega} - \underset{n}{\$}\frac{V_{\omega n} V_{n\omega'}(\omega)}{\omega_n - \omega} \qquad (3.11)$$

$$P_\omega(\omega) = P_\omega - \underset{n}{\$}\frac{P_n V_{n\omega}(\omega)}{\omega_n - \omega} \qquad (3.12)$$

$V_{nn'}(\omega)$ is the so called <u>reaction matrix</u> and $P_n(\omega)$ is a kind of effective dipole matrix element, determined by the reaction matrix. <u>The reaction matrix really is the key quantity to be calculated</u>. The photoionization amplitude then becomes

$$D(\omega) = P_\omega(\omega) / \left[1 + i\pi V_{\omega\omega}(\omega) \right] \tag{3.13}$$

(obtained from adding eqs.(3.10) and (3.7)) and we finally get the photoionization cross section as

$$\sigma(\omega) = 4\pi \alpha a_o^2 \omega \; Jm \; \alpha(\omega)$$
$$Jm \; \alpha(\omega) = \frac{1}{3} \frac{\ell+1}{2\ell+1} N_{4d} \pi |D(\omega)|^2, \quad (\ell = 2) \tag{3.14}$$

In principle, the reaction matrix integral equation (3.9) could be solved by matrix inversion techniques. In practice one can construct quite accurate approximate solutions in a very simple manner. In the present case of Ba we recall the total dominance of the 4d-4f transition in the φ^N basis. We therefore adopt the picture discussed in section 3.1 and treat the 4d-4f transition as a resonance interacting with a continuum background. We then first exclude the resonance from the sum over intermediate states, obtaining a reaction matrix for the remaining part of the channel

$$\overline{V}_{nn'}(\omega) = V_{nn'} - \oint_{n''\neq 4} \frac{V_{nn''} \overline{V}_{n''n'}(\omega)}{\omega_{n''} - \omega} \tag{3.15}$$

and likewise

$$\overline{P}_n(\omega) = P_n - \oint_{n'\neq 4} \frac{P_{n'} \overline{V}_{n'n}(\omega)}{\omega_{n'} - \omega} \tag{3.16}$$

This solves the problem in the absence of the n=4 resonance. This resonance is then reintroduced into the full reaction matrix in a renormalized form

$$V_{\omega\omega}(\omega) = \overline{V}_{\omega\omega}(\omega) - \overline{V}_{\omega 4}(\omega) \overline{V}_{4\omega}(\omega) / K(\omega) \tag{3.17}$$

$$P_\omega(\omega) = \overline{P}_\omega(\omega) - \overline{P}_4(\omega) \overline{V}_{4\omega}(\omega) / K(\omega) \tag{3.18}$$

where

$$K(\omega) = \omega_4 - \omega + \overline{V}_{44}(\omega) \tag{3.19}$$

Eq.(3.19) shows how the zeroth-order 4d-4f resonance has been shifted by an effective intrachannel interaction (renormalized 4d-4f interaction) to higher energy to a position given by the solution of $K(\omega) = 0$.

The solution of eq.(3.15) in the continuum can now be obtained by iteration which would not have been possible in the presence of the 4d-4f resonance. A much faster and very accurate method is, however, to use the Fredholm approximation

$$\overline{V}_{nn'} = V_{nn'} - \oint_{n''\neq 4} \frac{V_{nn''} V_{n''n'}}{\omega_{n''} - \omega} \Bigg/ \left[1 + \oint_{n\neq 4} \frac{V_{nn}}{\omega_n - \omega} \right] \tag{3.20}$$

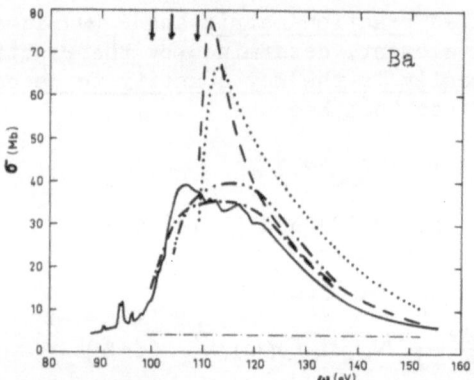

Fig.5. Photoionization cross sections for atomic Ba.
——— Experiment (vapour) [30]; ····· TDAE; — — — RPAE; —··— RPAE
+ relaxation of the 5p^6, 5s^2, 4d^{10} shells; —·— RPAE + relaxation of
the 5p^6, 5s^2, 4d^{10} shells + simulation of relativistic effects.
A constant (somewhat arbitrary) background —··— has been added to
the theoretical curves.

The resulting 4d-εf ^{1}P photoionization cross section for atomic
Ba in the TDAE is shown in fig.5 (dotted curve). I have also calculated the same cross section with the $\varphi^{N(LS)}$ basis using the V^{N-1} HF^{1}P
potential. The results should be identical and they actually came
out so close (difference ~ 1 percent) that the single curve well
describes both calculations (also see [19b,22,32]).

On the general question of numerical accuracy, there was some
discussion at the School concerning a suspected discrepancy between
the HF^{1}P results of myself and those of Fliflet et al. [19b] at energies above the continuum peak. More careful plotting of the curves
shows on the contrary nice agreement above the peak but some discrepancy below the peak instead. The origin of this discrepancy is not
known at present.

The RPAE ground state correlations in fig.2 can be introduced
in several ways. The simplest, slightly approximate way is to make
the substitution

$$\frac{1}{\omega_n-\omega} \rightarrow \frac{1}{\omega_n-\omega} + \frac{1}{\omega_n+\omega} = \frac{2\omega_n}{\omega_n^2 - \omega^2} \qquad (3.21)$$

in eqs.(3.11)-(3.20), i.e. the ground state correlations are introduced via negative frequency contributions to the polarizability.
This is exact if the TDAE matrix elements (3.2) are equal to the
pair excitation matrix elements. In practice the difference is ~10
percent and the effect on the cross section is often quite small.
For a discussion of this simplified version of the RPAE, see refs.
[4,16,17,29,31]. Refs. [29,32] also give a good account of the
method of Amusia and coworkers [2,3,15].

Another way is to include the ground state correlations as corrections to the direct intrachannel interaction [16,23,33]

$$V_{nn'} \rightarrow \quad \text{(a)} \quad + \quad \text{(b)} \quad + \quad \text{(c)} \quad + \dots \qquad (3.22)$$

$$P_n \rightarrow \quad \text{(a)} \quad + \quad \text{(b)} \quad + \quad \text{(c)} \quad + \dots \qquad (3.23)$$

A quite realistic approximation turns out to be

$$V_{nn'} \rightarrow V_{nn'} - \sum_{n''} \frac{VG_{nn''} \, VG_{n''n'}}{\omega_{n''} + \omega_{n'}} \qquad (3.24)$$

$$P_n \rightarrow P_n - \sum_{n'} \frac{P_{n'} \, VG_{n'n}}{\omega_{n'} + \omega_n} \qquad (3.25)$$

with the subsidiary conditions

$$\omega_n = \begin{cases} \omega_n, & \text{if } n > 4 \\ \omega_4 + V_{44}, & \text{if } n = 4 \end{cases} \qquad (3.26)$$

Eq.(3.26) represents inclusion to infinite order of diagrams like (3.22c) that are diagonal in n=4, resulting in a shifted energy denominator (cf. Kelly's lectures).

The results of using on one hand eq.(3.21) and on the other eqs.(3.22)-(3.26) is again represented by one single curve in fig.5 (dashed curve), the difference being about 2 percent at the top of the peak and about 1 percent at higher energies.

4. BEYOND THE RPAE

4.1. Introduction

The result in fig.5 clearly shows that the RPAE completely fails to describe the experimental cross section in a large region ($\sim$1 ryd) above the $N_{4,5}$ (4d) threshold in Ba. The reason is that the RPAE is based on HF-orbitals. One aspect of this is that the observed ionization energy (threshold) is considerably lower than in the RPAE. A second, and in my opinion even more important, aspect is that relaxation of the frozen HF-potential for excited states may lead to very conspicuous changes in the shape of the cross section.

The most obvious way of including relaxation effects is introducing the experimental (or calculated) binding energy in the RPAE (TDAE) equations (3.11) and (3.12) [2,3,4,17,19b,25]. However, one then neglects the change in the ionic potential seen by the photoelectron. The resulting cross section then is just the TDAE shifted to the experimental position without change in shape or the RPAE shifted with slightly changed shape (due to change in the ground state correlati-

ons). With this approximation the agreement with experiment can be
even worse than for the RPAE proper.

In principle, an approximation which only takes into account
relaxation via the energy shift of the core-hole is bound to be poor.
One then neglects that the photoelectron will feel a relaxed ion potential and also itself induce relaxation effects. One can expect
these effects to be relatively unimportant for deep inner shells and
also maybe for other transitions where the RPAE itself is unimportant.
However, when the RPAE is important there is a considerable degree of
localization of the photoelectron in the region of the core-hole for
low velocities. The photoelectron is then very sensitive to changes
of the potential around the core-hole due to relaxation effects. Also
a resonantly localized photoelectron will try to screen out the hole,
screening the change in effective charge. This makes the transition
more adiabatic, i.e. very smooth with little shake up.

Experimentally, the 4d-cross section becomes more and more peaked
as one goes through the series Xe, Cs, Ba and La. For Xe the RPAE is
quite good [2,15,16,17] ; for Cs there is considerable discrepancy in
the threshold region [3,35] ; for Ba there is very strong disagreement in the threshold region (fig.5) and for La the disagreement is
complete in the sense that the theoretical cross section does not
even predict any peak above threshold (fig.8). Simple shift of the
RPAE curves is clearly out of question and in this chapter I shall
attempt to formulate a systematic way to improve on the RPAE.

4.2. Influence of Core Relaxation on the Ionization Energy

The most straightforward way to determine the ionization energy
for a given electron is to measure the binding energy in an XPS (ESCA)
experiment. In this experiment the photoelectron energy is so high
that the sudden approximation is at least approximately valid. This
means that the photoelectron leaves the atom without further interaction with the core and the excitation spectrum becomes determined
simply by the eigenstates of the free ion (see the lectures by Åberg
and Krause). In diagrammatic language we then have the following
picture [2,17,31,36,38] (see also [5,37])

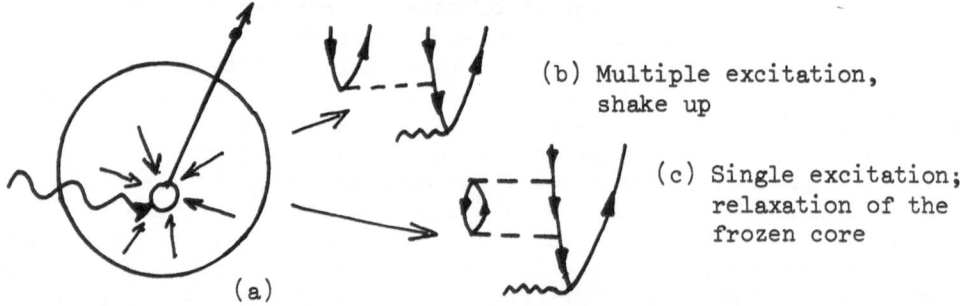

(b) Multiple excitation,
    shake up

(c) Single excitation;
    relaxation of the
    frozen core

(a)

Fig.6. Diagrams describing core-relaxation and shake up.

Fig.6c shows the diagram that essentially describes the relaxation of the froxen core. In order to make it appear as a correction to the HF-binding energy we must consider the infinite series

$$(4.1)$$

The self energy is defined by

$$\Sigma(\mathcal{E}) = \qquad = \sum_m \frac{V_m^2}{\omega_m + \mathcal{E}} \qquad (4.2)$$

and summation of the geometrical series (4.1) gives that

$$\frac{1}{\omega_n - \omega} \longrightarrow \frac{1}{\omega_n - \omega - \Sigma(\omega_n - \omega)} \qquad (4.3)$$

In the photoionization process the "end product" in single ionization is a free electron and a free ion (forgetting about Auger decay and similar things). The energy spectrum of the core-hole is nothing but the energy eigenstates of this free ion. Therefore, in an XPS experiment on a free atom the binding energy of a core-electron must be independent of the photon energy. In particular, if we set the photon energy ω equal to the ionization energy ω_I eq.(4.3) becomes

$$\left[-E_{4d} - \omega_I - \Sigma(-E_{4d} - \omega_I) \right]^{-1} \qquad (4.4)$$

We get onset of continuum absorption when the denominator in (4.4) is zero. Thus the relaxed ionization energy is obtained as the self-consistent solution of the the so called <u>Dyson equation</u>

$$\omega_I = \underbrace{(-E_{4d})}_{\substack{\text{Binding energy,} \\ \text{frozen HF-core} \ > 0}} - \underbrace{\sum(-E_{4d} - \omega_I)}_{\substack{\text{Relaxation} \\ \text{correction} \ > 0}} \qquad (4.5)$$

Actually there are also infinitely many other solutions that correspond to double excitations [34]. In order to apply the above results one should modify the reaction matrix eq.(3.11) and related quantities by introducing the self-energy correction according to eq.(4.3).

4.3. Systematic Modification of the RPAE to including Relaxation

As mentioned in section 4.1 the method of introducing experimental binding energies in the RPAE equations can be extremely bad when the photoionization cross section is strongly peaked near threshold.

In the case of 4d-ionization of Ba and La the photoelectrons of f-
symmetry become resonantly localized in the 4d-region. If the 4d-
hole is strongly influenced by relaxation of the frozen core this
must be true also for the escaping photoelectron at low velocity.
Crudely speaking, the 4d-hole will attract the surrounding electrons
while the resonant f-photoelectron will push them away so that the
net effect of rearrangement will be small. However, if we change the
photon energy away from resonance the photoelectron will leave the
core region much faster. The 4d-hole will then appear much less
screened and become much more affected by relaxation. If we think in
terms of the φ^N basis the 4d-4f resonance in the continuum will be
little affected by relaxation, remaining approximately at the "frozen
core position", while the flat continuum background will be starting
from the relaxed threshold [4,34] . This is certainly to oversimplify
the problem but it serves to illustrate a general conclusion, namely
that relaxation effects can not only shift ionization thresholds but
also strongly deform the shape of the cross section in the threshold
region.

Unfortunately (or perhaps fortunately!) there is no room for presenting any details so I shall confine myself to giving a rough
sketch of an attempt at a systematic treatment. The diagrams of crucial importance for relaxation (of shake up/off type; monopole relaxation) are shown in fig.7 [23,40,41]. We are working within the φ^N basis where the 4f-orbital is about as localized as the 4d. Therefore,
with n =n´ =n´´ =4 the sum of the diagrams in fig.7 is approximately
zero because a 4d- and a 4f-electron are about equally probable to
excite additional electrons, leading to destructive interference. It
is thus extremely important that one considers all of the diagrams
in fig.7 together on the same level of approximation. Modification
of the 4d-binding energy only includes the virtual shake process in
fig.7a and if one stops at that one has not made any meaningful approximation. On the other hand, for n, n´ and n´´ in the non-resonant

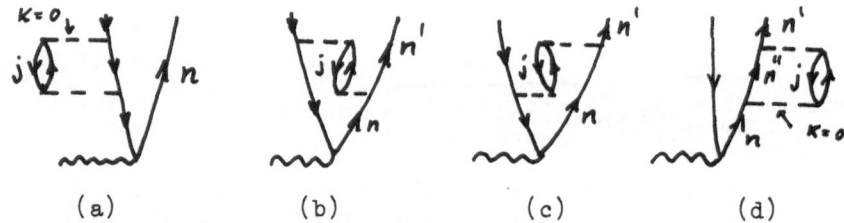

(a)   (b)   (c)   (d)

Fig.7. Modifications of the single-electron ionization process
due to relaxation (core rearrangement). (a) is a self-energy correction to the ionization energy, describing the relaxation of the free
ion. (b,c) describe lowest order screening of the electron-hole interaction. (d) represents an optical (self-induced) potential seen by
the photoelectron. (b)-(d) can be thought of as an effective intrachannel interaction that modifies the ordinary one discussed in the
TDAE and RPAE approaches.

continuum only the correction to the binding energy is really important. Finally there remain very important off-diagonal terms coupling
the 4d-4f resonance to the continuum via the relaxation processes,
e.g. n =4 and n´ = continuum in fig.7b,c. This is the reason why the
simple picture of shifting the continuum relative to the resonance
does not work in practice [4,34].

In summary, it is very important to treat all of the diagrams in
fig.7 on the same footing and this may be done in the following way
[23]: As discussed before fig.7a gives rise to a self-energy shift of
the one-electron binding energy. Fig.7b,c,d on the other hand, leads
to an <u>effective 4d-f intrachannel interaction</u>. The total intrachannel
interaction then approximately becomes

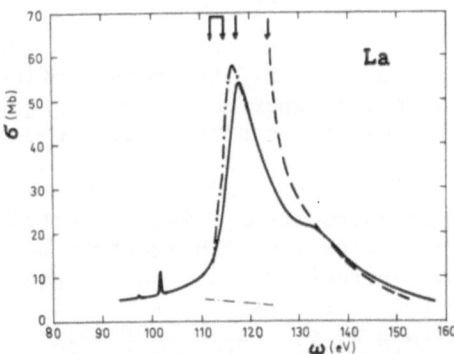

$$\text{(TOTAL)} \quad = \quad \text{(RPAE)} \quad + \quad \text{(RELAX)} \qquad\qquad (4.6)$$
$$\qquad\qquad\qquad (a) \qquad\qquad\qquad (b)$$

where (4.6a) is given by eq.(3.22) and explitly by eq.(3.24), whereas
the relaxation part (4.6b) is given by fig.7b,c,d. In the TDAE/RPAE
equations in section 3.2 we then simply introduce

$$V_{nn'} \longrightarrow \text{total intrachannel interaction}$$
$$\omega_n - \omega \longrightarrow \omega_n - \omega - \sum (\omega_n - \omega) \qquad\qquad (4.7)$$

In particular, the energy denominator of the modified 4d-4f resonance
(3.19) becomes

$$K(\omega) = \omega_4 - \omega - \sum (\omega_4 - \omega) + \bar{V}_{44}(\omega) \qquad\qquad (4.8)$$

Fig.8. Photoionization cross sections for atomic La. ———
Experiment (<u>metal</u>!) [30a,43]; — — —RPAE; — · — RPAE + relaxation of
the $6s^2$, $5p^6$, $5s^2$, $4d^{10}$ shells + simulation of relativistic effects.
— · — · — Assumed background (pure RPAE no background). Middle $N_{4,5}$
threshold: Relativistic effects omitted.

and here $\sum(\omega_4 - \omega)$ practically cancels the relaxation part of $\overline{V}_{44}(\omega)$.

The resulting $4d$-ℓf ^1P cross section for single ionization is shown in fig.5 for Ba. For La in fig.8 the curve is <u>not</u> shown but the corresponding threshold is marked and the cross section just rises like the RPAE towards this relaxed threshold. It must be kept in mind that the present results must not be made subject to any detailed comparison with experiment because they are really of an exploratory nature. The aim of the calculation has been to qualitatively demonstrate the importance of relaxation but there are many things to be tied up before one can say anything about the accuracy of the calculation. For instance, one has to include double excitations as shown in fig.3 plus other types discussed e.g. in [39,40,41] . Even so the qualitative behaviour could certainly be taken seriously:

(i) The high energy portion of the cross section is not much changed.

(ii) The oscillator strength in the RPAE (which <u>satisfies the f-sum rule</u>) from the HF-threshold and ~1 ryd upwards is <u>redistributed</u> over a region extending down to the relaxed threshold, reducing the peak height and increasing the width as compared with the RPAE.

(iii) The areas under the two curves, i.e. RPAE and RPAE + relaxation, are nearly equal, so that the improved scheme does not violate the f-sum rule in any serious way.

A more physical way to explain the broadening and the shift of the continuum resonance is to say that relaxation makes the core-hole float up a bit and become more diffuse so that the potential well seen by the photoelectron will be less narrow and also displaced. The f-resonance will then be less sharp and the photoionization peak appears considerably broadened as compared with the frozen core case.

4.4. Exploratory Inclusion of Relativistic Effects

The results for Ba in fig.5 look reasonable compared with experiment in what concerns the <u>shape</u> of the cross section (note that the complicated structure in the experimental Ba cross section must be due to more complicated excited configurations [4]). The reason for the difference in threshold energy is probably due to relativistic effects [4,5,23,34] which should amount to ~4 eV (average).

For La the relaxed calculation (fig.8) predicts that the peak should lie in the discrete part of the spectrum below the $N_{4,5}$ threshold. If this were the correct result then the peak in the experimental cross section must be due to solid state effects (La metal!). However, again there is a relativistic shift of ~ 4 eV of the threshold and included in a reasonable way this will cause the <u>atomic</u> cross cross section to peak <u>above</u> threshold (see fig.8).

A fully relativistic treatment belongs to the future and would also mean cracking a nut with a sledgehammer if applied to the present case. After all, $4d$-electrons in Ba and La are not relativistic in themselves but are influenced by the relativistic inner shell s- and p-electrons only through the self-consistent potential. I have there-fore made an exploratory simulation of relativistic effects by taking the difference between the Coulomb potential of a Hartree-Fock and a Dirac-Fock calculation [42]

$$\Delta V(r) = V_{DF}^{C}(r) - V_{HF}^{C}(r) \tag{4.9}$$

This is applied as an "external" perturbation, included to infinite order in the RPAE reaction matrix equations through

$\langle 4d \mid \Delta V \mid 4d \rangle$ included in Σ ; $\langle nf \mid \Delta V \mid n'f \rangle$ included in $V_{nn'}$

The calculation is not quite first principles since ΔV had to be scaled to give the correct relativistic shift but it is hoped to give a reasonable picture of the change in potential both for ground and excited states.

The final results for the $4d-\varepsilon f$ ^{1}P photoionization cross sec-tions are shown in fig.5 (Ba) and fig.8 (La). For Ba the difference is not very large but for La the effect is very important. The La curve is a statistically weighted superposition of two different cal-culations where ΔV has been scaled to simulate the N_5 and N_4 thres-holds resp. The result of the calculation suggests that the absorp-tion peak actually is an atomic property that should be observed in photoionization of La vapour as a collective giant dipole resonance in the continuum above the N_4 threshold

REFERENCES

1. D. Bohm and D. Pines, Phys. Rev. 92, 609 (1953)
2. M.Ya. Amusia, Invited lectures and progress reports, VIII ICPEAC, Beograd 1973
3. M.Ya. Amusia, Proc. IV Int. Conf. on Vacuum-UV Radiation Physics, Hamburg 1974 (Vieweg-Pergamon)
4. G. Wendin, Proc. IV Int. Conf. on Vacuum-UV Radiation Physics, Hamburg 1974 (Vieweg-Pergamon)
5. S. Lundqvist and G. Wendin, J. Electr. Spectr. Rel. Phenom. 5, 513 (1974)
6. D.J. Thouless, The Quantum Mechanics of Many-Body Systems (Acade-mic Press, New York and London, second edition 1972) p.113
7. G.E. Brown, Unified Theory of Nuclear Models and Forces (North Holland Publ. Co., 1971) p.49 (see also p.26)
8. D.L. Yeager and V. McKoy, J. Chem. Phys. 60, 2714 (1974)
9. F.S.M. Tsui and K.F. Freed, Chem. Phys. Lett. 32, 345 (1975)
10. a) D.J. Rowe, Rev. Mod. Phys. 40, 1531 (1968)
    b) L. Armstrong, J. Phys. B: Atom. molec. Phys. 7, 2320 (1975)
    c) E. Dalgaard, J. Phys. B: Atom. molec. Phys. 8, 695 (1975)

10b,c) treat the very interesting case of <u>RPAE for open shells</u>.

11. M.J. Jamieson, J. Phys. B: Atom. molec. Phys. <u>8</u>, 537 (1975)

12. T.N. Chang and U. Fano, Phys. Rev. A, to be published

13. U. Fano and J.W. Cooper, Rev. Mod. Phys. <u>40</u>, 441 (1968)

14. H.P. Kelly, Phys. Rev. <u>136</u>, B896 (1964)

15. M.Ya. Amusia, N.A. Cherepkov and L.V. Cernysheva, Soviet Physics-
    JETP <u>33</u>, 90 (1971)

16. G. Wendin, J. Phys. B: Atom. molec. Phys. <u>5</u>, 110 (1972)

17. G. Wendin, J. Phys. B: Atom. molec. Phys. <u>6</u>, 42 (1973)

18. H.P. Kelly and R.L. Simons, Phys. Rev. Letters <u>30</u>, 529 (1973)

19. a) A.W. Fliflet and H.P. Kelly, Phys. Rev. A<u>10</u>, 508 (1974)
    b) A.W. Fliflet, R.L. Chase and H.P. Kelly, J. Phys. B: Atom.
    molec. Phys. <u>7</u>,L443 (1974)

20. J.E. Hansen, J. Phys. B: Atom. molec. Phys. <u>5</u>, 1083, 1096 (1972)

21. D.J. Kennedy and S.T. Manson, Phys. Rev. A<u>5</u>, 227 (1972)

22. F. Combet Farnoux, Proc. of the Int. Conf. on Inner Shell Ioniza-
    tion Phenomena, Atlanta 1972 (ed. US AEC, Oak Ridge), Vol 2, p.1130

23. G. Wendin, Phys. Lett. <u>51A</u>, 291 (1975)

24. A.F. Starace, Phys. Rev. B<u>5</u>, 1773 (1972)

25. J.L. Dehmer and A.F. Starace, Phys. Rev. B<u>5</u>, 1792 (1972)

26. J.L. Dehmer, PhD thesis, University of Chicago (1971), unpublished

27. E.U. Condon and G.H. Shortley, The Thory of Atomic Spectra,
    (Cambridge University Press 1967)

28. A.F. Starace, Phys. Rev. A<u>2</u>, 118 (1970)

29. C.D. Lin, Phys. Rev. A<u>9</u>, 181 (1974)

30. a) P. Rabe, Internal report DESY F41-74/2
    b) P. Rabe, K. Radler and H.-W. Wolff, Proc. IV Int. Conf
    Vacuum-UV Radiation Physics, Hamburg 1974 (Vieweg-Pergamon)

31. G. Wendin, J. Phys. B: Atom. molec. Phys. <u>3</u>, 1080 (1971)

32. A.W. Fliflet, PhD thesis, University of Virginia (1975), unpubl.

33. A.F. Starace, J. Phys. B: Atom. molec. Phys. <u>7</u>, 14 (1974)

34. G. Wendin, Phys. Lett. <u>46A</u>, 101,119 (1973)

35. H. Petersen, K. Radler, B. Sonntag and R. Haensel, DESY report
    SR-74/14 (1974)

36. L. Hedin and A. Johansson, J. Phys. B: Atom. molec. Phys. <u>2</u>,
    1336 (1969)

37. H.W. Meldner and J.D. Perez, Phys. Rev. A<u>4</u>, 1388 (1971)

38. M.Ya. Amusia, N.A. Cherepkov and S.G. Shapiro, Soviet Physics-
    JETP <u>36</u>, 468 (1973)

39. a) T.N Chang, T. Isihara and R.T. Poe, Phys. Rev. Letters <u>27</u>, 838
    (1971)
    b) T.N. Chang and R.T. Poe, Phys. Rev. A<u>12</u>, 1432 (1975)

40. T.N. Chang, J. Phys. B: Atom. molec. Phys. <u>8</u>, 743 (1975)

41. J.-J. Chang and H.P. Kelly, Phys. Rev. A<u>12</u>, 92 (1975)

42. I. Lindgren, private communication

43. J.P. Connerade and coworkers are presently investigating the vapour
    spectrum of La and other rare-earth metals. At the School were pre-
    sented preliminary experimental data in the region of 5p-excita-
    tion and hopefully there will soon be data also in the 4d-region.

PHOTOIONIZATION CROSS SECTIONS AND AUGER RATES

CALCULATED BY MANY-BODY PERTURBATION THEORY*

Hugh P. Kelly

Department of Physics, University of Virginia

Charlottesville, Virginia 22901 U.S.A.

I. RAYLEIGH SCHRÖDINGER THEORY

In these lectures I will discuss methods for applying the many-body perturbation theory of Brueckner [1] and Goldstone [2] to atomic calculations with particular emphasis on calculation of photoionization cross sections and Auger rates. The derivation of the linked cluster diagrammatic many-body perturbation expansion was first given by Brueckner [1] and Goldstone [2] and has been discussed in review articles [3,4,5] and more recently in textbooks on the many-body problems such as those by Fetter and Walecka [6] and March, Young, and Sampanthar [7]. Rather than repeat this work, I would like to start by giving a feeling for diagrams by first presenting them for the particularly simple case of a one-particle system.

Consider the formulation of the time-independent, non-degenerate Rayleigh-Schrödinger perturbation theory as given, for example, by Messiah [8]. Let

$$H = H_o + H' , \tag{1}$$

where the Hamiltonian H has been split into an unperturbed part H_o and a perturbation H'. We are interested in a particular energy E_α and state $|\Psi_\alpha>$ which satisfy

$$H|\Psi_\alpha> = E_\alpha|\Psi_\alpha> . \tag{2}$$

In order to use perturbation theory we calculate the complete set of unperturbed states $|\psi_n^{(o)}\rangle$ which satisfy

$$H_o|\psi_n^{(o)}\rangle = E_n^{(o)}|\psi_n^{(o)}\rangle. \tag{3}$$

The exact energy E_α and state $|\psi_\alpha\rangle$ may be expanded as

$$E_\alpha = \sum_{n=0}^{\infty} E_\alpha^{(n)} , \tag{4}$$

and

$$|\psi_\alpha\rangle = \sum_{n=0}^{\infty} |\psi_\alpha^{(n)}\rangle . \tag{5}$$

Using the "intermediate normalization"

$$\langle\psi_\alpha^{(o)}|\psi_\alpha\rangle = 1 ,$$

and

$$\langle\psi_n^{(o)}|\psi_m^{(o)}\rangle = \delta_{nm} , \tag{6}$$

one readily finds that

$$E_\alpha^{(n)} = \langle\psi_\alpha^{(o)}|H'|\psi_\alpha^{(n-1)}\rangle , \tag{7}$$

$$|\psi_\alpha^{(n)}\rangle = (E_\alpha^{(o)} - H_o)^{-1} Q_o [(H' - E_\alpha^{(1)})|\psi_\alpha^{(n-1)}\rangle$$

$$- E_\alpha^{(2)}|\psi_\alpha^{(n-2)}\rangle \ldots - E_\alpha^{(n-1)}|\psi_\alpha^{(1)}\rangle] , \tag{8}$$

with

$$Q_o = 1 - |\psi_\alpha^{(o)}\rangle\langle\psi_\alpha^{(o)}| . \tag{9}$$

We now use diagrams to represent matrix elements of H' and eventually of $E_\alpha^{(n)}$. As shown in Fig. 1, H' is represented (arbitrarily) here by a dashed line terminated in a cross. Matrix elements of H' are drawn as shown in Fig. 1. The states $|\psi_n^{(o)}\rangle$ are denoted $|n\rangle$. The state $|\alpha\rangle$ is the unexcited state and all states $|n\rangle$ ($n \neq \alpha$) are called excited states. An occupied excited state is called a particle and an unoccupied unexcited state is called

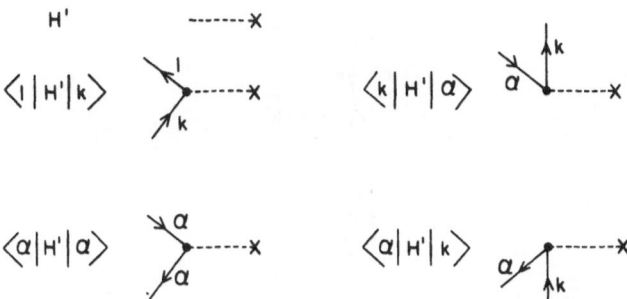

Fig. 1. Diagrammatic representations for H' and for matrix elements of H'. The lines drawn upward correspond to particle lines or occupied excited states. The lines labelled α with an arrow pointed down represent a hole or the absence of the electron from $|\alpha>$.

a hole. Particle lines are drawn with an arrow directed upward and hole lines are drawn with an arrow pointed down. For example, in Fig. 1 the matrix element $<\ell|H'|k>$ corresponds to scattering from excited state $|k>$ to excited state $|\ell>$. Diagrams corresponding to $E_\alpha^{(1)}$, $E_\alpha^{(2)}$, and $E_\alpha^{(3)}$ are shown in Fig. 2. In Fig. 2(a) is shown

$$E_\alpha^{(1)} = <\alpha|H'|\alpha> . \tag{10}$$

Note that the $<\alpha|H'|\alpha>$ matrix element of Fig. 1 has the two lines corresponding to $|\alpha>$ connected so that one now has a closed loop. Figure 2(a) is the diagram for

$$E_\alpha^{(2)} = \sum_{n \neq \alpha} \frac{<\alpha|H'|n><n|H'|\alpha>}{E_\alpha^{(o)} - E_n^{(o)}} . \tag{11}$$

In the diagrams, bottom to top corresponds to going from right to left in Eq. (7).

The general rule for drawing the nth order diagrams is the following: Draw n horizontal lines ------X. Connect these horizontal lines in all possible ways to get connected diagrams with no free lines at the bottom or top. (Note that there can be more than one hole line labelled α). From these rules, there are (n-1)! diagrams for $E_\alpha^{(n)}$.

The mathematical expression for each diagram is obtained by writing the energy denominators with the negative of the excitation

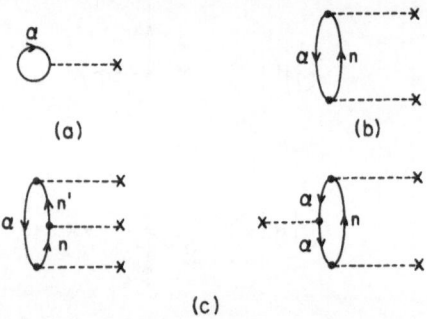

(a)   (b)

(c)

Fig. 2. Diagrams corresponding to (a) $E_\alpha^{(1)}$, (b) $E_\alpha^{(2)}$, (c) $E_\alpha^{(3)}$.

energy, and including a factor $(-1)^{h+\ell}$ where h is the number of hole lines and ℓ is the number of closed loops.

In Fig. 2(a) we have one hole line and one closed loop. In Fig. 2(b) there is also one hole line and one closed loop so $(-1)^{h+\ell}$ = +1. The energy denominators may be determined by thinking of a fictitious horizontal line between any two matrix elements of H'. For each hole line labelled   which is crossed, there is an energy contribution $E_\alpha^{(o)}$. For each particle line labelled n which is crossed, there is a contribution $- E_n^{(o)}$. In Fig. 2(b), between the two interactions there is one hole line α and one particle line n so the energy denominator is $E_\alpha^{(o)} - E_n^{(o)}$. Note that there is a sum over excited states $|n\rangle$.

In Fig. 2(c) are shown the diagrams corresponding to $E_\alpha^{(3)}$. These diagrams are readily seen to reproduce the mathematical expression for $E_\alpha^{(3)}$

$$E_\alpha^{(3)} = \langle\alpha|H' \frac{Q_o}{E_\alpha^{(o)} - H_o} (H' - E_\alpha^{(1)}) \frac{Q_o}{E_\alpha^{(o)} - H_o} H'|\alpha\rangle . \quad (12)$$

The expression for $E_\alpha^{(3)}$ in terms of matrix elements is obtained by inserting after each H' the identity operator

$$I = \sum_n |n\rangle\langle n| . \quad (13)$$

The left diagram in Fig. 2(c) corresponds to the case where the middle interaction in Eq. (12) is H'. Since there is one hole line and one closed loop, $(-1)^{h+\ell} = +1$. The right diagram in Fig. 2(c) corresponds to the case where the middle interaction in Eq. (12) is $-E_\alpha^{(1)}$. Note that $E_\alpha^{(1)}$ is obtained from the diagram because the middle interaction is $\langle\alpha|H'|\alpha\rangle = E_\alpha^{(1)}$. The sign is also given correctly because the right diagram has two hole lines and one closed loop so $(-1)^{h+\ell} = -1$.

For $E_\alpha^{(4)}$ there are $(4-1)! = 6$ diagrams as shown in Fig. 3. The particle lines have not been explicitly labelled. The expression for $E_\alpha^{(4)}$ is

$$E_\alpha^{(4)} = \langle\alpha|H' \frac{Q_0}{E_\alpha^{(0)} - H_0} (H' - E_\alpha^{(1)}) \frac{Q_0}{E_\alpha^{(0)} - H_0} (H' - E_\alpha^{(1)}) \frac{Q_0}{E_\alpha^{(0)} - H_0}$$

$$\times H'|\alpha\rangle - \langle\alpha|H' \frac{Q_0}{E_\alpha^{(0)} - H_0} E_\alpha^{(2)} \frac{Q_0}{E_\alpha^{(0)} - H_0} H'|\alpha\rangle . \tag{14}$$

The four diagrams of Fig. 3(a) – (d) corresponds to the first term on the right hand side of Eq. (14). Diagrams 3(e) + 3(f) correspond to the second term of Eq. (14) containing $E_\alpha^{(2)}$. This is readily checked by labelling the two particle lines and adding the expressions for the diagrams. For example, if the left particle line is labelled n' and the right line labelled n and we set $A = E_\alpha^{(0)} - E_n^{(0)}$ and $B = E_\alpha^{(0)} - E_n^{(0)}$, then the denominators of 4(e) and 4(f) are

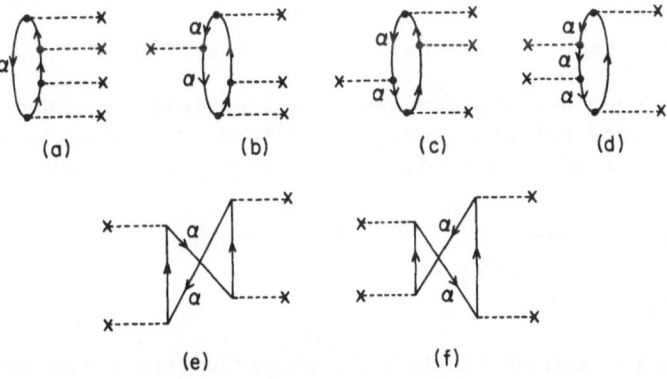

Fig. 3. Diagrams corresponding to $E_\alpha^{(4)}$.

$A^{-1}(A+B)^{-1}B^{-1}$ and $A^{-1}(A+B)^{-1}A^{-1}$ respectively. The sum of these is $A^{-2}B^{-1}$ (noting that the matrix elements and $(-1)^{h+\ell}$ are common for 3(e) and 3(f). Writing out the matrix elements and summing over the particle states, diagrams 3(e) plus 3(f) reproduce the second term of Eq. (14).

The diagrams corresponding to $|\psi_\alpha^{(n)}\rangle$ are obtained by removing the last interaction in the diagrams for $E_\alpha^{(n+1)}$ as is evident from Eq. (7).

It is sometimes of interest to sum certain classes of diagrams to all orders of perturbation theory. A simple example is demonstrated by considering diagram (b) of Fig. 2. This diagram may be considered to be modified by interactions on the particle and hole line as shown in Fig. 2(c). If we consider only the diagonal interactions on the particle line so that $n = n'$, then the ratio of the diagrams of 2(c) to 2(b) is

$$R = (\langle n|H'|n\rangle - \langle\alpha|H'|\alpha\rangle)/(E_\alpha^{(o)} - E_n^{(o)}) \ . \tag{15}$$

These interactions are repeated in fourth and higher orders. Each time, these diagrams in a given order are given by R times the next lower order result. One has then a geometric series which may be summed to give the second order diagram 2(b) times $(1-R)^{-1}$. Noting that 2(b) equals $\langle\alpha|H'|n\rangle\langle n|H'|\alpha\rangle/(E_\alpha^{(o)} - E_n^{(o)})$, we obtain the second-order diagram of Fig. 2(b) but with the denominator $(E_\alpha^{(o)} - E_n^{(o)})$ replaced by $(E_\alpha^{(o)} + \langle\alpha|H'|\alpha\rangle - E_n^{(o)} - \langle n|H'|n\rangle)$.

II.   MANY-BODY PERTURBATION THEORY

A.   Linked Cluster Expressions for Energy and Wave Function

We now consider the problem of N identical fermions interacting through two-body potentials v_{ij} in addition to any one-body potentials such as interaction with the nucleus. The Hamiltonian is

$$H = \sum_{i=1}^{N} T_i + \sum_{i<j}^{N} v_{ij} \ , \tag{16}$$

where T_i is the sum of the kinetic energy operator for the ith particle and all one-body potentials acting on the ith particle. For atoms of atomic number z,

$$T_i = -\frac{\nabla_i^2}{2} - z/r_i \ . \tag{17}$$

Atomic units will be used throughout this paper. In order to simplify the problem, Σv_{ij} is approximated by ΣV_i, where V_i is a Hermitian potential which may be arbitrarily chosen. We now have an approximate Hamiltonian

$$H_o = \sum_{i=1}^{N} (T_i + V_i) . \tag{18}$$

The unperturbed wave function Φ_o satisfies $H_o\Phi_o = E_o\Phi_o$ and is a determinant containing N single-particle solutions ϕ_n of

$$(T + V)\phi_n = \varepsilon_n\phi_n . \tag{19}$$

The states ϕ_n occupied in Φ_o are called unexcited states and all others are called excited states. Unoccupied, unexcited states are called holes and occupied, excited states are called particles just as in Section A.

One can systematically include correlation effects by using perturbation theory with

$$H' = \sum_{i<j}^{N} v_{ij} - \sum_{i=1}^{N} V_i . \tag{20}$$

According to the results of many-body perturbation theory [1,2],

$$\Psi_o = \sum_{L} \left(\frac{1}{E_o - H_o} H'\right)^n \Phi_o , \tag{21}$$

and

$$\Delta E = E - E_o = \langle\Phi_o|H'|\Psi_o\rangle , \tag{22}$$

where $\sum_L$ indicates that only "linked" terms are included [2].

The terms in the perturbation expansions of Eqs. (21) and (22) may be represented by diagrams analogous to those already considered in Section I. In Eq. (21) an unlinked part of a diagram is defined as any part of a diagram which is completely disconnected from the rest and has no external lines. A linked diagram is defined as one containing no unlinked parts [2]. The energy diagrams for Eq. (22) are connected diagrams and are linked when the last interaction H' is removed. First order corrections to Φ_o are shown in Fig. 4.

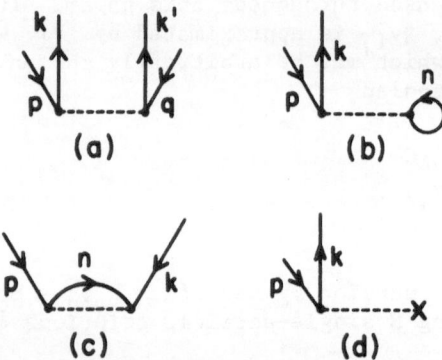

Fig. 4. First order corrections to Φ_0. (a) Two-body correlation in which electrons in states p and q are excited into k and k' through the interaction $<kk'|v|pq>$. (b) Electron in ϕ_p excited into ϕ_k through the interaction $<kn|v|pn>$ with the unexcited state ϕ_n. (c) Exchange interaction $-<nk|v|pn>$. (d) Interaction $-<k|V|p>$. Note that the symbol with the cross for the many-particle case indicates interaction with the perturbation $-V$. Diagrams (b), (c), and (d) sum to zero when V is the Hartree-Fock potential V_{HF}.

The diagram of Fig. 4(a) is given by

$$(\varepsilon_p + \varepsilon_q - \varepsilon_k - \varepsilon_{k'})^{-1}<kk'|v|pq>n_k^+ n_{k'}^+ n_q n_p \Phi_0 \ , \qquad (23)$$

where n_i^+, n_i are the usual Fermion creation or annihilation operators which operate on Φ_0. For example $n_k^+ n_{k'}^+ n_q n_p \Phi_0$ is a determinant in which orbitals p,q in Φ_0 have been replaced by orbitals k,k'. In Fig. 4(d) is shown a matrix element of interaction with the perturbation $-V$ of Eq. (20). When V is the Hartree-Fock potential V_{HF}, diagrams 4(b),(c), and (d) cancel.

In order to carry out the perturbation calculations, we calculate an essentially complete set of states from Eq. (19). Bound states are calculated to approximately n = 10 and then the n^{-3} rule [5] is used to include the remaining bound states.

Continuum states are calculated numerically at a sufficient number of k values to represent all the matrix elements. Typically, we calculate 30 continuum states for each ℓ.

In order to have an orthogonal set of single-particle states, all states of given ℓ should be calculated in the same Hermitian potential. It is possible to have great flexibility in potential choice by making use of the Silverstone-Huzinaga potential [9,10]

$$V = R + (1 - P)\Omega(1 - P) \tag{24}$$

where $P = 1 - \sum_{j=1}^{n} |i\rangle\langle j|$. The choice for Ω is arbitrary as long as it is Hermitian. As seen from Eq. (24), the occupied orbitals are calculated in the "potential" R. The excited orbitals are calculated in the "potential" $R + \Omega - P\Omega$ which may be chosen to give the best description of the physical situation. However, all orbitals are actually being calculated in the same general, nonlocal potential.

B. Frequency-Dependent Polarizability and the Photoionization Cross Section

Consider a perturbing electric field $F\hat{z} \cos \omega t$ so that we have a time-dependent perturbation

$$V_{ex}(\underset{\sim}{r},t) = F \cos \omega t \sum_{i=1}^{N} z_i . \tag{25}$$

The induced dipole moment $\underset{\sim}{P}$ is given by

$$\underset{\sim}{P} = \alpha(\omega)F\hat{z} \cos \omega t , \tag{26}$$

where $\alpha(\omega)$ is the frequency-dependent polarizability. In calculating $\alpha(\omega)$, we use the fact that $-\alpha(\omega)F^2$ is equal to the sum of all (time-independent) energy diagrams in which there are only two interactions with $F\sum_i z_i$ and any number of interactions with H'. Energy denominators between the two interactions with $F\sum_i z_i$ are shifted by $+\omega$ or by $-\omega$, with both cases contributing to $\alpha(\omega)$.

The lowest-order contribution to $\alpha(\omega)$ from the single-particle state ϕ_p occupied in ϕ_o is given by

$$-\sum_{k}{}' \; |\langle k|z|p\rangle|^2 \left(\frac{1}{\varepsilon_p - \varepsilon_k - \omega} + \frac{1}{\varepsilon_p - \varepsilon_k + \omega} \right) , \tag{27}$$

where the sum over k includes all excited states, bound and continuum.

We now derive the relationship between the photoionization cross section $\sigma(\omega)$ and $\alpha(\omega)$. In Eq. (27) if $|p\rangle$ and $|k\rangle$ represent exact eigenstates of $H_o + H'$ and z represents $\sum_i z_i$, then Eq. (26) gives $\alpha(\omega)$ exactly. Since $\varepsilon_p - \varepsilon_k + \omega$ may vanish, we add a small imaginary part $i\eta$ and note that

$$\lim_{\eta \to 0} (\varepsilon_p - \varepsilon_k + \omega + i\eta)^{-1} = P(\varepsilon_p - \varepsilon_k + \omega)^{-1} - i\pi\delta(\varepsilon_p - \varepsilon_k + \omega), \quad (28)$$

where P represents principal value integration. We replace Σ_k for continuum states by $(2/\pi)\int_0^\infty dk$ which assumes our continuum states are normalized according to

$$R_k(r) \to \cos[kr + \delta_\ell + (q/k)\ell n\ 2kr - \frac{1}{2}(\ell + 1)\pi]/r \qquad (29)$$

as $r \to \infty$, where $V(r) \to q/r$. Inserting Eq. (28) into Eq. (27) and replacing Σ_k by $2/\pi \int dk$, we find

$$\text{Im } \alpha(\omega) = (2/k)|<k|z|p>|^2 , \qquad (30)$$

where

$$k = (2\varepsilon_p + 2\omega)^{1/2} . \qquad (31)$$

The expression for $\sigma(\omega)$ is (in atomic units) [11]

$$\sigma(\omega) = (8\pi\omega/ck)|<k|z|p>|^2 \qquad (32)$$

taking account of the normalization of Eq. (28). Alternatively, we may write

$$\sigma(\omega) = (4\pi\ \omega/c)\ \text{Im } \alpha(\omega) . \qquad (33)$$

Upon consideration of the many-body diagrams for Im $\alpha(\omega)$, we find [12] that they are factored into a sum of diagrams to give the exact $<\Psi_k|\Sigma z_i|\Psi_o>$ times its complex conjugate plus corrections due to normalization diagrams [4]. In cases studied so far, we have found effects of the normalization diagrams to be 5% or less.

Diagrams contributing to $<\Psi_k|\Sigma z_i|\Psi_o>$ in which there is a $p \to k$ transition are shown in Fig. 5. The solid dot represents matrix elements of z. The other dashed lines represent coulomb interactions. The exchange diagrams are not explicitly shown but should be included. The lowest order diagram in Fig. 5(a) is $<k|z|p>$. The diagram of Fig. 5(b) is given by

$$\sum_{k'} \frac{<kq|v|pk'><k'|z|q>}{\varepsilon_q - \varepsilon_{k'} + \omega + i\eta} , \qquad (34)$$

and the diagram of Fig. 5(c) is given by

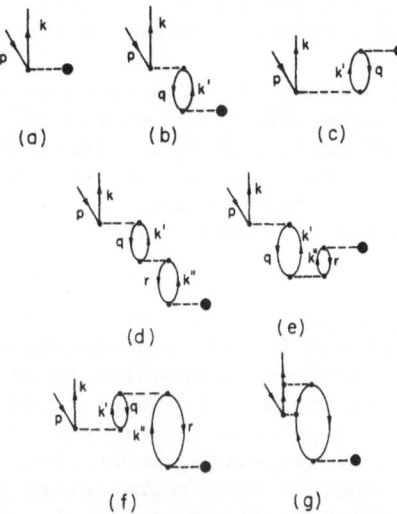

Fig. 5. Diagrams contributing to the matrix element $\langle \Psi_k | \Sigma z_i | \Psi_o \rangle$.
Solid dot indicates matrix element of z. Other dashed lines
represent Coulomb interactions. Exchange diagrams are also included.

$$\sum_{k'}' \frac{\langle q | z | k' \rangle \langle kk' | v | pq \rangle}{\varepsilon_p + \varepsilon_q - \varepsilon_k - \varepsilon_{k'}} \,. \tag{35}$$

The diagrams of Fig. 5 are read from bottom to top corresponding
to right to left in the mathematical expressions. In the diagrams,
Coulomb interactions (interactions with H') below the heavy dot
correspond to correlations in the initial state $| \Psi_o \rangle$ and Coulomb
interactions above the heavy dot correspond to correlations in the
final state $| \Psi_k \rangle$.

III. CALCULATIONS OF PHOTOIONIZATION CROSS SECTIONS

A. Iron

One of our first applications of many-body perturbation theory
to calculate $\sigma(\omega)$ was for FeI [12], a case of considerable astrophysical interest. The ground state of FeI is $(1s)^2 (2s)^2 (2p)^6$
$(3s)^2 (3p)^6 (3d)^6 (4s)^2 \, {}^5D$. In evaluating diagrams (particularly for
resonances) the multiplet structure of intermediate configurations
should be explicitly considered. This was done only in a very
crude way by considering the ground state of FeI to consist of a

half-filled shell of $3d^+$ electrons and one $3d^-$ electron with different orbital energies. Account was also taken of the $4s^\pm$ splitting. The calculation of $\sigma(\omega)$ included correlations among the $(4s)^2$ and $(3d)^6$ subshells which were found to be quite significant. There was a particularly large effect due to the $3d^+ \to 4p$ resonances. These resonances should, of course, show considerable multiplet splitting and these effects are being calculated now. In the original $\sigma(\omega)$ calculations for FeI, the resonance was included only by including the pole due to the diagram of Fig. 5(b) with $p = 4s^\pm$, $q = 3d^+$, and $k' = 4p$.

In additional work [13] on $\sigma(\omega)$ for FeI, account was taken of the process in which one photon is absorbed and one 4s electron is ejected while a second 4s electron is promoted to an excited bound orbital. When the second electron is promoted to an excited bound orbital of the same orbital angular momentum this is referred to as "shake-up". For FeI the lowest-lying bound orbital excitations are $4s \to 3d$ and $4s \to 4p$. The diagrams representing the lowest-order contributions to this process are shown in Fig. 6. These diagrams were calculated by Chang, Ishihara, and Poe [14] for the process $h\nu + Ne \to Ne^{++} + e + e$ and gave results in good agreement with experiment. Contributions from the diagrams of Fig. 6 were projected onto FeII multiplets of $3d^7$ and $3d^64p$. The resulting cross section was on the average approximately 20% as large as the one-electron cross section computed from the diagrams of Fig. 5.

B. Argon

For many years there had been a considerable discrepancy between

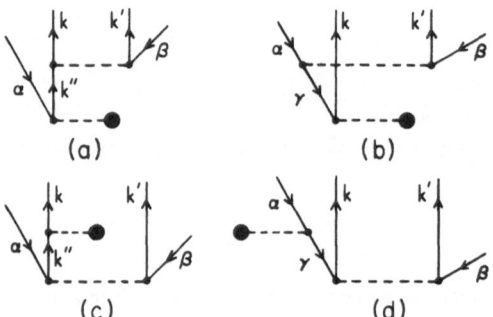

(a)   (b)

(c)   (d)

Fig. 6. Lowest-order diagrams contributing to photoionization with excitation. The final many-particle state contains two excited single-particle $|k\rangle$ and $|k'\rangle$, at least one of which is a continuum state.

the one-electron calculations and experiment [15] for the rare
gases. It was of great interest then when Amus'ya and coworkers
[16] reported a very impressive series of calculations on the rare
gases using the random phase approximation with exchange (RPAE).
In the RPAE one is summing to all orders in perturbation theory
those classes of diagrams which involve only the matrix elements
of creation or destruction of particle-hole pairs as shown in the
diagrams of Fig. 5(b) - (f) and their exchanges. For Xe Wendin
[17] concluded that one must include terms in perturbation theory
to infinite order in order to obtain accurate results; and he
interpreted this as evidence of a "collective oscillation."

It was of interest then to carry out many-body calculations
order by order to study these effects. It was found that the convergence of the perturbation expansion depended critically upon
the choice of potential. When the potential was chosen as that appropriate for the single determinant in which one occupied 3p orbital
is replaced by a continuum orbital, the potential has a negative
3p exchange term. In this case diagrams such as Fig. 5(b) or (d)
with p, q, r corresponding to 3p electrons of different m_ℓ and m_s
were found to be quite large (in fact, divergent). However, as
shown by Amus'ya et al. [16], the final state is $3p^5 k \,^1P$ and the
appropriate Hartree-Fock potential should be chosen accordingly,
i.e., from

$$\delta<3p^5k \,^1P|H|3p^5k \,^1P> = 0 \; , \tag{36}$$

where the continuum state k has $\ell = 0$ or $\ell = 2$. The state $|3p^5k \,^1P>$
now is a linear combination of determinants. The 3p exchange part
of the potential for the continuum state which results from Eq. (35)
is strongly repulsive. This changes greatly the character of the
continuum states, and the Hartree-Fock (using Eq. (36)) $\sigma(\omega)$ is
very different from that obtained when the exchange term in the potential is negative.

The Hartree-Fock length and velocity results are shown in
Fig. 7 in the curves labelled HFL and HFV respectively [21]. The
diagrams for the first-order corrections are shown in Fig. 5(b) and
(c). As pointed out by Amus'ya et al. [16] and essentially also
by Ishihara and Poe [18], when the Hartree-Fock potential is used
(according to Eq. (36)), diagram 5(b) is omitted when p and q refer
to orbitals of the same subshell. This is because there is a corresponding diagram involving interaction with the potential which
cancels 5(b). The first order corrections to $<k|z|p>$ then come only
from the ground state correlations within the 3p subshell as shown
by diagram 5(c). When this diagram is included the HFL and HFV
curves are in good agreement as shown in Fig. 7. In the range
approximately 30 - 45 eV, the diagram of Fig. 5(f) has some effect
in reducing the cross section and improving agreement with the

experimental values. The experimental values shown in Fig. 7 are
from Madden, Ederer, and Coding [19] below 37 eV and from Samson
[20] above 37 eV. It is very clear now from the results of Fig. 7
that the perturbation expansion in this case converges rapidly when
the appropriate Hartree-Fock potential is used for continuum states.
It would, of course, have been disturbing if the expansion in terms
of RPAE type diagrams did not converge rapidly since already in
second order there are many diagrams outside the RPAE which would be
omitted, and there is no reason for assuming they are small.

The window-like resonances in Fig. 7 are due to the basic resonance diagram of Fig. 5(b) with q = 3s, k' = np, and p = 3p. In
order to get the proper shape of the resonances, it is necessary to
include higher-order diagrams. The 3s → np denominators in Fig. 8

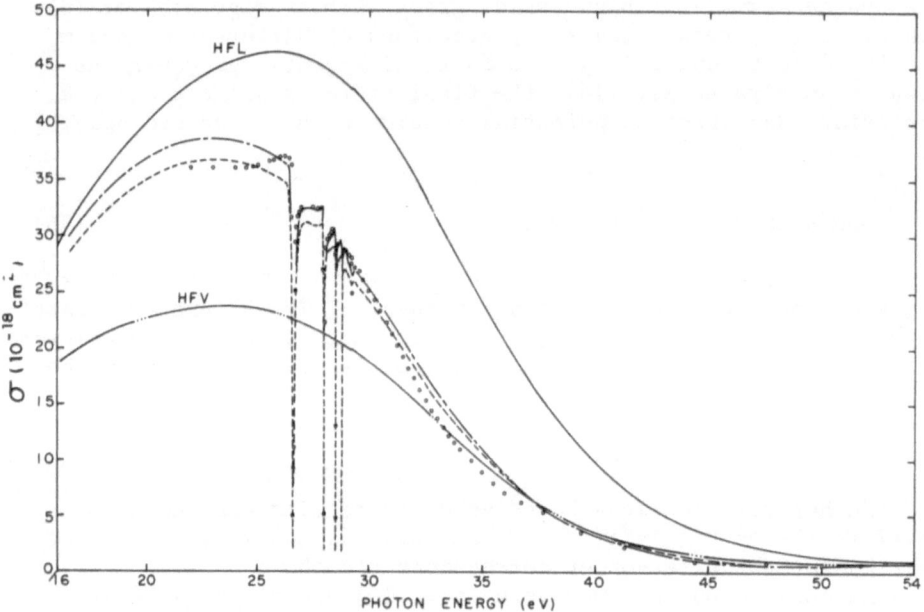

Fig. 7. Cross section σ for photoabsorption by ArI. HFL represents
Hartree-Fock length cross section. HFV represents Hartree-Fock velocity cross section. Dot-dashed line, calculated length cross section including higher-order terms. Dashed line, calculated velocity
cross section including higher-order terms. The circles represent
experimental data from Ref. [19] (below 37 eV) and from Ref. [20]
(above 37 eV). We have only shown the lowest 3s → np resonances,
there being an infinite number of them preceding the 3s threshold
at 29.24 eV.

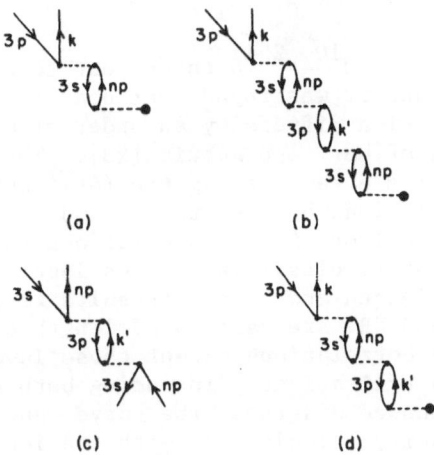

(a)   (b)

(c)   (d)

Fig. 8. Resonance diagrams. (a) basic resonance diagram; (b) next diagram in geometric series which results in shifted denominator of diagram (a); (c) segment by which (b) differs from (a); (d) diagram modifying dipole matrix element <np|z|3s>.

are given by $D = \varepsilon_{3s} - \varepsilon_{np} + \omega$. However, we note that diagrams 8(a) and (b) are the first two terms in a geometric series with ratio given by Fig. 8(c) divided by D. Summing this series results in a shift in D which becomes

$$D = \varepsilon_{3s} - \varepsilon_{np} + \omega + \Delta_n + i\Gamma_n/2 , \qquad (37)$$

where Δ_n is the negative of the real part of Fig. 8(c) and $\Gamma_n/2$ is the negative of the imaginary part of 8(c). It is also necessary for argon to consider correlation modifications of the dipole matrix element of the basic resonance diagram 8(a) as shown in 8(d). For argon these are important since the dipole matrix elements <kd|z|3d> are larger than <np|z|3s>. At resonance, 8(d) with the bottom denominator treated according to $-i\pi\delta(\varepsilon_{3p} - \varepsilon_{k'} + \omega)$ identically cancels the lowest order contribution given in Fig. 5(a). Since $\sigma(\omega)$ is large, this cancellation at resonance is more important than Fig. 8(a) and the result is the "window'like" resonances shown in Fig. 7 which are in good agreement with the experimental results of Madden et al. The resonance positions were "calculated" as $-[\varepsilon_{3s}(\exp) - \varepsilon_{np}]$ where $-\varepsilon_{3s}(\exp)$ is the experimental removal energy and ε_{np} is the single particle Hartree-Fock energy for $3s\ 3p^6 np\ ^1P$. The resonance energies can be calculated in a completely ab initio way but the main purpose of this work was to calculate $\sigma(\omega)$ [21].

C. Zinc

Calculations on ZnI $3d^{10}4s^2\ ^1S$ turned out to be particularly interesting [22] because it was found that near threshold the Hartree-Fock cross section differs by an order of magnitude from the experimental data of Marr and Austin [23]. The calculations showed large correlation effects among the $(4s)^2$ subshell and particularly large correlations between the 4s and 3d subshells. In Fig. 9 is shown a comparison of the calculations and the experimental results shown by circles. The curves labelled HFL and HFV are the Hartree-Fock length and velocity results respectively. The curves labelled NRL and NRV are calculated length and velocity results which include correlations except those involving resonance diagrams with 3d → np excitations. Including both non-resonance diagrams and the resonance diagrams, the curve labelled VCF was obtained for the velocity calculation, with the length curve being almost identical. These calculations were carried out by calculating all excited orbitals in the field of frozen core orbitals of ZnI but the configurations of ZnII. In order to estimate the relaxation effects, these calculations were also carried out with excited orbitals calculated in the field of ZnII. The velocity result is the curve labelled VCI, and the length result was very close. It is now very clear that for $\sigma_{4s}(\omega)$ of ZnI there are very important correlation effects both from resonance and non-resonance diagrams.

There is very interesting resonance structure due to 3d → np excitations. Also, for each n it is found that there are actually three resonances due to spin-orbit mixing. That is, one must consider $3d^9 4s^2 np\ ^1P_1, ^3D_1, ^3P_1$. These resonances are sufficiently close to interact among themselves. It was shown that these effects may be included by a modification of the procedure used to shift denominators as discussed for Argon. One considers the basic resonance diagram shown in Fig. 8(a) (except that for Zn 3p and 3s are replaced by 4s and 3d). Higher order diagrams are included as shown in Fig. 10. The horizontal line indicates that the denominator is treated according to $-i\pi\delta(D)$. Again, one forms a geometric series with the ratio being the segment above the horizontal line in Fig. 10(a) as shown explicitly in 10(c). In the 3d → mp excitations, one can now sum over mp and this accounts for the interaction between resonances np and mp. We also include the diagram of Fig. 10(b) and repeated interactions with the segment 10(c). In Fig. 10(d) is shown the segment which would have been used for the geometric ratio according to the summation discussed for argon in Section B. The summation using the segment of Fig. 10(c) reproduces the results given by Fano for the case of one continuum and many interacting resonances.

The intermediate-coupling energies and mixing coefficients for the $3d^9 4s^2 np\ ^1P_1, ^3D_1,$ and $^3P_1$ states were obtained by the method of Wilson [24]. The cross section $\sigma(\omega)$ is shown in the 4p resonance

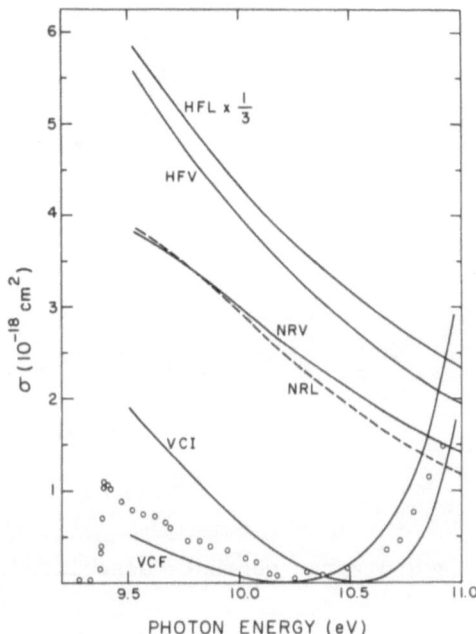

Fig. 9. Cross section $\sigma(\omega)$ for photoionization of the 4s subshell of ZnI near threshold. The curves HFL and HFV represent Hartree-Fock length and velocity cross sections using frozen-core (FC) orbitals. The curves NRL and NRV include correlations excluding resonance diagrams. The curves VCF and VCI represent velocity correlated 4s cross sections with FC and IC (ionic-core) orbitals respectively. The circles represent experimental data from ref.[23].

region in Fig. 11. There is reasonable agreement between the length and velocity calculations and the experimental results of Marr and Austin [23]. The absorption window between the two large resonances is obtained only when interactions between resonances are included. The diagrammatic resonance description used for Zn [22] is equivalent to that derived previously by Fano [25] for the case of one continuum and many interacting resonances.

D. Barium

There has been considerable interest in $\sigma(\omega)$ for the 4d subshell of BaI, particularly since the predicion of a collective resonance by Wendin [26]. It seemed desirable then to carry out Hartree-Fock calculations for $\sigma(\omega)$ along with inclusion of correlations to low order by perturbation theory. In this case, the

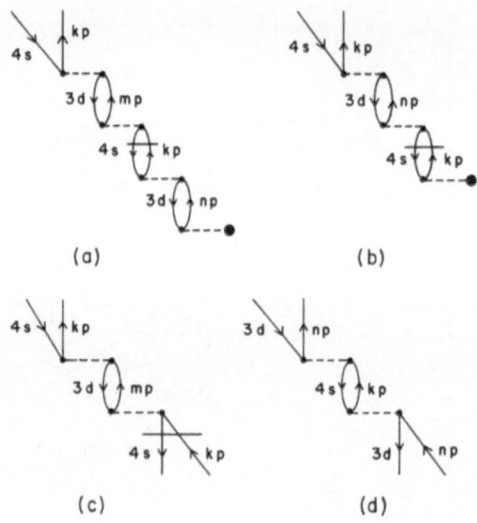

Fig. 10.   Diagrams and diagram segments associated with resonances.

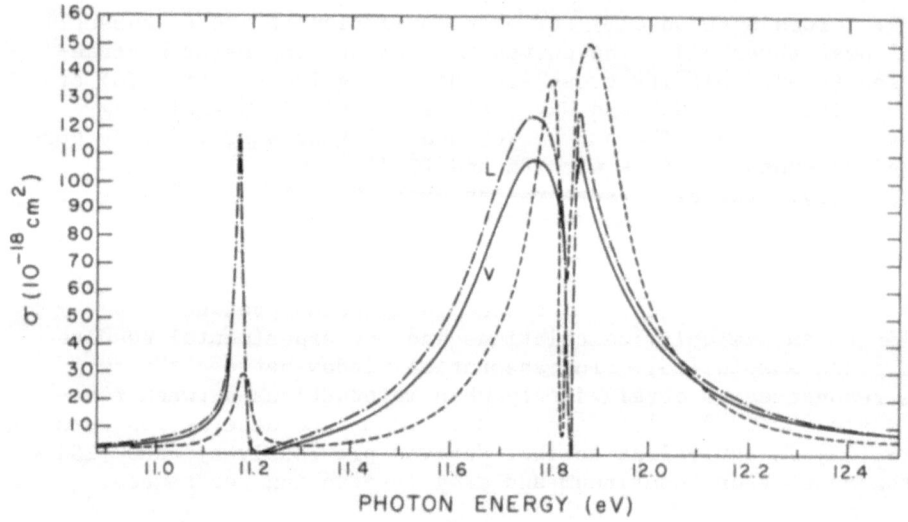

Fig. 11.   Cross section $\sigma(\omega)$ in the $3d^9 4s^2 4p$ resonance region.
Solid line, correlated velocity cross section; Dash-dot line, cor-
related length cross section; dashed line, experimental data from
Ref. [23].

excited states are calculated in a $V^{N-1}4d^9k\,{}^1P$ potential. The
experimental removal energy (99.28 eV) of Connerade and Mansfield
[27] was used in the calculations. The effect of 3 eV splitting
between the $4d^9\,{}^2D_{5/2}$ and ${}^2D_{3/2}$ cores [27] was estimated by
assigning the cross sections according to statistical weights.
The results of these calculations [28] are shown in Fig. 12. The
dashed curve includes the lowest order diagram (Fig. 5(a)) and the
diagram of Fig. 5(c) which accounts for ground state correlations
(GSC) to lowest order. The L and V results in this case are
close after Fig. 5(c) is included. The only effect of Fig. 5(c)
is to bring HF L and V closer. The dashed curve of Fig. 12 represents an average of L and V results. The dotted curve shows the
experimental results of Connerate and Mansfield [27] which has been
normalized to give an oscillator strength of 10. The solid curve
represents the results of an RPAE calculation [28]. Since the experimental removal energy was used in place of the Hartree-Fock
orbital energy the RPAE results for length and velocity are not
identical and the curve in Fig. 12 is an average of these results.
The dot-dashed curve is the calculation by Wendin of RPAE with

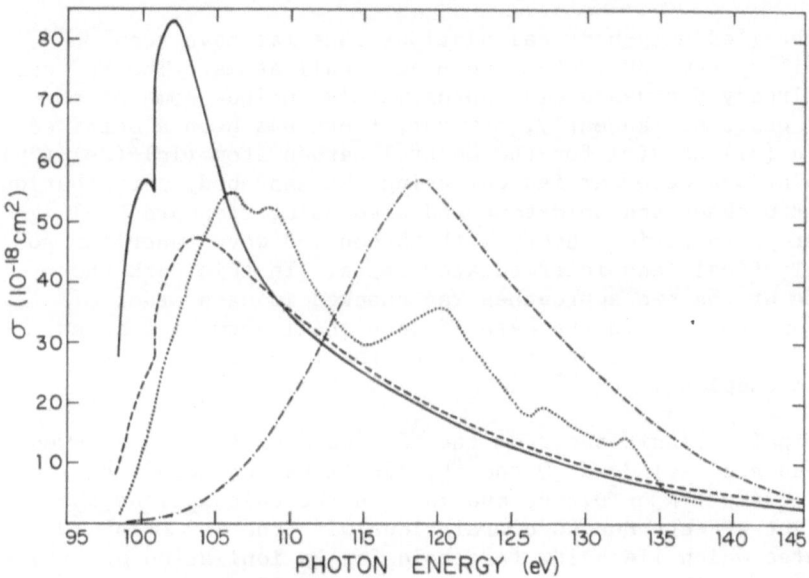

Fig. 12. Comparison of calculated and experimental results for σ_{4d}
of BaI:____, average of length and velocity form RPAE results including effect of the splitting between the $4d^9\,{}^2D_{5/2}$ and ${}^2D_{3/2}$
cores; ...., experimental data of Connerade and Mansfield [27] normalized to give an oscillator strength of 10;————, average of
length and velocity ground state correlation (GSC) results including spin-orbit splitting of cores;—·—·—, relaxed RPAE, Wendin [26].

relaxation. Recently Wendin [29] has also calculated $\sigma(\omega)$ using the RPAE. Measurements of $\sigma(\omega)$ for the 4d subshell of BaI vapor have also been reported by Rabe, Radler, and Wolff [30]. In their results the structure in $\sigma(\omega)$ particularly at 120 eV is less pronounced than in the data of Connerade and Mansfield [27].

When the second order RPAE diagrams such as Fig. 5(e) and (f) are included, the perturbation theory curve for $\sigma(\omega)$ approaches that of the RPAE [28]. It is interesting to note that the low order perturbation result (Hartree-Fock plus ground state correlations included to first order) clearly gives better agreement with experiment in this case than RPAE or RPAE with relaxation (Wendin, [26]). There is, of course, no justification for atoms in selecting the higher-order RPAE diagrams and omitting other types of higher-order diagrams. In this case it appears that there is significant cancellation among higher-order diagrams which is broken when one selectively includes only one class of diagrams such as in the RPAE.

E.   Carbon

The detailed many-body calculations thus far have been on closed-shell systems or systems such as alkali atoms. The FeI calculation already discussed did approximately include some of the open shell effects. Recently, however, there has been a detailed calculation [31] of $\sigma(\omega)$ for the neutral carbon atom $(1s)^2(2s)^2(2p)^2$ ^3P. Calculations were carried out using the many-body perturbation theory of Brueckner and Goldstone and also using standard Rayleigh-Schrödinger perturbation theory with LS-coupled wave functions for the initial, final, and intermediate states. In this work the equivalence of the two approaches was checked in each order of perturbation theory. In the case of open shell atom resonances, it is particularly important to use intermediate states of the appropriate coupling.

For dipole transitions from the ^3P ground state, the allowed final states are $(2s)^2 2pkd$ ^3D and ^3P; $(2s)^2 2pks$ ^3P; $2s2p^2$ ^4P kp ^3D, ^3P, and ^3S; $2s2p^2$ ^2P kp ^3D, ^3P, and ^3S. In the calculations there are important effects due to correlations with the $1s^2 2s 2p^3$ ^3D and ^3P states which lie below threshold for 2p ionization and therefore do not contribute to resonances. These excitations correspond to diagram 1(b) of Fig. 5 with p = 2p, q = 2s, k' = 2p.

All first-order perturbation terms were included as well as those second- and higher-order terms which were expected to be most significant.

Results for $\sigma(\omega)$ including correlations are shown in Fig. 13 with the dashed line for the velocity results and the solid line

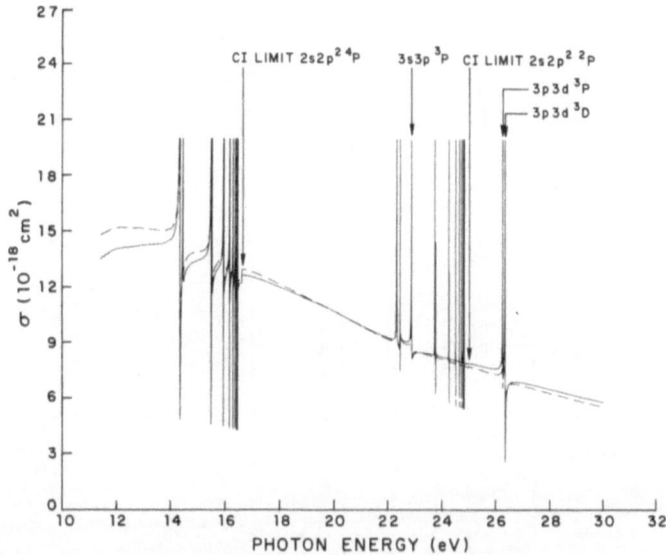

Fig. 13. Total ^{3}P carbon photoionization cross section. Solid line is length form. Dashed line is velocity form. Resonance lines are due to $2s2p^2$ (^{4}P, ^{2}P)np autoionizing configurations. Resonances from the $2s^2$ 3s3p and $2s^2$ 3p3d configurations are indicated. All resonance peaks have been truncated at 20 megabarns. Details of the cross section near threshold are shown in Fig. 14.

for length results. In the region 14 - 16 eV are shown resonances arising from $2s2p^2$ ^{4}P np ^{3}D and ^{3}P excited states with n $\geq$ 3. In the region 22 - 25 eV there is the series of resonances due to $2s2p^2$ ^{2}P np ^{3}D and ^{3}P excited states (n $\geq$ 3). Resonances due to the states $(2s)^2$3s3p ^{3}P, $(2s)^2$3p3d ^{3}D and ^{3}P are also shown.

After these calculations were completed, the existence of relative experimental data by Esteva, Mehlman-Ballofet, and Romand [32] in the range 14 - 16 eV was pointed out by Dr. D. L. Ederer. In Fig. 14 is shown (as an inverted mirror-image) a densitometer trace of their unnormalized carbon spectra and also the corresponding portion of the calculated cross section. The resonance line shapes for the $2s2p^2$ ^{4}P np ^{3}D series are clearly reproduced. Close to each ^{3}D resonance is also a less prominent $2s2p^2$ ^{4}P np ^{3}P resonance. The calculated energies are consistently larger than the experimental results primarily because they were evaluated according to $-(E_{2s} - \varepsilon_{np})$ where $-E_{2s}$ is the experimental [33] 2s removal energy and ε_{np} is the HF single-particle energy. In addition Δ_n of Eq. (37) was not included. The cross section for

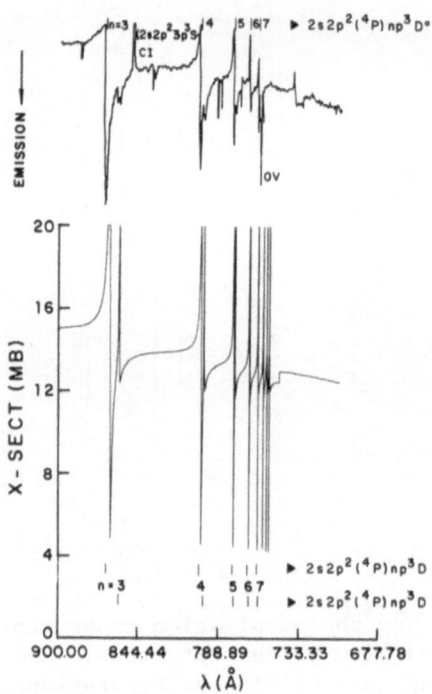

Fig. 14. Detail of velocity form curve from Fig. 13 in the resonance region of the $2s2p^2$ (^4P)np series. Unnormalized densitometer
trace is from Esteva et al. [32] and has been reproduced as an inverted, mirror-image.

ejection of two 2p electrons has also been calculated [34] and
has been found to be small, approximately 1% of the total. Preliminary results have also been obtained for the cross section for
simultaneous ejection of one 2s and one 2p electron. Here the results are larger, causing approximately a 5% increase in $\sigma(\omega)$ [34].

In summary, it is interesting to note that a careful perturbation calculation can successfully predict complex structure even
for open shell atoms.

F.   Other Atoms

In addition to the many-body calculations already discussed,
there have been other many-body calculations of photo cross sections for systems such as O^- where resonance structure due to
2s $\rightarrow$ 2p excitation was calculated [35] and good agreeement with
experiment [36,37] was obtained.

There have been two recent calculations of the 3s subshell of Na. In a calculation by Chang [38], the first-order correlation diagrams were calculated explicitly and the effect of higher order diagrams was approximated by use of a semiempirical polarization potential. After inclusion of the polarization potential, Chang's calculations agree well with experiment from threshold to approximately 10 eV where experiment increases [39] and all calculations predict a decrease. In another recent calculation [40] the second-order diagrams were calculated but the resulting cross section did not agree as well with experiment as that calculated with the semiempirical polarization potential.

For Si^+ calculations were carried out [41] which are similar to those of carbon described in the previous section.

Chang and Poe [42] have calculated the diagrams of Fig. 6 contributing to double electron photoionization from Ne. They have analyzed the contributions from the various diagrams and their total result (only the velocity form is given) is in good agreement with experiment [43,44,45]. They have also calculated the energy distribution between the two outgoing electrons for five different values of total kinetic energy of the outgoing electrons.

So far there have been few accurate molecular photoionization calculations. One case of considerable astrophysical interest is that of CH. Calculations [46] were carried out using a complete set of atomic carbon orbitals. Perturbation theory was used to correct for the effects of the extra electron and the potential due to the hydrogen nucleus. Correlation diagrams were not included and the result is an approximate Hartree-Fock cross section.

Amusia has given a recent review [47] of RPAE calculations by his group. These calculations have been on closed shell systems and he speaks of the "collective behavior of electron shells in the process of photoionization." Wendin has also reviewed his RPAE calculations in these same proceedings [29] and he states that "one has reason to talk about very strong collective effects and even collective giant dipole resonances, like those well known in nuclear physics."

IV. AUGER RATES

Many-body perturbation theory is also useful in calculating Auger rates and radiative decays. It is easily shown [48] that the Auger rate is proportional to the imaginary part of the energy when denominators D are treated according to $D \rightarrow D + i\eta$ and Eq. (28). One can also write down a perturbation expansion [49] for the Auger matrix elements as shown in Fig. 15. The lowest order diagram is shown in Fig. 15(a) and diagrams (b) - (i) represent effects of

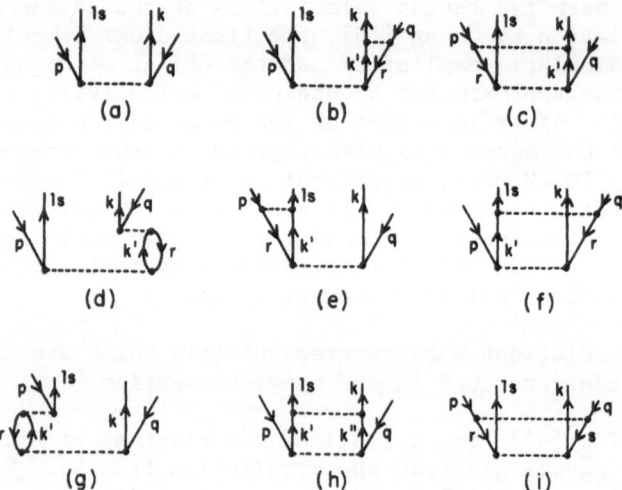

Fig. 15. Diagrams contributing to the Auger-rate matrix element.
(a) lowest-order diagram; (b), (c), (d), and (i) are correlation
correction diagrams equivalent to configuration mixing in the final
state.   Diagrams (e) – (h) involve correlations or configuration
mixing in the initial state. Diagrams (d) and (e) also represent
the diagrams obtained by including the exchange of the bottom matrix
element.

first-order correlation corrections.

Calculations of these diagrams have been carried out for
$Ne^+(1s^{-1})$, the neon atom with an initial 1s vacancy. The possible
final states in LS-coupling are: $(1s)^2(2p)^6$ ^1S, $(1s)^2(2s)(2p)^5$
^1P, $(1s)^2(2s)(2P)^5$ ^3P, $(1s)^2(2s)^2(2p)^4$ ^1S, and $(1s)^2(2s)^2(2p)^4$ ^1D.
Results of these calculations are given in Table I and compared
with experimental results by Mehlhorn, Stalherm and Verbeek [50].
These experimental results have been renormalized to a total rate
$(8.45 \pm 0.07) \times 10^{-3}$ a.u. obtained by Gelius et al. [51].

In these calculations there are significant contributions
from final state configuration mixing of $(1s)^2(2p)^6$ ^1S and
$(1s)^2(2s)^2(2p)^4$ ^1S. However, the other diagrams of Fig. 15 also
contribute large effects, particularly diagrams such as (b) and
(c) with 2p → 2s or 2s → 2p transitions on the hole line. For
example, in calculating the transition to $(1s)^2(2p)^6ks$ ^2S denoted
1s – 2s2s in Table I, the diagram (a) contributes 0.04197 a.u.
Diagrams (b) and (c) with r = 2p (and p,q = $2s^+$, $2s^-$) contribute
-0.00454 a.u. and -0.00151 a.u. respectively. Diagram (i) which
represents final state mixing of $(1s)^2(2s)^6$ ^1S and $(1s)^2(2s)^2(2p)^4$ ^1S

Table I. Auger Rates for Ne $(1s^{-1})$[a]

Transition[b]	HF[c]	CORR I[d]	CORR II[e]	Expt[f]
$1s-2s2s(^1S)$	0.9508	0.4879 0.7012[g]	0.4902	0.54 ± 0.11
$1s-2s2p(^1P)$	2.0335	1.3956	1.3670	1.48 ± 0.29
$1s-2s2p(^3P)$	0.7888	0.5015	0.4922	0.54 ± 0.1
$1s-2p2p(^1S)$	0.4560	0.8643 0.7067[g]	0.7707	0.85 ± 0.17
$1s-2p2p(^1D)$	5.6849	5.1983	4.9349	5.05 ± 0.102
Total	9.9140	8.4476	8.0550	$8.45 \pm .78$

[a]All results given in units of 10^{-3} a.u.
[b]See text for notation.
[c]Hartree-Fock or lowest-order results.
[d]Including correlation diagrams of Fig. 15.
[e]Including correlation diagrams in Fig. 15, estimates of higher-order diagrams, and small contributions from overlap and normalization factors.
[f]Experimental values from Ref. [50] renormalized to a total value of 8.45×10^{-3} a.u. using recent results of Ref. [51].
[g]Correlation effects only from mixing of $(1s)^2(2s)^2(2p)^4 {}^1S$ and $(1s)^2(2p)^6 {}^1S$.

contributes -0.00593 a.u.

In this case the $1s-2s2s$ rate is reduced by final state mixing of $(2s)^2(2p)^4 {}^1S$ and $(2p)^6 {}^1S$. However, as Z becomes smaller one has the possibility of $2s^2 - 2p^2$ mixing in the initial state for $Z \leq 6$. This mixing increases the $1s-2s2s$ rate. For example, for $Be^+(1s^{-1})$ the Auger rate in the Hartree-Fock approximation is 2.500×10^{-3} a.u. and 3.415×10^{-3} a.u. when the $(2s)^2-(2p)^2$ mixing in the initial state is included [52]. For $Be^+(1s^{-1})$, radiative decay is only possible when correlations are included. The allowed final states are $(1s)^2np$ [52]. A detailed calculation of radiative and Auger decays for Be with a 1s vacancy is given in Ref. [52].

REFERENCES

*Work supported by the National Science Foundation.
1. Brueckner, K. A., Phys. Rev. 97, 1353 (1955); Phys. Rev. 100, 36 (1955); The Many-Body Problem (John Wiley & Sons, Inc., New York, 1959).
2. Goldstone, J., Proc. Roy. Soc. (London) A 239, 267 (1957).
3. Brandow, B. H., Rev. Mod. Phys. 39, 771 (1967).

4.   Kelly, H. P., Advan. Chem. Phys. XIV, 129 (1969).
5.   Kelly, H. P., Advances in Theoretical Physics V. II (Academic, New York, 1968), p. 75.
6.   Fetter, A. L., and Walecka, J. D., Quantum Theory of Many-Particle Systems (McGraw-Hill, San Francisco, 1971).
7.   March, N. H., Young, W. H., and Sampanthar, S., The Many-Body Problem in Quantum Mechanics (Cambridge University Press, 1967).
8.   Messiah, A., Quantum Mechanics, Vol. II (North-Holland, New York, 1961).
9.   Silverstone, H. J., and Yin, M. L., J. Chem. Phys. $\underline{49}$, 2026 (1968).
10. Huzinaga, S., and Arnau, C., Phys. Rev. A $\underline{1}$, 1285 (1970).
11. Bethe, H. A. and Salpeter, E. E., Quantum Mechanics of One and Two-Electron Atoms (Academic, New York, 1957), p. 306.
12. Kelly, H. P., and Ron, A., Phys. Rev. A $\underline{5}$, 168 (1972).
13. Kelly, H. P., Phys. Rev. A $\underline{6}$, 1048 (1972).
14. Chang, T. N., Ishihara, T., and Poe, R. T., Phys. Rev. Letters $\underline{27}$, 838 (1971).
15. Fano, U., and Cooper, J. W., Rev. Mod. Phys. $\underline{40}$, 441 (1968).
16. Amus'ya, M. Ya., Cherepkov, N. A., and Chernysheva, L. V., Zh. Eksp. Teor. Fiz. $\underline{60}$, 160 (1971) [Sov. Phys. JETP $\underline{33}$, 90 (1971)].
17. Wendin, G., J. Phys. B $\underline{5}$, 110 (1972).
18. Ishihara, T., and Poe, R. T., Phys. Rev. A $\underline{6}$, 111 (1972).
19. Madden, R. P., Ederer, D. L., and Codling, K., Phys. Rev. $\underline{177}$, 136 (1969).
20. Samson, J. A. R., Advan. At. Mol. Phys. $\underline{2}$, 177 (1966).
21. Kelly, H. P., and Simons, R. L., Phys. Rev. Letters $\underline{30}$, 529 (1973).
22. Fliflet, A. W., and Kelly, H. P., Phys. Rev. A $\underline{10}$, 508 (1974).
23. Marr, G. V., and Austin, J. M., J. Phys. B $\underline{2}$, 168 (1972).
24. Wilson, M., J. Phys. B $\underline{1}$, 736 (1966).
25. Fano, U., Phys. Rev. $\underline{124}$, 1866 (1961).
26. Wendin, G., Phys. Lett. $\underline{46A}$, 119 (1973).
27. Connerade, J. P., and Mansfield, M. W. D., Proc. Roy. Soc. A $\underline{341}$, 267 (1974).
28. Fliflet, A. W., Chase, R. L., and Kelly, H. P., J. Phys. B $\underline{7}$, L443 (1974).
29. Wendin, G. in Vacuum Ultraviolet Radiation Physics (ed. E. E. Koch, R. Haensel, C. Kunz, Pergamon Vieweg, Braunschweig, 1974), p. 225.
30. Rabe, A., Radler, K., and Wolff, H.-W., in Vacuum Ultraviolet Radiation Physics (ed. E. E. Koch, R. Haensel, C. Kunz, Pergamon Vieweg, Braunschweig, 1974), p. 247.
31. Carter, S. L., and Kelly, H. P., to be published (1975).
32. Esteva, J. M., Mehlman-Ballofet, G., and Romand, J., J. Quant. Spectrosc. Radiat. Transfer $\underline{12}$, 1291 (1972).
33. Moore, C. E., Atomic Energy Levels, Nat'l. Bur. Std. Circ. No. 467 (U. S. GPO, Washington, D. C., 1949).

34. Carter, S. L., and Kelly, H. P., to be published (1976).
35. Chase, R. L., and Kelly, H. P., Phys. Rev. A $\underline{6}$, 2150 (1972).
36. Smith, S. J., in <u>Proceedings of the Fourth International Conference on Ionization Phenomena in Gases, Uppsala, Sweden</u> (North-Holland, Amsterdam, 1960), p. 219.
37. Branscomb, L. M., Smith, S. J., and Tisone, G., J. Chem. Phys. $\underline{43}$, 2906 (1963).
38. Chang, T. N., J. Phys. B $\underline{8}$, 743 (1975).
39. Hudson, R. D., and Carter, V. L., J. Opt. Soc. Am. $\underline{57}$, 651 (1967).
40. Chang, J. J., and Kelly, H. P., Phys. Rev. A $\underline{12}$, 92 (1975).
41. Daum, G. R., and Kelly, H. P., Phys. Rev. A, to be published.
42. Chang, T. N., and Poe, R. T., Phys. Rev. A $\underline{12}$, 1432 (1975).
43. Carlson, T. A., Phys. Rev. $\underline{156}$, 142 (1967).
44. Van der Wiel, M. J., and Wiebes, G., Physica $\underline{54}$, 411 (1971).
45. Samson, J. A. R., and Haddad, G. N., Phys. Rev. Letters $\underline{33}$, 875 (1974).
46. Walker, T. E. H., and Kelly, H. P., Chem. Phys. Letters $\underline{16}$, 511 (1972).
47. Amusia, M. Ya. in <u>Vacuum Ultraviolet Radiation Physics</u> (ed. E. E. Koch, R. Haensel, C. Kunz, Pergamon Vieweg, Branuschweig, 1974), p. 205.
48. Chase, R. L., Kelly, H. P., and Köhler, H. S., Phys. Rev. A $\underline{3}$, 1550 (1971).
49. Kelly, H. P., Phys. Rev. A $\underline{11}$, 556 (1975).
50. Mehlhorn, W., Stalherm, D., and Verbeek, H., Z. Naturforsch. A $\underline{23}$, 287 (1968).
51. Gelius, U., Svenson, S., Siegbahn, H., Basilier, E., Faxälv, A., and Siegbahn, K., Chem. Phys. Lett. $\underline{28}$, 1 (1974).
52. Kelly, H. P., Phys. Rev. A $\underline{9}$, 1582 (1974).

PHOTOABSORPTION SPECTROSCOPY OF ATOMS IN THE EXTREME ULTRAVIOLET (XUV) SPECTRAL REGION

K. Codling

The University of Reading, Physics Department

Whiteknights, Reading, U.K.

It is not the object of this short review to attempt to cover all experiments in photoabsorption over the past 40 years. The reader is referred to the reviews of Garton (1), Samson((2) and Marr (3) and the compilations of Hudson and Kieffer (4), Hubbell (5) and Veigele (6). Rather it is intended to look anew at some older experiments in the light of recent theoretical advances, to present more recent data and to pinpoint the deficiencies in both.

On reviewing the literature, it is clear that absolute photoabsorption cross section data in all but the inert gases is sparse indeed, that is for neutral atoms in the free state. Photodetachment of negative ions (e.g. Kasdan and Lineberger (7)) will not be discussed and the photoabsorption of positive ions will be mentioned only briefly. Measurement of mass absorption coefficients for atoms in thin film form is relatively easy in the XUV and one gains valuable information on atomic properties (8). However, in determining the precise role of electron correlation effects, one wishes to avoid the complications of the solid state. Also one wishes to obtain, if possible, accurate atomic ionization potentials (binding energies for the various inner and subshell electrons).

Interaction of XUV radiation with an atom is expected to produce ionization, except in those cases where quasi-stable states are excited and even then autoionization (or Auger decay) is more likely than radiative decay. One is thus determining photoionization cross sections in an XUV atomic absorption experiment.

The intensity of radiation, $I(\lambda)$, transmitted through a gas of length ℓ is given by the expression

$$I(\lambda) = I_0(\lambda) \exp \left[- \sigma(\lambda) n\ell \right] \qquad (1)$$

where $I_0(\lambda)$ is the incident intensity at wavelength λ, $\sigma(\lambda)$ the total absorption cross section and n the number of atoms per cc.

Expression (1) can be rewritten in the form :

$$\sigma(\lambda) = \frac{T}{273} \cdot \frac{760}{p\ell} \cdot \frac{1}{N} \cdot \log_e \left[\frac{I_0(\lambda)}{I(\lambda)} \right] \qquad (2)$$

where N is Loschmidt's number, T the absolute temperature and p the pressure. Cross sections are usually quoted in megabarns ($1Mb=10^{-18}$ cm^2). Alternatively the absorption is given as a linear absorption coefficient, k, in cm^{-1} at S.T.P. and relates to σ by the expression $k = N\sigma$. One may also find the continuum cross section given as an oscillator strength distribution $\frac{df}{d\epsilon}$, where $\sigma = \frac{\pi e^2}{mc} \cdot \frac{df}{d\epsilon}$.

EXPERIMENTAL DETAILS

A typical photoabsorption experiment (9) is shown in figure 1, where the absorption cell is placed behind the exit slit of the monochromator. The monitor photomultiplier (PM) effectively monitors $I_0(\lambda)$. In photographic spectroscopy, the cell is placed between light source and entrance slit, but to measure absolute cross sections with an ionization chamber (10) or partial cross sections by photoelectron spectroscopy (11) the former arrangement is essential.

Over the years, a number of light sources have been developed whose aim is to produce a source of pure continuum in the XUV. Examples are

(i) The ' Hopfield ' He_2 continuum (12) ...
(ii) The ' Lyman ' continuum ... for example, the Garton (13) flash tube
(iii) The BRV source (14).

None of these sources produces a totally pure continuum, the first has problems of differential pumping and limited range, the others produce large amounts of debris that can harm optical components.

The most interesting and useful source at the present time is synchrotron radiation (15) which spans the entire spectral range from infra-red to the x-ray region, is free from spurious emission lines and is intense,

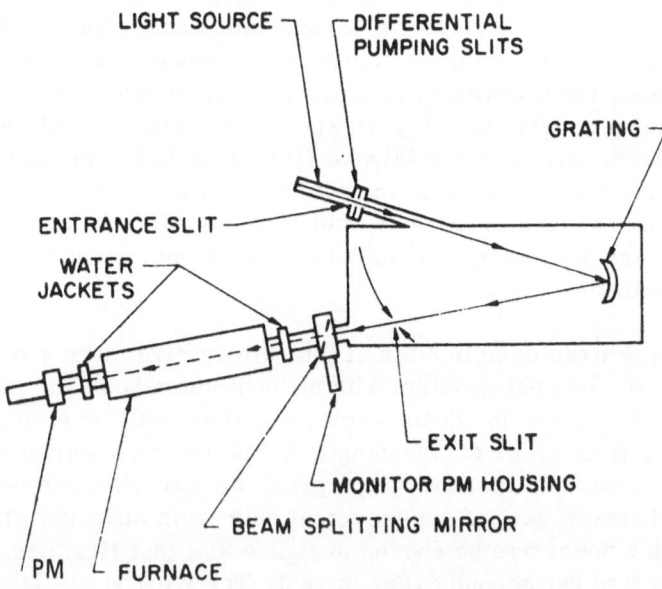

Figure 1. Experimental arrangement to determine cross sections with 2.2m. monochromator (9).

highly collimated and clean. Two additional properties - the high degree of polarization and the pulsed nature of the radiation (particularly in a storage ring) - have yet to be fully exploited.

The intriguing possibility of a tunable laser for the XUV has yet to be realised, but if it is, most other ources will presumably become redundant.

Monochromators

These should be divided into various categories. Firstly one uses a grating monochromator up to about 600eV (20Å) and crystal monochromators beyond. We concentrate here on grating monochromators, which can be further subdivided into normal incidence (up to 40eV) and grazing incidence (30eV-1keV) types. Various monochromators have been devised in the past either to maintain good resolution over an entire spectral range or to allow light source or experiment (or both) to remain stationary as the wavelength is scanned (2).

Recently a number of monochromators have been built to take

advantage of the high collimation of synchrotron radiation and take account
of the immovability of the source. In normal incidence, Skibowskii and
Steinman (16) have built a simple, high aperture monochromator but in
grazing incidence, the problems are more severe, particularly if high
resolution is being sought (17). Figure 2(a) shows schematically a low
resolution monochromator of the Miyake (18) – type built specifically for
the NINA synchrotron. Light from the synchrotron falls on the plane grating,
G, and then to one of two mirrors M_1 and M_2 with different radii of curva-
ture. The mirrors when placed in one of two positions, focus the virtually
parallel light onto the exit slit.

The reason for the complication of four alternative mirror positions
is to overcome the biggest problem with all continuum sources, that of
order overlap. For some incidence angle, α , there will be emitted into
angle β not only first order of wavelength λ but second order of wavelength
$\lambda/2$. This problem, and the stray light problem, are often the cause of
discrepancies between sets of experimental data from different groups.
The angles of incidence are so chosen in figure 2(a) that first order is
efficiently reflected but second order is not. The West et al (19) mono-

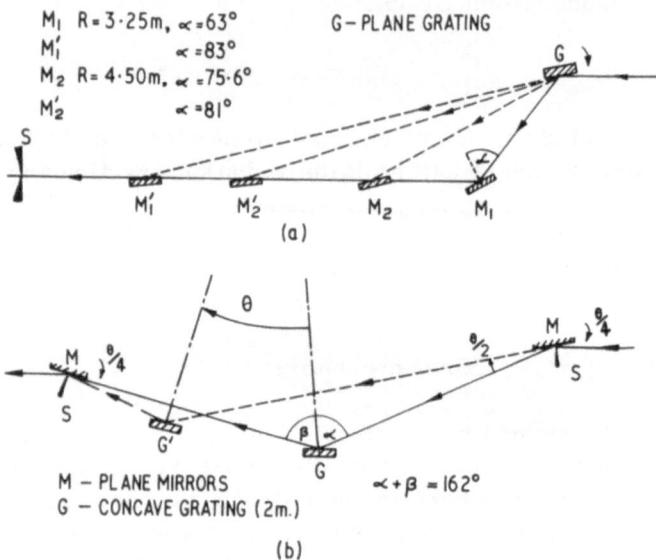

Figure 2(a). A plane grating monochromator for use with synchrotron
            radiation (18) ;
        (b) A concave grating, Rowland circle, constant deviation mono-
            chromator (20).

chromator achieved good order sorting over the range 30-300eV.

Figure 2(b) shows a monochromator using a Rowland circle mounting with constant deviation and slit-width-limited resolution (20). Slits are replaced by mirror-slits and as the grating rotates through θ, both mirrors must rotate through $\theta/4$ to maintain constant (or zero) deviation. The monochromator can be used in positive or negative order.

In monochromators of fixed slit width and hence constant wavelength band pass, the energy resolution worsens as the inverse square of the wavelength. It becomes increasingly more difficult therefore to analyse resonances in the continuum at high energy. This problem of determining true cross sections when the spectrometer bandwidth is greater than the width of the structure under investigation has been discussed in detail (9) (21).

At higher energies, crystal monochromators are used, coupled with x-ray bremsstrahlung sources. The immovability of the synchrotron source is a problem here also, and novel multi-crystal monochromators are being built (22). The highly directional property of synchrotron radiation can be used to good advantage and for an equivalent resolution, orders of magnitude in intensity are to be gained over conventional bremsstrahlung sources in certain types of x-ray experiment.

Absorption Cells

The problems here divide into two types, depending upon whether an inert gas is to be studied or a reactive metal vapour. With inert gases, one can define the path length, ℓ , with great accuracy and errors come mainly from pressure measurement and gas flow problems. West and Marr (23) have recently remeasured the inert gases He through Kr to an accuracy of better than 5% using a monochromator (19) free from order-overlap. Up to that time, the most accurate data was obtained with an ionization chamber absorption cell (24). One assumes, quite reasonably, that absorption leads to ionization with a yield of 100% and cross sections are determined by collecting the ions produced.

The determination of metal vapour cross sections is considerably more difficult, due primarily to vapour containment problems. To avoid coating slits, mirrors etc. with the vapour under investigation, a buffer gas is used to contain it. Ditchburn et al (25) devised a furnace to study the alkalis just above their ionization potentials and other groups (9),(26) have continued to use it. A schematic diagram is shown in figure 3(a).

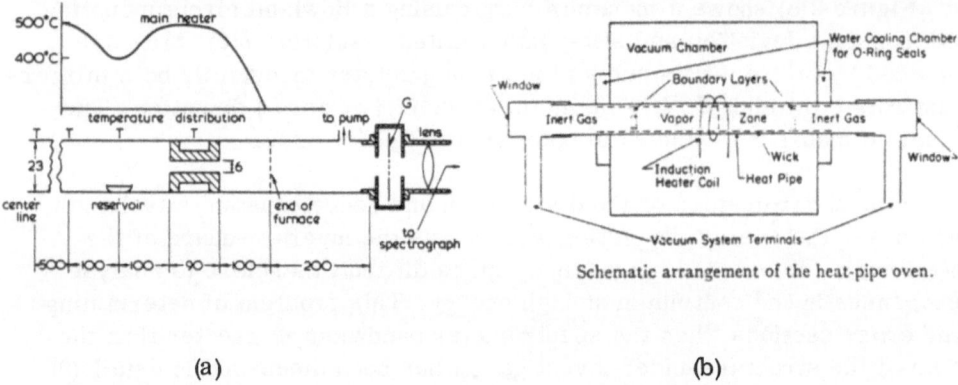

Schematic arrangement of the heat-pipe oven.

(a)   (b)

Figure 3 (a). Metal vapour absorption cell (3), right–hand side only and
furnace omitted. Dimensions in millimeters ; and (b) The
" heat–pipe " absorption cell (27).

The problem with such a cell is to determine the product "pl". In
fact one must determine the integral $\int pdl$ over the length of the tube and
accurate pressure measurement is limited by the knowledge of vapour
pressure versus temperature curves. Hope for the future may lie with the
" heat-pipe " (27) shown schematically in figure 3 (b). One should be able
to define a path length accurate with such a furnace and the pressure of
metal vapour is given effectively by the pressure of the inert gas buffer.
However, it is always advisable to confirm the column density of atoms
by an independent experiment.

Figure 3 (a) shows a lens, 3 (b) windows, none of which could be used
below 1100Å, the LiF cut-off. Here, differential pumping is required,
possibly in combination with thin films of zapon or Al to maintain the
monochromator below 10^{-3} torr while the furnace may be at a pressure
of 1 torr or more. A further problem occurs at 504Å, the ionization
potential of He, because now the buffer gas itself absorbs strongly and
one must choose carefully the inert gas with the lowest cross section for
the region under study. Situations may occur where the absorption cross
sections of the metal vapour is less than that of the buffer gas and the
buffer regions must be of minimal length.

By far the majority of results on metal vapour absorption have been
obtained using high resolution photographic techniques. One is usually
interested in accurate energy determination of resonances and their analysis
in terms of one- and two-electron excitations. In such circumstances, a

large range of absorption cells have been used. Beutler (28) used a wire wound furnace with rapidly circulated inert gases at each end to hinder diffusion of the vaporized species. At higher temperatures, Garton (1) and others (29) have used a King furnace. More recently Connerade and Mansfield (30) have used a cell with no buffer gas ; problems of inert gas absorption and differential pumping are thus circumvented. One requires only that the vapour remains in the furnace for the duration of an exposure (possibly minutes or even seconds). Garton and co-workers have used the Tomkins and Ercoli (31) induction furnace, with some important improvements, to observe structure in the rare earths at even higher temperatures.

Although particularly suitable for observation of absorption spectra of ionized species, Mehlman and Esteva (32) have used a double -BRV source (14), fired sequentially, to obtain neutral atom XUV absorption spectra. Other techniques for obtaining vaporized metal atoms involve the use of shock tubes, flash pyrolysis (1) and high-powered lasers.

LIMITATIONS OF THE PHOTON ABSORPTION EXPERIMENT

It is in the nature of photon absorption experiments that only certain excited states can be reached from the ground state. For example with the inert gas ground state configuration $^1S_0$, only those states will be observed with a $^1P_1^0$ component. In cases where various levels are populated in the ground state, for example $p^2 \ ^3P_{2,1,0}$, the possibilities are numerous but the limitations remain. (No such limitations occur in electron loss (33) experiments.) To overcome this fundamental limitation one wishes to populate in substantial numbers the excited states of atoms. Bradley et al (34) and Carlsten et al (35) have used tunable dye lasers to populate intermediate states from which absorption is observed. Bradley et al (34) have measured the photoionization cross section in the region of the $3p^2 \ ^1S_0$ resonance, by observing the absorption process $3s3p \ ^1P_1^0 \rightarrow 3p^2 \ ^1S_0$. The upper state is clearly forbidden from the ground state ($3s^2 \ ^1S_0$).

In figure 4 (a), the same Nd - glass laser was used to excite two dye lasers, one to produce the $3s3p \ ^1P_1^0$ state population, the other to produce photoionization. The metal vapour oven is in fact an ionization chamber and the electrons produced were monitored at each pulse of the laser. Figure 4 (b) shows the experimentally observed cross section in the region of the $3p^2$ resonance.

In discussing photoabsorption out of excited states, one should mention

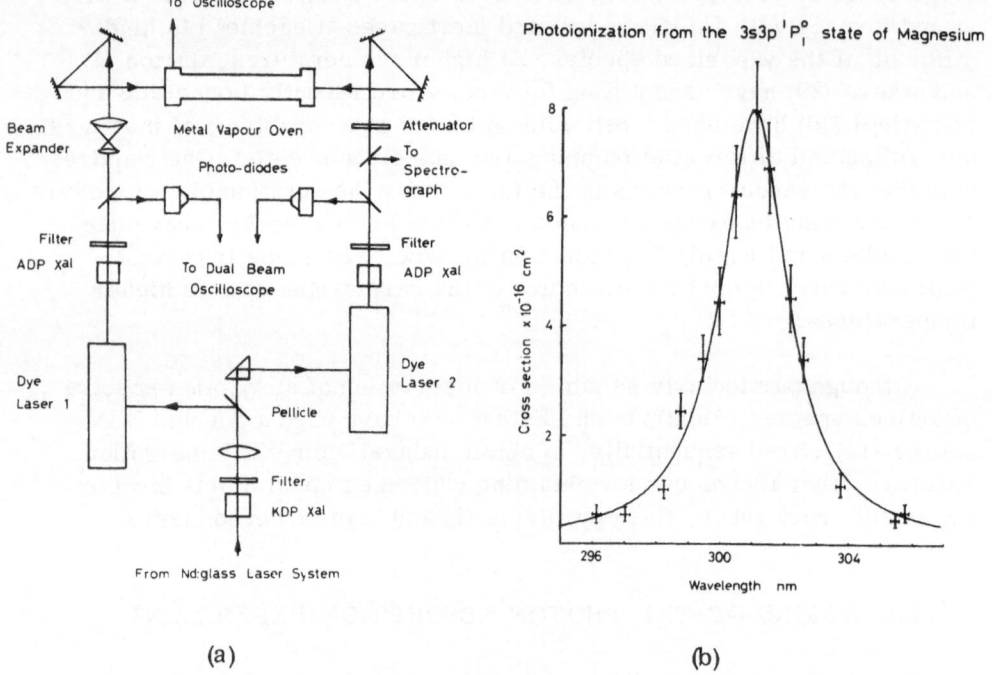

(a)   (b)

Figure 4 (a) Experimental arrangement with dye laser 1 providing the
              selectively exciting flux and laser 2 the photoionizing flux ;
         (b) The relative cross section obtained by tuning laser 2.

the work of Stebbings et al (36) who also used a tunable laser and a source
of metastable helium atoms produced by electron bombardment.

COMPLEMENTARY TECHNIQUES

The present review is limited basically to the straightforward
absorption experiment. However, the ionization chamber has already been
introduced as a method of measuring absorption cross sections. Mention
should be made, therefore, of the complementary techniques of photoionization mass spectrometry (37) and photoelectron spectroscopy.

Mass spectroscopy can be useful in two ways. Firstly, when atoms
and molecules co-exist in the same vapour, for example in the alkalis ;
secondly, when the photon energy is sufficiently high that multiple ionization occurs. Mass spectrometry can be used to determine the ratio of
multiply to singly ionized species (37). One advantage of this technique and

of photoelectron spectroscopy is clear. The use of atomic beams removes
the problems of metal vapour containment and of buffer gas absorption.
The advantages of photoelectron spectroscopy are discussed elsewhere (11).
Such experiments do depend heavily upon cross sections determined by the
photoabsorption techniques discussed here to place the " partial " cross
section measurements on an absolute basis.

EXPERIMENTAL RESULTS

This section will quite arbitrarily be separated into two parts. The
first will treat discrete resonance in the continuum and by this is meant
the manifestations of quasi-stable states due to two-electron excitation or
the excitation of an inner or subhell electron, with widths typically 0. 1eV
or less. The second will deal with gross structure stretching over many
eV (sometimes 20-30eV). In as much as both are manifestations of interchannel (or intrachannel) interaction and since discrete states and continuum are part of the same channel, this separation may appear rather
specious.

Discrete States in the Continuum

Absorption spectra exhibit numerous Rydberg series of lines, each
converging to an excited state of the positive ion. Most lines lie above the
first ionization limit and thus correspond to transitions from the ground
state to discrete states which are stationary only in the independent-elec-
tron model. Interchannel interaction, in this case between the discrete
states of one channel and the continuum of another channel of equivalent
energy causes the discrete states to autoionize with lifetimes of the order
of 10^{-13} seconds. (This short lifetime compared with a radiative lifetime
of $\sim 10^{-8}$ seconds is the reason why such states are probed by absorption
rather than emission spectroscopy).

Since He is the simplest neutral atom capable of exhibiting electron
correlation effects, its absorption spectrum is of considerable theoretical
interest (38,39,40). The spectrum of He observed in the 210-160Å region
is now well-known (41). Asymmetric resonances were observed, due to
two-electron excitation such as $1s^2$ $^1S_0 \rightarrow$ 2s2p $^1P_1^0$; 3s3p $^1P_1^0$, etc.

These resonances could be parameterised in terms of the Fano formalism (42), the cross section in the region of the 2s2p $^1P_1^0$ resonance
being given by :

$$\sigma = \sigma_A \frac{(q+\varepsilon)^2}{1+\varepsilon^2} \tag{3}$$

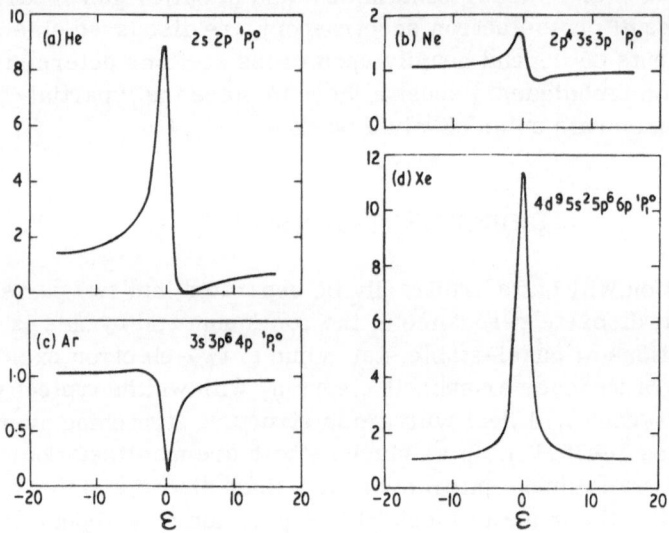

Figure 5. Resonance profiles in the inert gases (a) He, with q=-2.8 and
Γ =0.038eV ; (b) Ne, with q=-2.0, Γ =0.01eV and ρ^2=0.17 ;
(c) Ar, with q=-0.2, Γ =0.08eV and ρ^2=0.86 ; (d) Xe, with
q~200, Γ =0.11eV and ρ^2 ~ 0.0003.

where σ_A is the cross section in the wings of the resonance, q is a line
profile index involving 3 matrix elements and ε an energy variable in
units of the half-width of the resonance $\Gamma/2$. In as much as the 2s2p $^1P_1^0$
resonance is the simplest example of a single state interacting with a
single set of continuum states (1s εp) the resonance should have a zero
minimum in the cross section.

Figure 5(a) shows the line profile for He 2s2p $^1P_1^0$ and the zero
minimum. From equation 3 it can be seen that q can take positive or
negative values, and in the case of the 3s3p $^1P_1^0$ resonance, q=+2.0.
Moreover, this resonance is much less obvious than the 2s2p $^1P_1^0$
resonance even though the width is larger. One must introduce a further
parameter, ρ^2, in the situation of a single resonance and more than one
continuum, here the 1sεp, 2sεp, 2pεs and 2pεd continua. In this somewhat more general case, the cross section is given by

$$\sigma = \sigma_A \frac{(q+\varepsilon)^2}{1+\varepsilon^2} + \sigma_B \qquad (4)$$

where σ_A is the part of the continuum that does, σ_B that which does not, interact with the discrete state and

$$\rho^2 = \frac{\sigma_A}{\sigma_A + \sigma_B} \tag{5}$$

where ρ is the autoionization-to-dipole correlation coefficient or overlap integral (43). Dhez and Ederer (44) have parameterised the 3s3p $^1P_1^0$ resonance, giving $\rho^2 = 0.012$ and $\Gamma = 0.13\,eV$. Autoionization leads to the n=2 continua, direct ionization to the n=1 continuum ; hence the overlap is small and ρ^2 is small.

Returning to the series approaching the n=2 limit, one observes a strong (+) and a weak series (-) of resonances. A physical interpretation of the series in terms of the correlated in-phase and out-of-phase motions of the two excited electrons suggested by Cooper et al (45) still appears to have some validity although the detailed analysis has changed considerably (39).

In Ne and Ar, the excited states due to one-electron excitation, nsnp^{6}mp $^1P_1$, and two-electron excitation states, ns^2 np^{4}mℓ m'ℓ', occur in the same energy range. In the case of Ne (46), the 2s^2 2p^4 3s3p $^1P_1^0$ resonance at 45eV has $\rho^2 = 0.17$ (overlap of 41%) see figure 5(b), whereas in the 2s2p^{6}3p $^1P_1^0$ resonance the final state overlap via the two alternative channels is 84%. By comparing the strengths of the two-electron and one-electron excitation states one can gauge the importance of two-electron excitation processes (correlation effects), something that could be confirmed by photoelectron spectroscopy (11).

Figure 5(d) shows the Xe 4d $\rightarrow$ 6p resonance at 65eV and is an example of the kind of resonance profile generally seen at higher photon energies. Since autoionization leads to doubly excited states of the Xe ion, and direct photoionization to singly excited ones, again ρ^2 is very low and no minimum in the cross section is observed. The profile is Lorentzian (47) with q very large (~ 200) and ρ^2 almost zero. This situation and that observed in Ar in figure 5(c) can best be understood if equation 4 is rewritten in a more instructive way :

$$\sigma = (\sigma_A + \sigma_B)\left[1 - \frac{\rho^2 q^2}{1+\varepsilon^2} - \frac{\rho^2}{1+\varepsilon^2} + \frac{2q\varepsilon\rho^2}{1+\varepsilon^2}\right] \tag{6}$$

The second term in the square brackets represents the Lorentz profile,

the third term the effect of spectral repulsion and the last term the interference which averages out over the whole profile. When q~0, as in Ar, a
" window " resonance is observed (48) since absorption through the discrete
state fails to compensate the effect of spectral repulsion.

The simple situations described so far rarely occur in practise due
to intermixing of series approaching the same (or only slightly displaced)
ion limits. But when they do, certain useful relationships emerge. The
line profile indices q and ρ^2 remain constant within a series (43). The
width of autoionizing series reduces as the inverse cube of the effective
quantum number, n^* , and the resonances get closer at the same rate.
(At higher photon energies, the width of resonances is governed by the
Auger process and widths are constant). The oscillator strength per unit
energy range remains constant through the Rydberg series into the continuum
and one is able to evaluate the absorption " jump " at the opening of a new
channel by parameterising the Rydberg series. This " jump " may even be
negative, as in the case of the Ar " window " series discussed above.

Usually, however, one cannot assume resonances to be isolated and
according to well known selection rules, configuration interaction (mixing)
will occur. Mies (49) extended the Fano (42) theory to consider the problem
of overlapping resonances which are coupled through the underlying continuum. If the widths of adjacent resonances are of the order of their
distance apart, and the so-called " overlap matrix " is large, then the
effect on the absorption profiles can be profound. The $3d^9 4s^2 4p$ $^1P_1$, $^3P_1$,
$^3D_1$ resonances in the photoionization continuum of zinc (50) were analysed
using the Mies theory.

A unified treatment of perturbed series and continua has been developed
recently by Fano and co-workers (51). It is based on the multi-channel
quantum defect theory of Seaton (52). Lu (53), for example, has analysed
the photoabsorption spectrum of Xe approaching the $^2P_{\frac{1}{2},\frac{3}{2}}$ limits at ~12eV.
Brown and Tilford (54) have used the Lu-Fano graphical technique to extend
the analysis of the absorption spectrum of SiI. The theory makes no distinction between interacting series of the same J-value and parity and therefore
all reference to term symbols, series classifications and correlation of
series with different ion limits has (unfortunately) to be abandoned.

Gross Features in the Photoionization Continuum

Absolute XUV absorption cross section data on all but the rare gases
is very incomplete and where it exists is often suspect and needs repeating
and extending to higher energies. In many instances relative cross sections
only are measured ; some are put on an absolute basis by application of

the sum rule (55). In others the absorption spectrum is obtained from thin film transmission measurements, where the perturbation due to the atomic environment is a complicating factor. Nevertheless, by the early 1960's it was apparent that, although the hydrogen-like treatment of photoionization was adequate for the x-ray region, accumulating data on atoms and particularly on the rare gases showed that it was inadequate for the XUV region.

From an experimental standpoint, one can view the systematic gathering of experimental data in two ways. One can either measure the cross sections for a group of atoms of similar electron configuration or alternatively, one can study a set of absorption features in a particular energy region as a function of increasing atomic number. In this way such phenomena as the onset of shell collapse can be investigated.

The rare gases. These have been the testing ground for many theoretical calculations partly because their absolute cross sections are known to high accuracy and partly because they are conveniently interspersed throughout the periodic table.

Cooper (56) used a single electron model with realistic central fields to predict the gross spectral shape of the cross sections of the rare gases from threshold to 150eV. He explained the " Cooper " minimum in the Ar $3p \rightarrow \varepsilon d$ photoionization and showed why the near-threshold behaviour of Ne would be expected to be quite different. Although the Ar calculations predicted a threshold cross section that was too high and a fall-off in cross section that was too abrupt, Cooper pointed out that correlation effects would be expected to smear out the rapid drop to give better agreement with experiment. Subsequently, more sophisticated calculations have born out this prediction (57, 58, 59). In figure 6(a) is shown the experimental total cross section data of West and Marr (23) which is in fact a weighted mean of all data taken in the 15-45eV region. Shown for comparison are the RPAE calculation of Amusia (58), and a recent R-matrix calculation of Burke and Taylor (60). The experimental and R-matrix curves are not drawn in the region of the series of " window " resonances, but the theory provides the parameters q, Γ and ρ^2 for the first two members that are in fair agreement with experiment (61).

Another favourite for theoretical calculations has been the 4d partial cross section of Xe. When the cross section was first measured (62) (63), a broad absorption feature $\sim$ 35eV in half-width was observed centred some 30eV above the 4d threshold. There appeared to be very little or no absorption " jump " at threshold. The oscillator strength due to transitions $4d \rightarrow \varepsilon f$ seemed to be held back in some way and the lack of discrete struc-

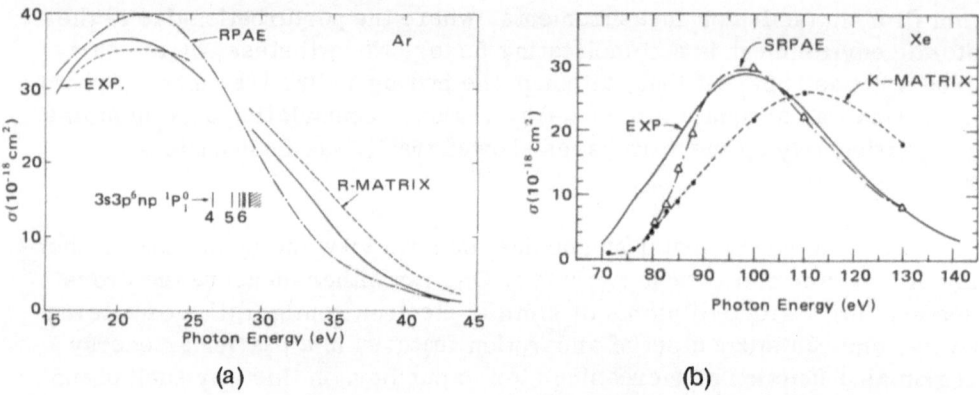

Figure 6(a). The total absorption cross section of Ar (23). RPAE by
            Amusia et al (58) ; R-matrix by Burke and Taylor (60). The
            energy locations of the " window " resonances are shown.
        (b). The 4d → εf cross section in Xe (67). SRPAE by Lin (59) ;
            K-matrix by Starace (57).

ture approaching the limit due to 4d → nf confirmed this (64). Here was
an example of a potential barrier causing a delayed onset of oscillator
strength. Starace (57) pointed out the reason for failure of the Hartree-
Slater model to reproduce the height, position and shape of the experimental
curves - the strong exchange interaction between excited electron and the
remaining $4d^9$ electrons. The intrashell interaction caused a " shape "
resonance near threshold. Hartree-Fock (65) and RPAE calculations (58),
(66) have since found fair to good agreement with experiment. Figure 6(b)
compares the experimental data of Haensel et al (67) obtained at the DESY
Synchrotron with the K-matrix result of Starace (57) and the SRPAE calculation of Lin (59). The agreement in the region of the peak absorption may
be somewhat fortuitous since double and triple electron excitation or ionization are not specifically included.

Deslattes (68), using a conventional x-ray bremsstrahlung source and
double crystal monochromator, measured the absorption spectrum of Xe
around the $M_{IV, V}$ edge at 700eV. At the time it was surprising that at such
high energies, non-hydrogen-like behaviour should occur. Again a centrifugal barrier was the reason but in this case the intershell spectrum was
reasonably well reproduced with a Hartree-Slater model calculation (65).

The alkalis.   The oscillator strength of the valence electron of the
alkalis is known to be concentrated in the lowest member of the principal
series. The intensity of successive lines decreases rapidly until at the

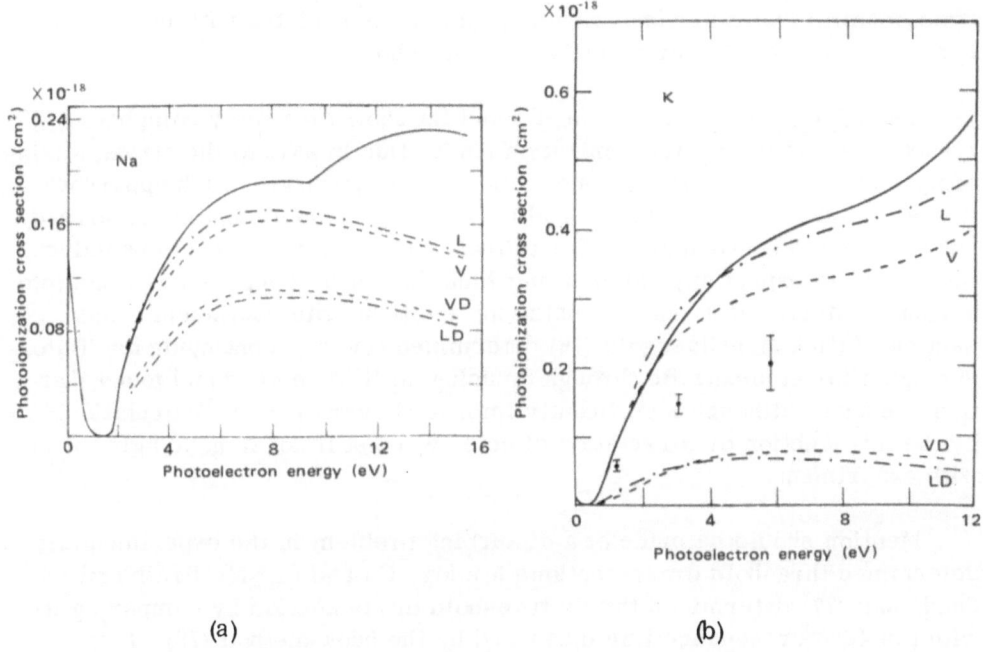

(a)   (b)

Figure 7(a). The absorption cross section of Na from Hudson and Carter
            (70) compared with the theory of Chang (73) ;
        (b). The cross section of K-Hudson and Carter (71), I Marr and
            Creek (74). Dashed lines from the theory (73).

limit the cross section is very low. Hartree-Fock calculations of Seaton
(69) showed that the photoabsorption goes through a near-zero minimum
a few eV above threshold before rising again at higher photon energies.
The experimental data of Hudson and Carter (70) (71), using a Ditchburn-
type furnace show a zero minimum in the cases of Na and K.

Many theoretical calculations have been performed on the alkalis (72)
but in figure 7(a) and 7(b) are shown the recent calculations of Chang (73).
It appears that the effect of virtual excitation to the inner p shell has a
drastic effect on the photoionization cross sections of the outer s electrons.
The lower curves include the intershell correlation effects.

Although the rise in K at 10eV can find explanation in the close proximity of resonances due to excited states $3p^54snℓ$, in Na the equivalent
structure is at much higher energy and the cause of the disagreement at
10-16eV is not known. The cross section should be remeasured and extended

to higher energies, to overlap the work of Wolff et al (75) using the DESY
Synchrotron. There are known to be problems associated with molecular
species in the alkalis particularly near threshold.

The alkaline earths.   Ca, Sr and Ba show extremely complex
structure close to the first ionization limit. Due in part to the strong mixing
of np^2 and $(n-1)d^2$ in the ground state ns^2 configuration, such apparent
two-electron transitions as $(n-1)d \, mp$ show up strongly against an already
weak threshold continuum similar to that of the alkalis. It would be naive,
therefore, to expect any calculation which did not take such transitions into
account to arrive at a sensible value for the threshold continuum cross
section. Altick and Glassgold (76) determined line and continuum oscillator
strengths for elements Be through Sr using an RPA method and found that
their results, although significantly improved over previous calculations
by the introduction of correlation effects, were still not in good agreement
with experiment.

Mention should be made of a disturbing problem in the experimentally
determined threshold cross sections for Mg, Ca and Sr. Mc Ilrath and
Sandeman (77) determined the Ca threshold cross section by comparing it
with the 4226Å resonance line measured by the hook method (78). They
found the Carter et al (79) value, already increased by a factor 2.8 over
earlier furnace measurements because of improved vapour pressure data,
was still out by a factor of 2.2. By determining the oscillator strength of
the Rydberg series by the hook methods, Parkinson et al (80) find similar
factors of 2 and 1.9 in Mg and Sr. Presumably inaccurate vapour pressure
data is still the major cause of the discrepancy. There was no sign of
molecular absorption in the furnace measurements of Ditchburn and Marr
(81) on Mg.

A large effort (82,83,84) has recently been expended to determine the
absorption spectrum of Ba above 15eV. Two regions are of particular
interest ; firstly, the 15-25eV region, where complex structure is observed
due to excitation of the 5p electron and secondly, the region between 90 and
150eV, where excitation of the 4d electron occurs. Fano (85) comments
that the continuous absorption below the 5p excitation structure is more
intense than that above the 5p ionization threshold. The implication is that
electron correlation effects have redistributed much of the $5p^6$ oscillator
strength into the $6s^2$ continuum. The 4d electron excitation spectrum
consists of discrete structure below the 4d ionization threshold and a broad
feature some 25eV in half-width, above it. Hansen et al (86) disagrees
with the Ederer et al (84) interpretation of the overall structure, saying
in particular that the excited state $4d^9 4f \ ^1P_1^0$, assumed to be responsible
for the broad feature at 110eV, lies below the ionization limit.

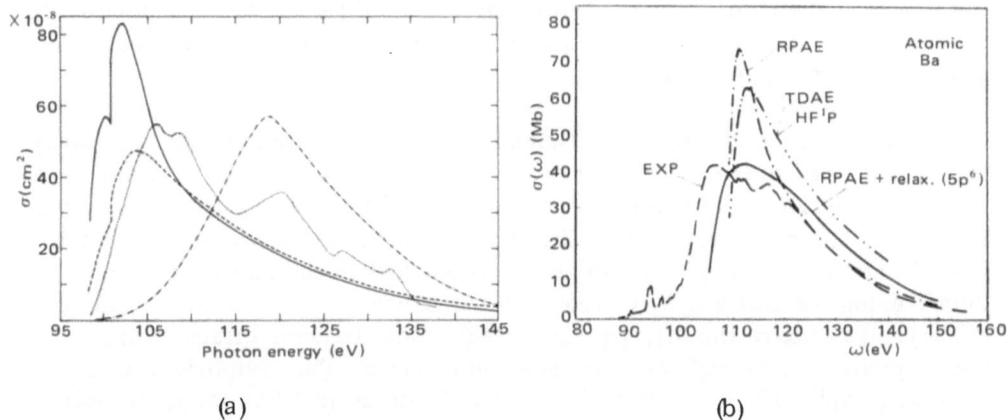

(a)   (b)

Figure 8(a). The absorption cross section of Ba. Dotted line is experiment
            (82). For meaning of other curves, see Fliflet et al (87) ;
        (b). The absorption cross section of Ba (83) and various theoreti-
            cal models (89).

Figure 8(a) shows the spectrum obtained with the Bonn Synchrotron
(82) compared with theoretical calculations of Fliflet et al (87) and Wendin
(88), while figure 8(b) shows the curve obtained using the DESY Synchrotron and a heat-pipe (83) compared with the RPAE calculation of Wendin
(89) including relaxation effects. Undulations in both experimental curves
are due to two-electron excitation states. In both cases the relative cross
section data were put on an absolute basis by application of the sum rule,
assuming the broad feature to be due exclusively to 4d-shell excitation.
This is not a safe procedure since some of the oscillator strength will be
associated with outer electron excitation, via intershell interaction.

    The rare earths.   Zimkina et al (90) made a systematic study of the
rare earth absorption spectra in the 50-500eV region. They worked with
thin films but this did not obscure the basically atomic nature of the spectra
obtained, in particular the large absorption features around 150eV due to
the excitation of the 4d electrons. Later work (91) led Dehmer et al (92)
to reconsider the original suggestion of Fomichev et al (93) that the structure observed was due to transitions $4d^{10}4f^{N} \to 4d^{9}4f^{N+1}$ in the triply
ionized atom. It was suggested that the large exchange interaction between
the 4f electrons and 4d hole spreads the levels of these upper state configurations over 20eV or more, so that the higher levels autoionized into the
$4d^{9}4f^{N}\varepsilon f$ continuum. This interpretation was supported by the evidence
that the strength of the absorption feature decreased in relation to the
number of 4f vacancies until at Lu ($4d^{10}4f^{14}$), no structure was observed.

Scaling factors were used (94) (95) in order to achieve good agreement on the discrete and continuous structure observed. Starace (96) has performed a recent ab initio RPAE calculation for La and achieved equally good agreement with experiment.

The transition elements.   Dehmer et al (92) pointed out that the ideas involved in interpreting the rare earth spectra might well apply to any optical transitions where the final state depended critically upon a centrifugal barrier. Earlier, the thin film transmission spectrum of the transition elements Ti through Ni had been studied (97) in the energy range 30-200eV using the DESY Synchrotron and the surprise was the large width of the structure above the $M_{II, III}$ edge. Centrifugal barrier effects were again important, as well as exchange interaction. The transitions involved here were $3p^6 \, 3d^N \rightarrow 3p^5 \, 3d^{N+1}$. Combet Farnoux (98) has made extensive calculations which appear to agree well with experimental data.

CONCLUSIONS

One must differentiate between two types of experiment performed in the XUV, those that attempt to determine absolute cross sections and those that simply photograph the atomic absorption spectra. In the latter type of experiment, many beautiful spectra have been obtained and subsequently analysed to reveal the existence of correlation effects. Such spectra are essential in order to separate the atomic and solid state effects. In the former experiment, it is disturbing that cross sections for those elements that have received much attention in the past, the alkaline earths, are in error by large factors near threshold. It may be that other elements suffer from similar problems and that relating continuum cross sections to discrete state oscillator strengths determined by an independent technique is always advisable.

Even when absorption cross sections are known with great confidence they cannot tell the whole story. At any photon energy, one is determining the response from all of the atomic electrons in their various shells. Evidence is fast accumulating from absorption spectroscopy that inter-electron correlations can redistribute the continuum oscillator strength in unusual ways. Calculations show (99) that partial cross sections may be reduced in certain energy ranges, enhanced in others, due to intershell interaction. In Ar, for example, the 3s partial cross section is lowered close to threshold due to interaction with the 3p outer shell. In Xe, on the other hand, the 5s and 5p cross sections are expected to be enhanced in the region of the 4d excitation. Such phenomena cannot, of course, be explored by absorption spectroscopy, which in Xe determines the sum of the partial

4d, 5s and 5p cross sections and the double and triple excitation and ionization processes.

Photoelectron spectroscopy (PES) can examine such effects because the 5s electron can be differentiated from the 4d electron by energy analysis. Experiments have recently been performed in Ar (100) and Xe (101), giving evidence of the intershell interaction mentioned above. Mass spectrometry is required as a complementary technique. In the future, with the possibility of using dedicated storage rings as sources of intense XUV radiation, emphasis is likely to shift to the more sophisticated PES and mass spectrometry techniques. One can, for example, envisage experiments on determinations of the angular distribution of electrons from the interaction of polarised atoms with polarised radiation.

REFERENCES

(1) W.R.S. Garton, Adv. Atom. Molec. Phys., ed. D.R. Bates and I. Esterman 2, 93 (1966).

(2) J.A.R. Samson, " Techniques of Vacuum Ultraviolet Spectroscopy ", Wiley : New York (1967).

(3) G.V. Marr, " Photoionization Processes in Gases ", Academic Press : New York (1967).

(4) R.D. Hudson and L.J. Kieffer, Atomic Data 2, 205 (1971).

(5) J.H. Hubbell, Atomic Data 3, 241 (1971).

(6) W.J. Veigele, Atomic Data 5, 51 (1973).

(7) A. Kasdan and W.C. Lineberger, Phys. Rev. A10, 1658 (1974).

(8) R. Haensel, this volume.

(9) R.D. Hudson, V.L. Carter and P.A. Young, Phys. Rev. 180, 77 (1969).

(10) J.A.R. Samson, J. Opt. Soc. Am. 54, 6 (1964).

(11) M.O. Krause, this volume.

(12) Y. Tanaka, R.E. Huffman and J.C. Larrabee, J. Quant. Spect. Rad. Trans. 2, 451 (1962).

(13) W.R. Garton, J. Sci. Instrum. 36, 11 (1959).

(14) G. Balloffet, J. Romand and B. Vodar, C.R. Acad. Sci. (Paris) 252, 4139 (1961).

(15) K. Codling, Rep. Prog. Phys. 36, 541 (1973).

(16) M. Skibowskii and W.S. Steinmann, J. Opt. Soc. Am. 57, 112 (1967).

(17) C. Kunz, Int. Symp. for Synchrotron Radiation Users, Daresbury 1973, ed. G.V. Marr and I.H. Munro.

(18) K.P. Miyake, R. Kato and H. Yamashita, Sci. Light 18, 39 (1969).

(19) J.B. West, K. Codling and G.V. Marr, J. Phys. E 7, 137 (1974).

(20) K. Codling and P. Mitchell, J. Phys. E 3, 685 (1970).

(21) D.L. Ederer, Appl. Optics 8, 2315 (1969).

(22) J.H. Beaumont and M. Hart, J. Phys. E 7, 823 (1974).

(23) J.B. West and G.V. Marr, J. Phys. B, Proc. Ray. Soc. (London) (to be publ.).

(24) J.A.R. Samson, Adv. Atom. Molec. Phys., ed. D.R. Bates and I. Esterman, 2, 177 (1966).

(25) R.W. Ditchburn, J. Tunstead and J.G. Yates, Proc. Roy. Soc. (London) A181, 386 (1943).

(26) G.V. Marr and J.M. Austin, Proc. Roy. Soc. (London) A310, 137 (1969).

(27) C.R. Vidal and J. Cooper, J. Appl. Phys. 40, 3370 (1969).

(28) H. Beutler, Z. Physik 93, 177 (1935).

(29) C.M. Brown, S.G. Tilford, R. Tousey and M.L. Ginter, J. Opt. Soc. Am. 64, 1665 (1974).

(30) J.P. Connerade and M.W.D. Mansfield, Proc. Roy. Soc. (London) A335, 87 (1973).

(31) F.S. Tomkins and B. Ercoli, Appl. Optics 6, 1299 (1967).

(32) J.M. Esteva and G. Mehlman, Astrophys. J. 193, 747 (1974).

(33) P.G. Burke, Adv. Atom. Molec. Phys., ed. D.R. Bates and I. Esterman 4, 173 (1968).

(34) D.J. Bradley, P. Ewart, J.V. Nicholas, J.R.D. Shaw and D.G. Thompson, Phys. Rev. Lett. 31, 263 (1973).

(35) J.L. Carlsten, T.J. McIlrath and W.H. Parkinson, J. Phys. B. 7, L244 (1974).

(36) R.F. Stebbings, F.B. Dunning, F.K. Tittel and R.D. Rundel, Phys. Rev. Lett. 30, 815 (1973).

(37) R.B. Cairns, H. Harrison and R.I. Schoen, Adv. Atom. Molec. Phys. ed. D.R. Bates and I. Esterman 8, 131 (1972).

(38) K.T. Chung and I-h Chen, Phys. Rev. Lett. 28, 783 (1972).

(39) C.D. Lin, Phys. Rev. A10, 1986 (1974).

(40) D.R. Herrick and O. Sinanoglu, Phys. Rev. A11, 97 (1975).

(41) R.P. Madden and K. Codling, Astrophys. J. 141, 364 (1965).

(42) U. Fano, Phys. Rev. 124, 1866 (1961).

(43) U. Fano and J.W. Cooper, Phys. Rev. 137, 1364 (1965).

(44) P. Dhez and D.L. Ederer, J. Phys. B. 6 L59 (1973).

(45) J.W. Cooper, U. Fano and F. Prats, Phys. Rev. Lett. 10, 518 (1963).

(46) K. Codling, R.P. Madden and D.L. Ederer, Phys. Rev. 155, 26 (1967).

(47) D.L. Ederer and M. Manalis, J. Opt. Soc. Am. 65, 634 (1975).

(48) R.P. Madden, D.L. Ederer and K. Codling, Phys. Rev. 177, 136 (1969).

(49) F.H. Mies, Phys. Rev. 175, 164 (1968).

(50) G.V. Marr and J.M. Austin, J. Phys. B 2, 107 (1969).

(51) U. Fano, J. Opt. Soc. Am. 65, 979 (1975).

(52) M.J. Seaton, Proc. Phys. Soc. (London) 88, 801 (1966).

(53) K.T. Lu, Phys. Rev. A4, 579 (1971).

(54) C.M. Brown and S.G. Tilford, J. Opt. Soc. Am. 65, 385 (1975).

(55) U. Fano and J.W. Cooper, Rev. Mod. Phys. 40, 441 (1968).

(56) J.W. Cooper, Phys. Rev. 128, 681 (1962).

(57) A.F. Starace, Phys. Rev. A2, 118 (1970).

(58) M.Ya Amusia, N.A. Cherepkov and L.V. Chernysheva, Soviet Phys.
      J.E.T.P. 33, 90 (1971).

(59) C.D. Lin, Phys. Rev. A9, 181 (1974).

(60) P.G. Burke and K.T. Taylor, J. Phys. B16, 2620 (1975).

(61) H.P. Kelly and R.L. Simons, Phys. Rev. Lett. 30, 529 (1973).

(62) D.L. Ederer, Phys. Rev. Lett. 13, 760 (1964).

(63) A.P. Lukisrskii, I.A. Britov and T.M. Zimkina, Opt. Spectrosk.
      17, 438 (1964).

(64) K. Codling and R.P. Madden, Phys. Rev. Lett. 12, 106 (1964).

(65) D.J. Kennedy and S.T. Manson, Phys. Rev. A5, 227 (1972).

(66) G. Wendin, J. Phys. B 6, 42 (1973).

(67) R. Haensel, G. Keitel, P. Schreiber and C. Kunz, Phys. Rev. 188,
      1375 (1969).

(68) R.D. Deslattes, Phys. Rev. Lett. 20, 483 (1968).

(69) M.J. Seaton, Proc. Roy. Soc. (London) A208, 418 (1951).

(70) R.D. Hudson and V.L. Carter, J. Opt. Soc. Am. 57, 651 (1967).

(71) R.D. Hudson and V.L. Carter, J. Opt. Soc. Am. 57, 1471 (1967).

(72) J.C. Weisheit, Phys. Rev. A5, 1621 (1972).

(73) T.N. Chang, J. Phys. B 8, 743 (1975).

(74) G.V. Marr and D.M. Creek, Proc. Roy. Soc. (London) A304, 233
      (1968).

(75) H.W. Wolff, K. Radler, B. Sonntag and R. Haensel, Z. Physik 257,
      353 (1972).

(76) P.L. Altick and A.E. Glassgold, Phys. Rev. 133, 632 (1964).

(77) T.J. McIlrath and R.J. Sandeman, J. Phys. B 5, L217 (1972).

(78) F.P. Banfield, M.C.E. Huber, W.H. Parkinson and E.F. Tubbs,
      Appl. Optics 12, 1279 (1973).

(79) V.L. Carter, R.D. Hudson and E.L. Breig, Phys. Rev. A4, 821
      (1971).

(80) W.H. Parkinson, E.M. Reeves and F.S. Tomkins, J. Phys. B.

(81) R.W. Ditchburn and G.V. Marr, Proc. Phys. Soc. (London) A66,
      655 (1953).

(82) J.P. Connerade and M.W.D. Mansfield, Proc. Roy. Soc. (London)
      A341, 267 (1974).

(83) P. Rabe, K. Radler and H. Wolff, Proc. Int. Conf. on VUV Radiation
      Physics, Hamburg 1974 ed. E.E. Koch, R. Haensel and C.Kunz, p.247.

(84) D.L. Ederer, T.B. Lucatorto E.B. Saloman, R.P. Madden and J.
      Sugar, J. Phys. B 8, L21 (1975).

(85) U. Fano, Comments on Atom. Molec. Phys. $\underline{4}$, 119 (1973).

(86) J.E. Hansen, A.W. Fliflet and H.P. Kelly, J. Phys. B $\underline{8}$, L127 (1975).

(87) A.W. Fliflet, R.L. Chase and H.P. Kelly, J. Phys. B $\underline{7}$, L443 (1974).

(88) G. Wendin, Phys. Lett. $\underline{46}$A, 119 (1973).

(89) G. Wendin, Phys. Lett. $\underline{51}$A, 291 (1975).

(90) T.M. Zimkina, V.A. Fomichev, S.A. Gribovskii and I.I. Zhukova, Soviet Phys. - Solid St. $\underline{9}$, 1128 (1967).

(91) R. Haensel, P. Rabe, B. Sonntag and C. Kunz, Solid St. Commun. $\underline{8}$, 1845 (1970).

(92) J.L. Dehmer, A.F. Starace, U. Fano, J. Sugar and J.W. Cooper, Phys. Rev. Lett. $\underline{26}$, 1521 (1971).

(93) V.A. Fomichev, T.M. Zimkina, S.A. Gribovskii and I.I. Zhukova, Soviet Phys.- Solid St. $\underline{9}$, 1163 (1967).

(94) J. Sugar, Phys. Rev. B $\underline{5}$, 1785 (1972).

(95) J.L. Dehmer and A.F. Starace, Phys. Rev. B$\underline{5}$, 1792 (1972).

(96) A.F. Starace, Phys. Rev. B$\underline{5}$, 1773 (1972).

(97) B. Sonntag, R. Haensel and C. Kunz, Solid St. Commun. $\underline{7}$, 597 (1969).

(98) F. Combet Farnoux, Physica Fennica $\underline{9}$S, 80 (1974).

(99) M. Ya Amusia, Proc. Int. Conf. on VUV Radiation Phys., Hamburg 1974 ed. E.E. Koch, R. Haensel and C. Kunz, p.205.

(100) R.G. Houlgate, J.B. West, K. Codling and G.V. Marr, J. Phys. B$\underline{7}$, L470 (1974).

(101) J.B. West, P.R. Woodruff, K. Codling and R.G. Houlgate, J. Phys. B, to be publ.

PHOTOELECTRON SPECTROMETRY : EXPERIMENTS WITH ATOMS

Manfred O. Krause

Transuranium Research Laboratory, Oak Ridge National

Laboratory, Oak Ridge, Tennessee 37830, USA

INTRODUCTION

These lectures will present a brief review of the technique of Photoelectron Spectrometry and outline its capacity and potential to delineate
the electronic structure and dynamics of atoms. Experimental results and,
in particular, those photoelectron spectrometric experiments that give
evidence of many-electron interactions in atoms will be discussed. I shall
emphasize experiments that are especially instructive or promising for
future work in the area of electron-electron correlation effects. This
should introduce the student to the characteristics of the technique and the
manifestations of many-electron interactions in photoelectron spectra, and
at the same time, will familiarize him with the latest experimental results,
since the most relevant data are of recent origin.

While the term many-electron interaction, or electron correlation, is
defined in different ways by different authors, and may undergo changes
as time goes on, it will be taken loosely in the context of these lectures
as a description of all effects that are not described adequately, or cannot
be described at all, by the best relativistic Hartree-Fock independent,
single-particle model of the atom. Electron correlation (EC) would then
comprise all effects that require the explicit consideration of the interaction of two electrons or, in a somewhat more general sense, of two configurations or states.

Electron correlation effects exhibited in photoelectron spectrometry
may be strong and direct, such as the excitation of two or more electrons

in the photoionization process, generally known as shake-up and shake-off ;
or they may be weak and indirect such as the shift of an inner shell energy
level by a small amount, for example by less than 1eV for the Ne 1s level
that has a binding energy of 870eV. The EC effects may give rise to new
spectral features, such as shake-up and configuration interaction satellites,
or they may be embedded in " single-electron " properties such as the energies, natural widths and photoionization cross sections of a single-electron
level characterized by its quantum numbers $n\ell j$.

Study of EC is of interest from the point of view of providing a deeper
understanding of atomic structure and its response to external perturbations.
Similarly, investigation of EC effects is of " practical " significance, having
immediate consequences in diverse areas of chemistry, solid state physics
and radiation technology. Consider the following two examples. First, the
outer electrons of mercury responsible for chemical bonding and the solid
state behavior of this element have not only s-electron character, as a
simple description of these electrons as 6s electrons might indicate, but
also p and d character due to an admixture of p and d symmetric configurations present in the atomic state. An alteration and increased complexity
of the behavior of mercury in compounds follows as a natural consequence.
Second, the total and, especially the partial photoionization cross sections
may be changed, sometimes drastically, by the presence of various many-
electron interactions. The strong depression of the 3s partial photoionization cross section of argon near 40eV photon energy is such a manifestation
of EC in photoeffect. Thus, reliable calculations for various radiation-in-
matter processes can only be made if EC effects are recognized and properly taken into account. Similarly, the full potential of various analytical methods that are based on atomic phenomena (e.g. ESCA, Auger
Electron and X-Ray Spectrometry) can only be realized if EC is included.

Electron correlation is not a phenomenon that has just appeared on the
scene, but has come into focus in recent years through the availability of
more refined experimental and theoretical tools. While electron spectrometry is not the only technique suitable to demonstrate the existence of EC
and to investigate EC, it is one of the most powerful techniques since, by
virtue of the photoelectric effect, it is capable of providing a very direct
and detailed view of the electronic structure and dynamics. In electron spectrometry we can " see " and inspect the atom, electron by electron, configuration by configuration and process by process. Figure 1 is a good illustration of this ; it shows the photoelectron spectrum of mercury obtained
with the monochromatized AlKα X-rays as a radiation source giving a partial view of electronic structure and dynamic features of this atom.

General treatises on electron spectrometry, usually with emphasis on photoelectron spectrometry, have been given by Siegbahn et al. (1-3), Sevier (4), Krause (5), Rudd and Macek (6), and Carlson (7) ; reviews and papers of specific interest to the subject of this school have been presented by Carlson (8), Spears et al. (9), Krause (10, 11), Krause and Wuilleumier (12), Wuilleumier and Krause (13), Gelius (14), Carlson et al. (15), Kowalczyk et al. (16), Houlgate et al. (17), Berkowitz et al. (18), Sützer and Shirley (19), and others (20).

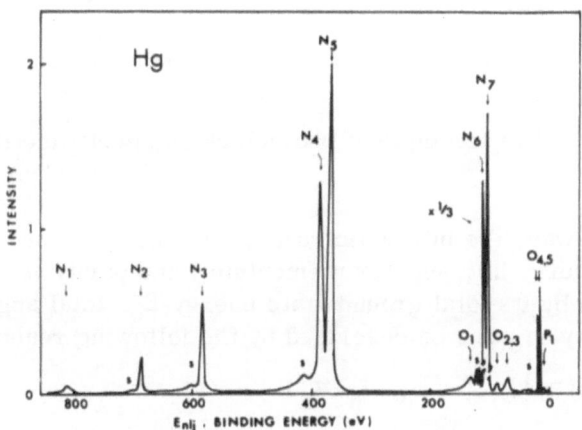

Fig. 1. Photoelectron spectrum of mercury obtained in the gaseous phase and with monochromatized AlKα radiation (K. Siegbahn, Ref 3).

PRINCIPLE OF PHOTOELECTRON SPECTROMETRY AND BASIC RELATIONS

The photoeffect expressed in the Einstein Relation

$$h\nu = E_{e,kin} + E_{n\ell j} \qquad (1)$$

forms the basis of photoelectron spectrometry. As illustrated in Fig. 2, a photon of energy $h\nu$ strikes an atom, often called the converter, and imparts the kinetic energy $E_{e,kin}$ to the photoelectron that is ejected from the level $n\ell j$ with the energy $E_{n\ell j}$. It is then that a measurement of the photoelectron energy $E_{e,kin}$ with a suitable apparatus that allows us to determine either $E_{n\ell j}$ provided $h\nu$ is fixed and known, or $h\nu$ provided $E_{n\ell j}$ is fixed and known. In either case, the determined quantity, $E_{n\ell j}$ or $h\nu$, can be correlated with the various properties of the atom under study. In these lectures only one of the two possible modes of photoelectron spectrometry, that in which $E_{n\ell j}$ is determined, shall be considered. The other mode, photoelectron spectrometry for the analysis of X-rays (PAX), is described elsewhere (21).

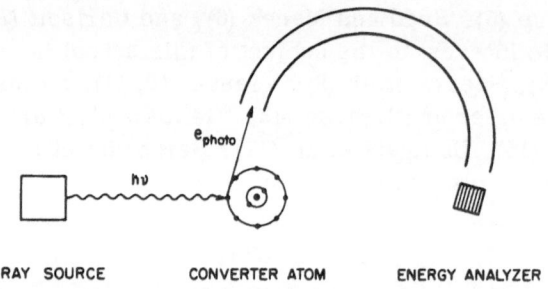

X-RAY SOURCE   CONVERTER ATOM   ENERGY ANALYZER

Fig. 2. The principle of photoelectron spectrometry.

In a general way, the interaction of a photon or quantum γ characterized by its energy $h\nu$, angular momentum j and parity π with an atom A characterized by its total ground state energy E_o, total angular momentum J_o and parity π_o can be described by the following reaction :

$$\gamma (h\nu, \text{ j=1}, \pi\text{=-1}) + A (E_o, J_o, \pi_o) \longrightarrow$$

$$A^+ (E_f, J_f, \pi_f) + e^- (E_{e,kin}, \ell sj, \pi_e = (-1)^\ell) \quad (2)$$

whereby the ionized atom is characterized by the final state parameters E_f, J_f and π_f and the outgoing photoelectron by its energy $E_{e,kin}$, its quantum numbers ℓsj and parity π_e.

Energy, angular momentum and parity are conserved, and the following relations hold :

(a) the energy balance, for which we rewrite Eq. 1 by substituting $E_o - E_f$ for $E_{n\ell j}$:

$$E_{e,kin} = h\nu - (E_o - E_f), \quad (3)$$

(b) the angular momentum balance :

$$\vec{J_o} + \vec{j_\gamma} = \vec{J_f} + \vec{s} + \vec{\ell}, \quad (4)$$

(c) the parity balance :

$$\pi_o \cdot \pi_\gamma = \pi_f \cdot \pi_e = -\pi_o = (-1)^\ell \pi_f \quad (5)$$

Until recently, photoelectron spectrometry has benefited principally from the application of Eq. 3 and has acknowledged Eqs. 4 and 5 only in their simplest interpretations. However, a few experimental (22) and theoretical (23) studies in recent years have utilized explicitly Eqs. 4 and 5, indicating that much may be gained in the future by exploiting the momentum and parity balances.

THE EXPERIMENTAL ARRANGEMENT

The experimental arrangement follows the principle depicted in Fig. 2, and consists of a photon source, a source chamber containing the photoelectron source, an analyzer as the selective element and a detector.

Electrical discharge lamps and x-ray tubes either with or without associated monochromators, as well as synchrotrons or storage rings with monochromators, may serve as photon sources supplying a photon beam of the desired discrete energy. Frequently used sources with outstanding characteristics are the He I resonance line ($h\nu = 21.22\text{eV}$; $\Gamma = 3\text{meV}$ FWHM), the Zr Mς line ($h\nu = 151.4\text{eV}$; $\Gamma = 0.75\text{eV}$), MgKα ($h\nu = 1.2\text{KeV}$; $\Gamma = 0.65\text{eV}$), AlKα ($h\nu = 1.5\text{KeV}$; $\Gamma = 0.85\text{eV}$), monochromatized AlKα x-rays ($\Gamma \approx 0.2\text{eV}$) and monochromatized synchrotron radiation ($h\nu \lesssim 200\text{eV}$; $\Gamma \approx 1\text{eV}$ in most cases). While other line sources in the regime of discharge and x-ray tubes are available, few of them possess the characteristics needed for detailed investigations of EC effects. Further improvements of synchrotron radiation sources should extend their utility beyond 200eV and thus cover continuously a large part of the energy range in which EC is especially important.

Little need be said about the source chamber, except that it should insure a high density of atoms, equivalent to 1–10 Pa, a minimum loss of photoelectrons on their way to the analyzer, and the possibility of a change of angle between photoelectron and photon beam propagation directions.

The analyzer may be of the integral or differential type, use magnetic or electrostatic deflection or time-of-flight separation, and focus in one or two directions. Electrostatic dispersive devices with double focusing have come into wide use because of their versatility, good resolution, sensitivity and easy protection against ambient magnetic fields. Most of these instruments, for example the type depicted in Fig. 3, have a resolution, $\Delta E/E$, in the order of 10^{-4}, a large étendue, a defined and simple intensity response, individual electron detection, and a large focal plane, and can be equipped with multipurpose source chambers, multiple detector systems, diverse excitation sources and differential pumping between source and analyzer.

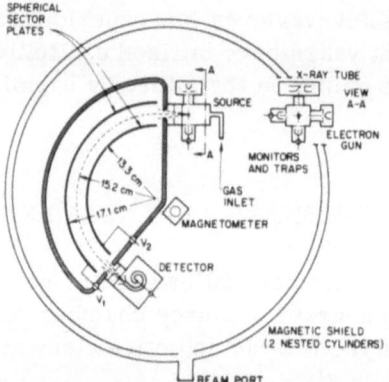

Fig. 3. Typical electron spectrometer of the electrostatic type. Fixed source chamber replaceable by turret chamber (M.O. Krause, Ref. 21).

The preferred detectors are the discrete-dynode and continuous electron multipliers, with the latter being available as single channel or as multiple channel (array) detectors. With these detectors electrons can be counted individually and with a probability that varies but slightly with energy over a wide range.

CHARACTERISTICS OF AN ELECTROSTATIC ENERGY ANALYZER AND BASIC, PRIMARY INFORMATION CONTAINED IN A PHOTOELECTRON SPECTRUM

Photoelectron spectrometry spans many orders of magnitude in energy, from less than 1eV electron energy to about 100keV ; however, the often used electrostatic spectrometer is best suited for energies below a few keV, which is   the region of greatest interest in the present context. Photoelectron spectrometry encompasses most of the Periodic Table, from helium to einsteinium, as demonstrated in Fig. 4. Of course, only a few elements are available for study as free atoms or can readily be so prepared with most others requiring a greater and possibly a consuming effort.

The energy of an electron is measured in terms of the potential V applied to the deflector plates of the analyzer, namely

$$E_{meas} = fV \tag{6}$$

where f is a constant of the apparatus. If precision in the energy determi-

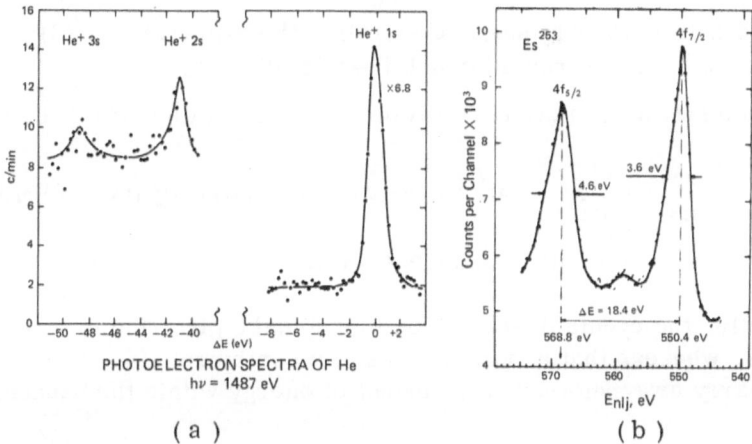

Fig. 4. Illustration of the range of photoelectron spectrometry. (a) the helium spectrum (T.A. Carlson et al., Ref 15) ; (b) the 4f doublet of einsteinium (M.O. Krause et al., unpublished data).

nation is desired (to be better than 10^{-3} at 1keV for example), Eq. 6 must be replaced by the relativistic formula

$$E_{meas} = fV \frac{E_{meas}}{E_{meas} + 2 E_o} \tag{7}$$

with E_o being the rest mass energy of the electron. Except for a small additive constant (2,5), E_{meas} is equal to $E_{e,kin}$ of Eq. 3. In accordance with Eqs. 2 and 3, the energy positions of the features exhibited in a photo-electron spectrum can then be correlated with a series of different events characterized by ($E_o - E_f$). They are the following :

a) Single-electron emission from the shell $n\ell j$ with

$$E_o - E_f = E_{n\ell j} = A^o \text{ (neutral) } - A^+ \text{ (} n\ell j \text{ ionized)}, \tag{8}$$

where $E_{n\ell j}$ has a single value for a closed shell atom, or several values (multiplet) for an open shell atom. It is often fruitful to equate E_o and E_f with initial and final state configurations, rather than using the binding energies $E_{n\ell j}$.

b) Two-electron excitation, with an $n\ell j$ electron ionized and an $n'\ell'j'$ electron excited to an $n''\ell''j''$ state :

$$E_o - E_f = E_{n\ell j} + E^*_{n'\ell'j' \to n''\ell''j''} = A^o \text{ (neutral) } - A^{+*} \text{ (} n\ell j, n'\ell'j' \to n''\ell''j'' \text{)}, \tag{9}$$

where the asterisk indicates that the energy is that of a transition in an atom with a hole in the $n\ell j$ shell. Generally, the expression $E^*(Z) \approx E(Z+1)$ holds quite well for the transition $n'\ell'j' \rightarrow n''\ell''j''$.

c) Two-electron excitation with two electrons, $n\ell j$ and $n'\ell'j'$, being ionized.

$$E_o - E_f = E_{n\ell j} + E^*_{n'\ell'j'} = A^o \text{ (neutral)} - A^{++} (n\ell j, n'\ell'j') \qquad (10)$$

with $E^*(Z) \approx E(Z+1)$ for $n'\ell'j'$, the energy of the " second " hole.

Evidently, the events described by Eqs. 8 and 9 give rise to discrete photopeaks, whereas that of Eq. 10 leads to a continuum since two electrons can each carry away an arbitrary amount of energy within the boundaries given by

$$E_{e,kin}(n\ell j) + E_{e,kin}(n'\ell'j') = h\nu - E_{n\ell j} - E^*_{n'\ell'j'} \qquad (11)$$

according to Eqs. 1 to 3.

The number of electrons of energy $E_{e,kin}$ ejected by $N(h\nu)$ photons into the direction given by the angle Φ between photon and electron propagation directions and transmitted by the analyzer is given by

$$N(e)dE = K \, E_{e,kin} \left. \frac{d\sigma_{n\ell j}}{d\Omega} \right|_\Phi d\Omega N(h\nu)dE, \qquad (12)$$

provided an electrostatic analyzer with equal entrance and exit aperture and no preacceleration of the electrons is used (5,21). The factor K is generally slowly varying with energy and is the product of a geometry factor G (which includes electron source volume, solid angles, etc.), the particle density

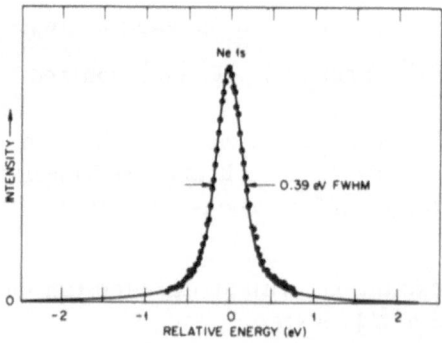

Fig. 5. The Ne 1s photoline. The K level was recorded with monochromatized AlKα x rays ($\Gamma = 0.2eV$), so that its Lorentzian wings show clearly. (U. Gelius Ref. 14)

N, the transmission $T = T(h\nu)$ of the window between x-ray and electron source, the detector response $\eta = \eta(E_{e,kin})$ and the loss $\epsilon = \epsilon(E_{e,kin}, N)$ of escaping electrons in secondary collision processes. If all factors and $N(h\nu)$ are known or have been determined, a measurement of $N(e)dE$ leads to an <u>absolute</u> determination of the partial photoionization cross section $d\sigma_{n\ell j}/d\Omega$. Generally, however, only the energy dependences of the various parameters are known which suffices for a <u>relative</u> determination of $d\sigma_{n\ell j}/d\Omega$, dispensing with the formidable task of determining G and N for example. The relation

$$N(e) \propto E_{e,kin} \; d\sigma/d\Omega \Big|_{\Phi} , \tag{13}$$

in which the indices $n\ell j$ are omitted, is a useful approximation of Eq. 13 to remember.

The contour V of a photoline, as depicted for example in Fig. 5, is a convolution of the Gaussian function G that is usually representative of the spectrometer window function and the Lorentzian function L that represents the ideal natural shapes of x rays and levels. Hence,

$$V = \int G(x - \zeta) \, L(\zeta) \, d\zeta . \tag{14}$$

The natural width of a level, $\Gamma_{n\ell j}$, can then be extracted from a measured photoline by means of Eq. 14 keeping in mind that all Gaussian widths that may occur add up quadratically and all Lorentzian widths arithmetically. Since the width $\Gamma_{n\ell j}$ is inversely proportional to the lifetime of a hole created in the $n\ell j$ level and proportional to the transition rates of the processes filling this hole, lifetimes and transition rates can be extracted from the width of an observed photoline. We note here a contact point between photoelectron spectrometry and the Auger and x-ray spectrometries.

Assuming an unpolarized target and restricting ourselves to dipole interactions, the angular distribution of photoelectrons, i.e. the differential photoionization cross section, is given by

$$\frac{d\sigma}{d\Omega} = \frac{\sigma}{4\pi} (1 + \beta P_2(\cos\theta)) \tag{15}$$

for polarized photons, and by

$$\frac{d\sigma}{d\Omega} = \frac{\sigma}{4\pi} \left(1 - \frac{\beta}{2} P_2(\cos\Phi)\right) \tag{16}$$

for unpolarized photons, whereby the indices $n\ell j$ for σ and β are omitted in these formulae and in the following paragraphs. The angle θ is the angle between polarization vector and photoelectron direction, and Φ the angle between photon and photoelectron directions. Furthermore,

$P_2 = \frac{1}{2}(3\cos^2\theta - 1)$; and $\beta = \beta(R_{\ell\pm1}, \ell, \delta)$ is the asymmetry parameter which is a function of the matrix element $R_{\ell\pm1}$, the quantum number ℓ, and the phaseshift δ. The parameter β may vary within the limits $-1 \leq \beta \leq 2$. For a quick survey, the functions of Eqs. 15 and 16 are plotted in Fig. 6 using the special values $\beta = 2, 0, -1$ and an intermediate value of β, chosen arbitrarily to be $+1$. Note that in Fig. 6 the ordinates are given in units of $\sigma/4\pi$. Several practical considerations follow from Fig. 6 representing Eqs. 15 and 16 : (i) for $\beta = 0$ we have $d\sigma/d\Omega = \sigma/4\pi$; (ii) at an angle of 54° 44' for θ in the case of polarized radiation or Φ in the case of unpolarized radiation, we obtain $d\sigma/d\Omega$ $\sigma/4\pi$ independent of the actual value of β; and (iii) for different lines, the variation of the signal strength ($d\sigma/d\Omega$) at 90° can be large for polarized radiation depending upon the values of β, but never exceeds a factor of two for unpolarized radiation. According to the above, experiments should be carried out at 54° 44', the " magic angle ", instead of the customary right angle, if the numerical value of β does not enter into the consideration of a particular study.

When multipoles other than dipole become important at higher photon energies, e.g. $h\nu \gtrsim$ 2keV, the Legendre polynomials P_1 and P_3 enter into Eqs. 15 and 16 in addition to P_2. If partially linearly polarized light is used, the resulting angular distribution is the sum of two distributions of the form of Eq. 15 using the appropriate fractional amplitudes for each

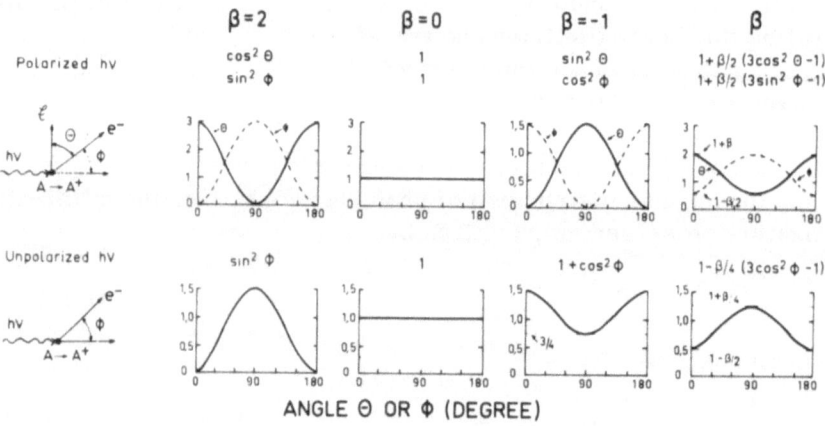

Fig. 6. Angular distributions of photoelectrons ; Plots of Eqs. 15 and 16 for different values of β, including a general case for which $\beta = 1$ arbitrarily.

of these distributions that are functions of the angles θ_x and θ_y, respectively (24). For elliptically polarized light, expressions for $d\sigma/d\Omega$ have been obtained by Schmidt (25) and by Samson and Starace (24). These expressions of greater complexity than Eq. 15 are applicable to the analysis of data obtained with synchrotron radiation.

The variation of $d\sigma/d\Omega$ with Φ (or θ) must be known if we are to take full advantage of the intensity measurements that are possible with an electron spectrometer; but, more specifically, a determination of $\beta = \beta(n\ell j)$ by means of a measurement of $d\sigma/d\Omega$ allows us to probe, in conjunction with theoretical predictions, the symmetry of an orbital, s, p, d or f, and certain dynamic effects, as for example the autoionization resonance structure (23) observed in xenon (22) and mercury (26).

CORRELATION EFFECTS IN SINGLE-ELECTRON PROPERTIES

In this section, the manifestations of electron correlation in single-electron properties will be presented. Binding energies, level widths, line shapes and initial-state configuration interaction will be discussed, but partial photoionization cross sections will not be treated until after the discussion of two-electron processes.

Within the independent-particle model, the best value of the binding energy $E_{n\ell j}$ is obtained with a relativistic Hartree-Fock calculation of the difference between the total energies for the initial, neutral state of the atom and for the final state of the atom that is ionized in the $n\ell j$ level. In addition, various quantum electrodynamic effects, such as the Lamb shift, need to be taken into account. Going beyond the independent-particle model, pair correlations between the electron in the $n\ell j$ shell and other electrons in that shell and neighboring shells need to be explicitly considered in both the initial and final states. The major contribution to EC is made by the electrostatic energy. Generally, the EC effects on the binding energy are in the order of 1eV and, therefore, often very small compared with the total binding energy. Energies of photolines, such as those displayed in Figs. 1 and 5, can be measured with an accuracy of better than 0.1eV, so that even small energy shifts due to EC can be probed reliably. Values of other single-electron properties, among them transition energies, level widths, branching ratios, and photoionization cross sections are modified by EC, sometimes by a small amount as in transition energies, and sometimes drastically as in branching ratios.

The role EC plays in general is shown in Table I by specific numerical examples (27-32). Listed are various properties that are customarily

Table I. Significance of Correlation Effects in One-Electron-Like Properties
Exemplified by Neon.

level or line	HF	HF*+Corr	Expt.
A) Energies (eV)[a]			
1s	868.6	870.53 870.5 [b]	870.32(2)[c]
1s-2s2s(^{1}S)	747.0	748.15	748.4 (1)[d]
1s-2p2p(^{1}D)	806.0	804.51	804.55(2)[c]
B) Level Width (eV)[a]			
1s	0.27	0.22	0.23(2)[e]
C) Auger Rates (eV)[a]			
1s-2s2s(^{1}S)	0.026	0.013	0.014 [d]
1s-2p2p(^{1}D)	0.155	0.134	0.139 [d]
D) Partial Photoionization Cross Section (Mb) at $h\nu$ = 130 eV			
2s	0.5 [f]	0.35 [g]	0.27 [h]

a) HF and HF*+Corr according to H.P. Kelly (Ref. 27), the asterisk indicates
   that values of Ref. 27 include relativistic and quantum-electrodynamic
   effects.
b) D.R. Beck and C.A. Nicolaides (to be published).
c) T.D. Thomas and R.W. Shaw, Jr. (Ref. 28).
d) M.O. Krause et al. (Ref. 29), normalized to (c) and (e) respectively.
e) U. Gelius (Ref. 14).
f) D.J. Kennedy and S.T. Manson (Ref. 30).
g) P.G. Burke and K.T. Taylor (Ref. 31).
h) F. Wuilleumier (Ref. 32 and Ref. 13).

considered to be one-electron-like ; EC can be seen to have a relatively
small influence on the energies and possibly a rather strong one on the
dynamic properties. It is apparent that the accurate measurements possible with photoelectron spectrometry and the reliable, complete calculations
possible with modern theoretical apparatus are essential to demonstrate
EC. Note that in Table I the HF values are nonrelativistic ; relativistic
effects contribute nearly 1eV to the Ne1s energy leaving slightly less than
1eV for the EC correction. However, the relativistic correction is small
for widths, rates and cross sections. In these, the improvement of theory
and its excellent accord with experiment is due largely to the inclusion of
EC.

Level widths can be determined with an accuracy of a few hundredths
of an eV (Fig.5), so that the EC in this property can be probed reliably.
This accuracy is realistic in terms of the decay rates, or the lifetime of
an innershell hole, only if broadening by hidden multiplet structure can be
removed or is absent as in closed shell atoms. Level widths should be

rather mildly affected by EC, since they are a measure of the total decay rates and do not respond to shifts within a decay branch. By contrast, branching ratios may be altered severely as shown in Table I by the Auger rates of neon. Electron correlation in Auger and x-ray spectra, however, are treated in detail in the lectures by W. Mehlhorn and J. P. Briand.

Line Shape Distortion

That sharp, discrete photolines of the type shown in Figs. 1 and 5 can be observed in photoelectron spectra is due to the fact that the initial process of photoionization is independent of the subsequent decay processes. The only manifestation of the decay process is contained in the width of the photoline (Eq. 14), which is proportional to the decay rate, assuming a negligible line broadening by photon energy-breadth and experimental apparatus. This rule of separability, fortunate for both experimental and theoretical considerations of photoelectron-, Auger- and x-ray spectrometries, is not absolute, however, but may be broken by strong EC under certain conditions. As an illustration, Fig. 7 shows the xenon photoelectron spectrum exhibiting a feathered $4p_{3/2}$ level and a $4p_{1/2}$ level that seems to have flowed apart. Simultaneously, the $4p_{3/2}$ binding energy seems to be shifted toward lower energies by about 10eV according to a calculation that makes reasonably accurate predictions of other binding energies (14).

The reason for this unusual behavior of the 4p, especially $4p_{1/2}$, level is the near-degeneracy of the $4p_{1/2}$ binding energy and the energy of the $4d^2$ configuration that can be reached by the Coster-Kronig transition $4p \rightarrow 4d4d$. But energy degeneracy alone is not sufficient, since nothing unusual happens with the 3p level of krypton, which has nearly the same energy as the $3d^2$ configuration. Required, also, is a strong overlap between the 4p and 4d wavefunctions to make the $4d^2$ state easily accessible. Again, this necessary requirement is not sufficient, since the Kr 3p and 3d wavefunctions also overlap well and, in another example, the 4p wavefunction of mercury overlaps strongly with that of the 4f level leading to nothing more than a lifetime broadening of the Kr 3p and the Hg 4p levels due to intense Coster-Kronig transitions (33). Finally, wide channels for the escaping electron are necessary, which in the case of xenon are the ϵf continuum channel and the nf bound channels. If all three conditions, energy degeneracy, strong overlap and wide exit channel are met, a resonance situation occurs (34), which transforms the photoline of single-electron emission into a broad asymmetric energy distribution with the possibility of a superimposed discrete multipeak structure. This structure is attributed to discrete final states involving the 4f level (14).

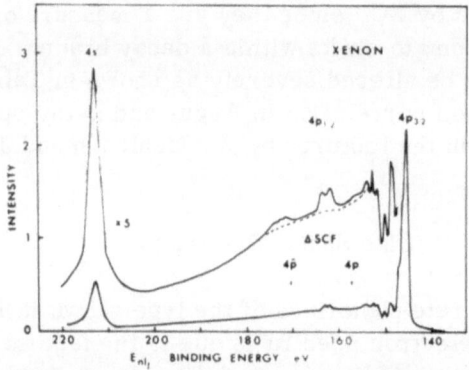

Fig. 7. The diffuse $4p_{1/2}$ level in xenon, suggesting an interaction between decay and excitation processes. (U. Gelius, Ref. 14).

Inspecting the trend of the 4p/4d wavefunction overlap, the access to the 4f or ϵf channel, and the energies of 4p and $4d^2$ states through the periodic system, we expect resonance conditions to exist for the range $50 \lesssim Z \lesssim 70$. Kowalczyk et al (16) have systematically followed up this point ; their observations and those of Gelius (14) are displayed in Fig. 8 showing that the 4p photolines start to spread with Z=48 (Cd) and are not restored until Z=73 (Ta) is reached. It is immaterial in the context of this discussion that the elements, the spectra of which are reproduced in Fig. 8, were investigated in the solid phase. However, spectra of greater clarity could have been recorded with free atoms as a target.

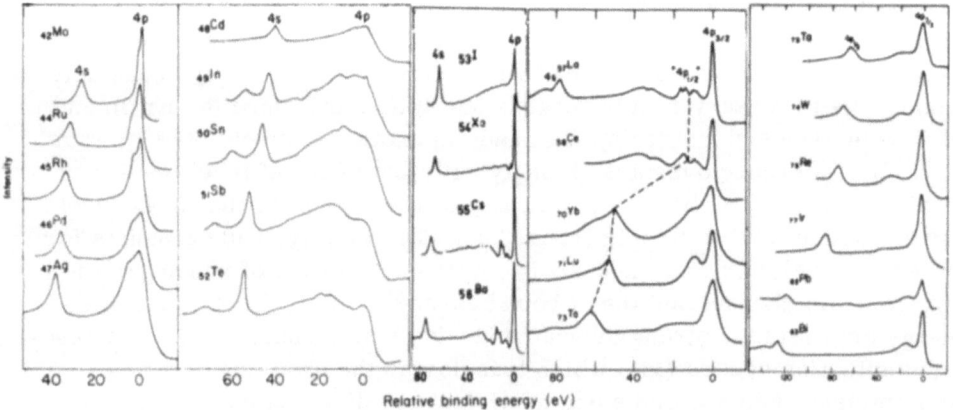

Fig. 8. The behavior of the 4p photolines as a function of atomic number for $42 \leq Z \leq 83$. (S. P. Kowalczyk et al., Ref. 16 ; and U. Gelius, Ref. 14).

Ground-State Configuration Interaction

Although it is the usual practice to speak of electron orbitals and of wavefunctions representing these orbitals, it is more realistic to think in terms of states or configurations. Then it becomes evident that all states of a given symmetry, occupied or unoccupied, need to be considered to find the wavefunctions that give a minimum total energy of the atom. For example, the wavefunction for the ground state of Cd would not have the simple form

$$\Psi(\,^1S\,) = |\,4d^{10}5s^2\,;\,^1S\,\rangle \tag{17}$$

according to the orbital structure (Kr core is inferred), but rather

$$\Psi(\,^1S\,) = a\,|\,4d^{10}5s^2\,;\,^1S\,\rangle + b\,|\,4d^{10}5p^2\,;\,^1S\,\rangle + \ldots, \tag{18}$$

indicating an admixture to the $5s^2(\,^1S\,)$ state of $5p^2(\,^1S\,)$ and possibly other (^1S) states with higher quantum numbers. The mixing coefficients, a and b, are chosen to yield a minimum total energy. This type of EC is known as multi-configuration mixing in the ground or initial state. As a consequence of this description of the atom, photoelectron lines of Cd, if excited by HeI photons, should be observable that would correspond to the 4d (doublet), the 5s, and the 5p (doublet) states and possibly others such as 4f, 5f, 6p etc. Indeed, the 4d, 5s, and 5p states are seen in the photoelectron spectrum of Cd (19, 35) displayed in Fig. 9. The intensity of the 5p photolines, labelled as ^2P of Cd^+, is smaller than that of the 5s photoline. To draw a quantitative conclusion in regard to the values of the mixing coefficients would be premature because the observed intensities are weighted according to Eq. 12 by the respective differential photoionization cross

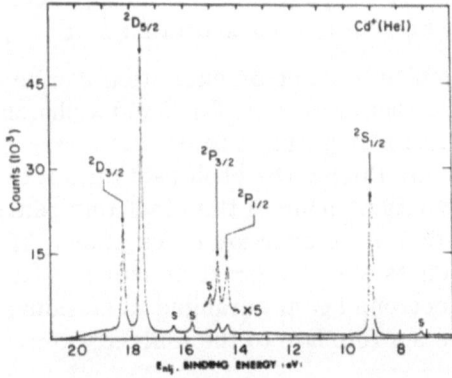

Fig. 9. The cadmium atom ionized by 21.22eV photons. Initial state correlation (configuration mixing) leads to the appearance of the ^2P photopeaks (Süzer and Shirley, Ref. 19).

sections. However, as our inventory of $\sigma_{n\ell_j}$ and $d\sigma_{n\ell_j}/d\Omega$ values increases in the future, spectra of the kind shown in Fig. 9 will allow the experimental determination of the mixing coefficients. Presence of ground state configuration interaction (CI) has been demonstrated recently for several atoms other than Cd, namely for Zn (35), Hg (18,36), Ca (35), Ba (35-37), Eu (35), Yb (35) and Pb (38). Admixture of more than one state to the nominal ground state has been shown to occur in Ca, Sr, Ba, Eu and Yb. In Ba, a great number of states are populated at a photon energy of 21.2eV by way of a complex excitation mechanism (37) yet to be fully interpreted. Using Hg as an example, it was shown (18) that the ^2P states of Hg^+ can be reached either by a single-electron transition from the initial configuration represented, in analogy to Cd, by

$$\Psi(\,^1S\,) = a\,|\,5d^{10}6s^2\,;\,^1S\,\rangle \;+\; b\,|\,5d^{10}6p^2\,;\,^1S\,\rangle \;+\,\ldots \qquad (19)$$

or by a two-electron transition of the conjugate shakeup type to be discussed in the next section. The detailed interpretation of the Hg photoelectron spectrum led to the conclusion, that CI is responsible for 80% and the conjugate shakeup, $5d^{10}6s^2 \rightarrow 5d^{10}6p\,\epsilon s$, for 20% of the observed intensity. Both mechanisms can be expected to be active in other elements ; and their contributions have presumably proportions similar to thosé in Hg.

TWO-ELECTRON PROCESSES

Electron correlation manifests itself directly in the simultaneous excitation of two or more electrons. Two-electron excitation is the most probable and constitutes the excitation to discrete levels, a process which is treated by K. Codling, or the excitation of one electron and the ionization of another (Eq. 9), a shakeup or $n\ell$, $\epsilon\ell$ process, or the ionization of both electrons (Eq. 10), a shakeoff or $\epsilon\ell$, $\epsilon'\ell'$ process (39).

The occurence of multiple electron excitation can be visualized in a simple manner. Imagine that atom A of Eq. 2 and a photon are at a large distance ; then all electrons of A are in their stationary orbits, the eigenstates of the neutral atom. During the photon-atom interaction, which corresponds to a perturbation, none of the electrons remain in these eigenstates by the very fact that an interaction takes place. After the perturbation has ceased and the products are at a great distance apart, the resulting ionized atom has its electrons again arranged in stationary states, which are then the eigenstates appropriate to the ionized atom.

Taking the neon atom as an example, this means that a photon of energy somewhat greater than the K electron binding energy can throw the atom into a number of excited states, namely those corresponding to the ejection of the 1s electron with the remaining electrons in the eigenstates of the K ionized atom, or the ejection of the 1s electron and the excitation of a 2p electron into the 3p state with the remaining electrons in their new eigenstates, or any other combination of excitations that are energetically possible and, in addition, are compatible with the selection rules. How the various accessible states are populated will depend on the quantum mechanical probability distribution.

As a consequence of this general view of the photon-atom interaction, multiple electron excitation is inherent in atomic dynamics and is as " natural " as single-electron excitation. That we, nevertheless, customarily separate single from multiple excitation processes has its roots in the difficulty of treating the many-body problem and the fact that one-electron processes are generally predominant and, as such, are described rather successfully by the independent single-particle model. Under specific circumstances, it is even possible, as shown below, to separate a two-electron process artificially into two single-electron processes, one following the other, although the excitation of two electrons in reality occurs simultaneously.

Dipole Selection Rules

Two-electron processes are governed by selection rules that were derived in 1931 by Goudsmit and Gropper (40) in the dipole approximation for transitions to discrete states. These are the following (Note that Ref.40 also treats three-electron transitions) :

a) One electron changes only its n, the other changes its n by an arbitrary amount and its ℓ by ± 1 ; in brief, $(n_1 \ell_1 ; n_2 \ell_2) \longrightarrow (n'_1 \ell_1 ; n'_2 \ell_2 \pm 1)$, whereby the indices can be interchanged between pairs.

b) Both electrons change their n's by arbitrary amounts, one changes its ℓ by δ, the other its ℓ by $\varepsilon \pm 1$ with the condition $\varepsilon + \delta =$ even. In brief, $(n_1 \ell_1 ; n_2 \ell_2) \longrightarrow (n'_1 \ell_1 + \delta ; n'_2 \ell_2 + \varepsilon \pm 1)$.

Hence, for the total atomic system (Eq. 2), the following rule applies : $\Delta L = \pm 1$ (dipole process), and $\Delta M = \Delta S = \Delta J = 0$.

Transitions that conform with rule (b) require the change of the angular momentum quantum number for both electrons and are therefore less intense than those that conform with rule (a). For rule (a), transitions of the type $n'_1 = n_1 + 1$ are the most probable with n'_2 running from n_2 to ∞, for rule (b), transitions satisfying $\varepsilon + \delta = 0$ are the most probable. Absorption measurements bear out this expectation (41).

Rule (a) contains the well known shakeup events as well as the recently observed conjugate shakeup event (12, 13, 18). In shakeup it is the outgoing electron that changes its ℓ by ± 1, in conjugate shakeup it is the excited electron that changes its ℓ by ± 1, while the other electron changes only its n in either event. The above dipole selection rules also cover the two-electron transitions that arise from configuration interactions (9, 14, 42).

MECHANISMS OF TWO-ELECTRON EXCITATION

In a theoretical treatment of double photoionization in the L shell of neon, Chang and Poe (43) identified the various mechanisms responsible for double excitation (39) and assessed quantitatively their relative importance. Using the many-body perturbation theory, they found the following effects to be the main contributors : core rearrangement, ground-state correlation and virtual Auger process. Other possible EC effects were found to be of lesser importance.

The main mechanisms of two-electron excitation are symbolized in Fig. 10. Although these diagrams are specific to double ionization of rare gases, they can be generalized to serve as a suitable framework for categorizing two-electron processes. For example, in the absence of a fundamental difference between excitation processes to bound and to continuum states, the diagrams describe also simultaneous excitation and ionization or $n\ell$, $\epsilon\ell$ processes. Formally, the k_1 designating the continuum channel must then be replaced by an n', which is greater than the connecting n. Figure 10 also contains the final-state configuration interaction (CI) to discrete states, which photoelectron spectra have indicated to be an important pathway in $n\ell$, $\epsilon\ell$ events. The diagram for the virtual Auger process represents this route if k_2 is replaced by n or n'. In spectroscopic

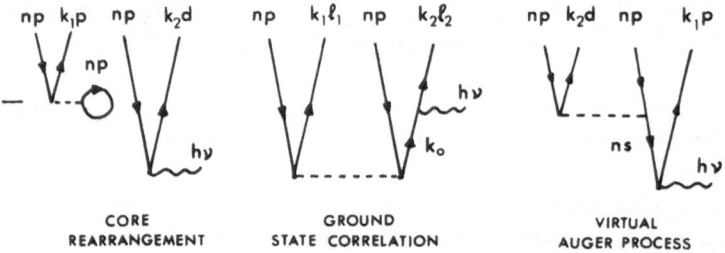

CORE REARRANGEMENT   GROUND STATE CORRELATION   VIRTUAL AUGER PROCESS

Fig. 10. The most important interactions contributing to double ionization in the L shell of neon or the outer shells of the other rare gases (T. N. Chang and R. T. Poe, Ref. 43).

terms, this would read as

$$3s3p^6(\,^2S\,) \longrightarrow 3s^23p^43d(\,^2S\,) \tag{20}$$

for CI in the arbitrarily chosen case of argon, instead of

$$2s2p^6 \longrightarrow 2s^22p^4\epsilon d \tag{21}$$

for the virtual Auger process in neon (k_2 replaced by ϵ).

In neon, all important processes involve a single shell, the L shell, whereas in heavier elements substantial contributions may come from interactions between principal shells. Such intershell EC can be fitted into the pertinent diagrams of Fig. 10 by appropriate relabeling.

Contributions from the different EC effects are not additive (43), and their relative importance may vary greatly from case to case. Often one effect is dominant, so that it is possible and fruitful to categorize the various two-electron excitation events under the rubrics suggested by the diagrams of Fig. 10.

Core Relaxation

The showpiece of the core relaxation (CR) mechanism, the Ne 1s photoelectron spectrum, is presented in Fig. 11. It is a high-resolution recording (14) of the spectrum, which was among the first spectra that gave direct

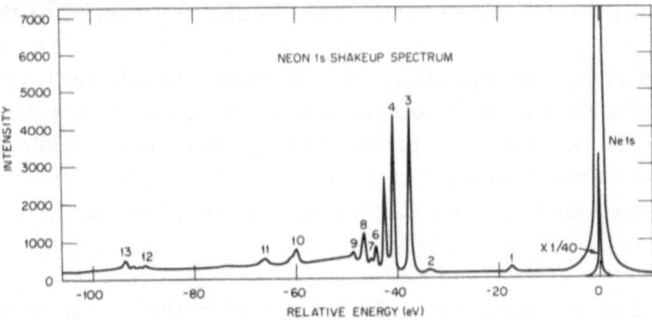

Fig. 11. The Ne 1s photoelectron spectrum, showing the single-electron emission photopeak, the shakeup peaks involving (a) a 1s, 2p electron pair (lines 2-9), (b) a 1s, 2s pair (10, 11), and (c) the $1s, 2p^2$ electrons (12, 13), and the KL double-electron ionization continuum (below $\Delta E=-47eV$). Excitation by monochromatized AlKα x rays with a bandwidth of 0.2eV (U. Gelius, Ref. 14).

evidence of two-electron excitation in photo-effect (10, 44, 45). The spectrum shows the Ne 1s photoline of single-electron ionization, the discrete shakeup or $n\ell$, $\epsilon\ell$ peaks and the underlying shakeoff continuum with onset at $\Delta E = -47$ eV. Peaks 3 and 4 are due to the transition $2p \rightarrow 3p$ in the presence of a 1s hole leading to a doublet because of the spin-spin interaction between two partially filled shells. All processes conform with the dipole selection rule (46). The shakeup peaks obey rule (a) of the two-electron selection rules, and in fact allow further differentiation of this rule. Comparison with optical spectroscopy data shows that peaks 3-9 correspond to shakeup processes in which the 2p electron retains its ℓ and the 1s electron changes its ℓ by +1. Similarly, peak 2 corresponds to a conjugate shakeup process in which the 2p electron changes its ℓ by -1 and the 1s electron retains its ℓ. That essentially all intensity of the $n\ell$, $\epsilon\ell$ processes comes from shakeup and little from conjugate shakeup suggests that CR is the major contributing mechanism.

Energy separations between shakeup and Ne 1s lines are in good accord with HF calculations and with estimates that use the optical transition energies of the Z+1 element (Eq. 9). The relative intensity of the sum of shakeoff continuum and shakeup discrete structure is in satisfactory agreement with predictions of the shakeoff theory (11, 15). The calculation of the intensity of the shakeup component is a difficult task and only now are detailed calculations appearing (47).

Core relaxation is equivalent to the phenomenon referred to by the shakeoff theory. This mechanism can be described in the language of the multi-body perturbation theory as coming from the incomplete cancellation of the term that contains the pertinent electron-pair interactions by the term that represents the interactions included in the single-particle potential of the V^{N-1} type (43). Also, CR can be viewed in the language of the shakeoff theory as resulting from a change in potential, or screening, which the L electrons suddenly experience when the 1s electron is photo-ejected. On the basis of the latter view, the two-electron process can be decomposed into two sequential one-electron processes : first, ionization of a 1s electron, and second, excitation (shakeup) or ionization (shakeoff) of an L, especially a 2p, electron.

The probability of exciting the electrons $n\ell j$ in the second step can be conveniently calculated in the sudden approximation by

$$P_{n\ell j} \approx 1 - |\int \Psi^*_{n\ell j} (A^+) \Psi_{n\ell j} (A) d\tau|^{2N}, \qquad (22)$$

where $\Psi^*(A^+)$ and $\Psi(A)$ represent the orbitals $n\ell j$ in the initial state and final state of the atom ionized in a given subshell, and N is the number of $n\ell j$ electrons.

The basis and the limits of the shakeoff concept are discussed by T. Åberg (48) ; extensive calculations of shakeoff probabilities have been given by Carlson and Nestor (49) ; and comparisons of theoretical with experimental results can be found in Refs. 8, 11, 14, and 15.

Core relaxation is the physical effect that is predominant when the two electrons come from different principal shells, except for heavy atoms in which two subshells with different n may be in close proximity. The reason is that small or negligible EC exists between electrons in shells sufficiently far apart.

The Ne 1s spectrum can be described by a momentum function (50):

$$E_a I_a = E_o I_o + \sum_{i \geq 1} E_i I_i + \int_{E_{c,o}}^{o} E_c I_c dE, \tag{23}$$

where E_o is the energy of the 1s line, E_i that of the shakeup lines, E_c that of the shakeoff continuum and I the respective intensities. For precise interpretation, especially at low photon energies, it should be noted that the integral in Eq. 23 contains two components (10, 45) (see also Eq. 11).

The energy E_a was shown (50) to be identical with the Koopmans energy, the eigenvalue ϵ of the orbital. It usually is close to E_o because of the large weight I_o. The difference $E_R = E_o - E_a$ is the relaxation energy which is defined as $E_R = \epsilon - (E_A - E_{A^+})$, where E_A and E_{A^+} are the exact total energies of the neutral and ionized atoms. Eq. 23 can aid in the apportionment of shakeup and shakeoff if the intensity of only one component is known, and in the estimation of the importance of EC effects other than CR. Finally, the I-weighting-factors can be interpreted as partial photoionization cross sections (51).

Ground-State Correlation

This mechanism rarely is the sole contributor to two-electron excitation, but may play a significant role among the various EC effects. Its relative importance is great when excitation involves electrons of the same shell and when that shell is a closed shell. Under these conditions, CR and final-state CI are of lesser significance.

Double photoionization in the L shell of Ne, reported in detail by V. Schmidt, is a case in which ground-state correlation is of primary importance (52). Fig. 12 illustrates (43) to which degree the various effects depicted in Fig. 10 account for the observed probability of double ionization.

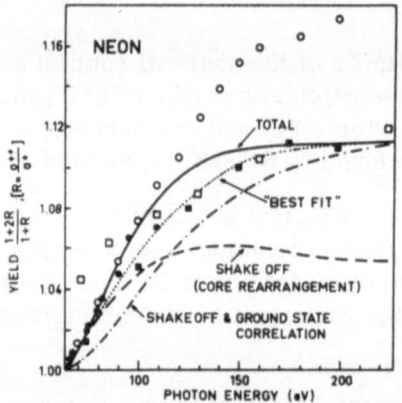

Fig. 12. Magnitude and energy dependence of double photoionization probability in L shell of neon (" best fit " curve) and of the contributing physical effects. (Ref. 43)

The role of ground-state correlation is even more pronounced in He (45, 52) ;
there this EC effect accounts for the bulk of the double ionization events.
This was demonstrated theoretically by using a correlated ground-state
wavefunction (53), an approach which quickly becomes difficult with increasing Z.

In a study of He it was shown (12, 54) that, in analogy to $\epsilon l, \epsilon' l'$ events,
the $n l, \epsilon l$ events are also dominated by ground-state correlation. However,
as is the usual in all cases of this category, other EC interactions cannot
be neglected if a full accounting for the observation is to be achieved.

Final-State Configuration Interaction

This mechanism comes into play when states of the same symmetry
are present ; it assumes a dominant role when these configurations are
closely spaced in energy. Under these circumstances, CI can drain much
intensity from a state that corresponds to single-electron excitation, and
transfer it to a state that corresponds to two-electron excitation. Equation
20 gives the configurations involved for the outer shell ionization of Ar,
and for Kr and Xe if n=3 is raised to n=4 and n=5. The corresponding Ar
and Xe spectra are displayed in Fig. 13. In both spectra, lines due to CR
can be found but their intensities are much lower than those due to CI. The
state np^4nd of Eq. 20 could be interpreted as arising from the simultaneous
ionization of an np electron, $np \rightarrow \epsilon p$, and the excitation of another to nd
with $l' = l + 1$. As such it would be a conjugate shakeup process ; however,

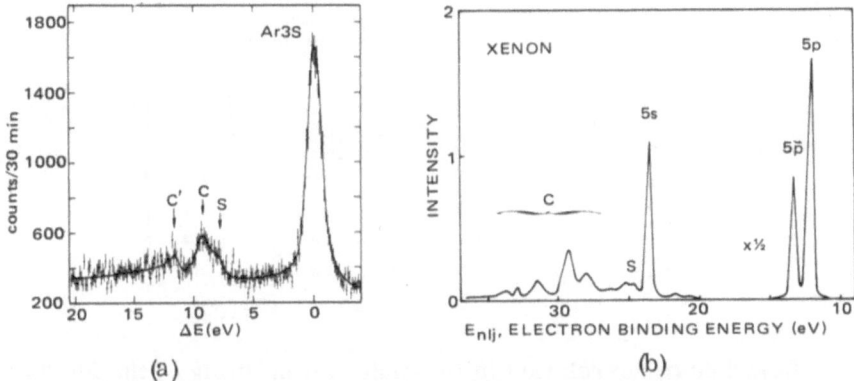

(a)   (b)

Fig. 13. Configuration interaction (C) and shakeup (S) satellites occurring in photoionization in the outer shells of (a) argon and (b) xenon. [(a) D. P. Spears et al., Ref. 9 ; and (b) U. Gelius, Ref. 14 .]

theoretical considerations and experimental evidence indicate that this state is populated almost exclusively via the CI route. For example, the diagram x-ray line $L_{\eta,\ell}$(2p → 3s) was found to be accompanied by a strong satellite that is attributable (55) to the same CI in the final state as represented by Eq. 20.

An interesting example of CI is offered by the Mn 3s photo-electron spectrum, Fig. 14a, which was recorded (16) in a compound, MnF_2. Because of the partially filled d-subshell, the Mn 3s photoline is split into two components, ^7S and ^5S, in the single-particle model (Fig. 14b,top). With CI included, the ^5S component undergoes further splitting (Fig. 14b, bottom) involving the following interaction :

$$3s^1 3p^6 3d^5 \,(^5S) \quad \longrightarrow \quad 3s^2 3p^4 3d^6 \,(^5S) . \qquad (24)$$

Remarkable in the Mn 3s spectrum, and common to CI events, is the shift of the principal peak, here the ^5S, to lower binding energy, so that the baricenter of the CI spectrum coincides with the Koopmans position in analogy to the formula of Eq. 23.

Ambivalence of the Two-Electron Selection Rule

Rule (a), the more common of the two rules derived for two-electron dipole transitions, does not specify which electron changes its ℓ by ±1 and which retains its ℓ. However, the ambivalence of the rule can be

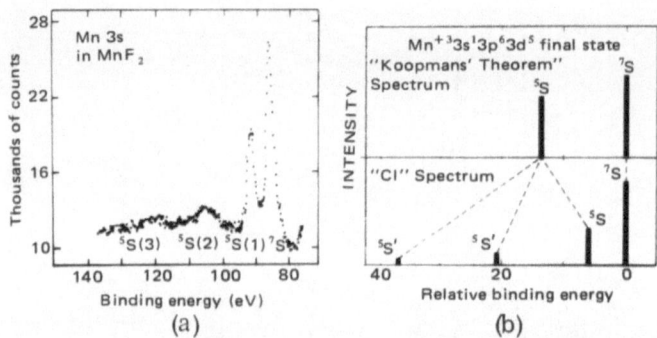

Fig. 14. Evidence of correlation in the multiplet splitting of the Mn 3s level in MnF_2. (a) Observed spectrum, (b) Calculated spectrum with and without configuration interaction. (S. P. Kowalczyk et al. , Ref. 16).

resolved experimentally by measuring the angular distribution of the out-going electron of a shakeup or $n\ell, \epsilon\ell'$ process. For He, the following transitions are allowed :

$$1s^2 \begin{cases} \epsilon p, 2s ; & \text{shakeup} \\ \epsilon s, 2p ; & \text{conjugate shakeup} \end{cases}$$

(25)

(26)

Although the two final states are degenerate (Fig. 4a), the two tran-sitions can be separated by an angular distribution measurement (12). Since the shakeup state $\epsilon p, 2s$ has the same symmetry as $\epsilon p, 1s$, reached by single ionization $1s^2 \to \epsilon p, 1s$, the ejected ϵp photoelectron must have have the asymmetry parameter $\beta_s = 2$, hence, a $\sin^2$-distribution in either case. Conversely, the ϵs photo-electron of conjugate shakeup has a β_p, which in general is different from β_s and smaller than 2, leading to a distribution of the type $A + B \sin^2\Phi$ (see Fig. 6). It follows then, that the intensity ratio $I(\epsilon p, 2s)/I(\epsilon p, 1s)$ is independent of Φ, and the ratio $I(\epsilon s, 2p)/I(\epsilon p, 1s)$ varies with Φ.

In He, the intensity ratio of the two-electron peak to the normal peak, was found to depend on Φ indicating the presence of conjugate shakeup events. Comparison of experimental (12) and theoretical (54) results shows that shakeup occurs only 33% of the time at threshold and dominates at high photon energies. Similar conclusions could be drawn for Ne 2p (13). By contrast, conjugate shakeup amounts to only 1% of all simultaneous excita-tion and ionization events in the Ne 1s spectrum (peak 2 of Fig. 11).

The exact correlation of these results with the various mechanisms of two-electron excitation has not been established so far, except for the requirement that conjugate shakeup be absent for CR in the validity range of the sudden approximation (48).

PARTIAL PHOTOIONIZATION CROSS SECTION

In principle, the partial photoionization cross section $\sigma_{n\ell j}$ can be determined with the aid of Eqs. 12 and 16. In practice, this is a difficult task and it is simpler to determine $\sigma_{n\ell j}$ on a relative basis for <u>all</u> processes, single and multiple , and normalize the results to the total photoionization cross section σ_{tot}. The procedure can be stated in brief as follows:

$$
I_{x,rel} \propto d\sigma_x/d\Omega_{rel} \xrightarrow{\beta_x} \sigma_{x,rel} \xrightarrow{\sigma_{tot}} \sigma_{x,abs} , \qquad (27)
$$

where $I_{x,rel}$ is the relative intensity of the peaks and continua of a photoelectron spectrum, and x designates the various subshells and subshell combinations.

The scheme of Eq. 27 was used to partition the σ_{tot} of Ne into its components over a wide energy range (13), and thus obtain absolute values for the $\sigma_{n\ell j}$. The result, displayed in Fig. 15, demonstrates the influence the multiple-electron processes have on the correct evaluation of the single-electron subshell cross sections, since, for example, the rates for KL transitions are about 3 times greater than those for L transitions at 1.5keV. In the same vein, at 50eV neglect of the $n\ell, \epsilon\ell$ processes would entail a considerable error in σ_{2s}. A comparison of $\sigma_{n\ell j}$ of Ne with the theoretical

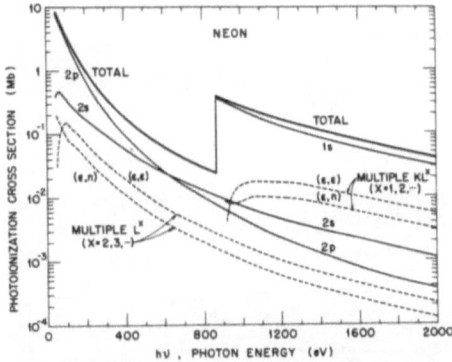

Fig. 15. Partition of the photoionization cross section of neon into its components of single and multiple ionization in the different subshells (F. Wuilleumier and M.O. Krause, Ref. 13).

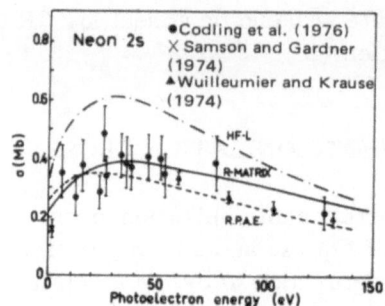

Fig. 16. Partial photoionization cross section for single-electron emission from the 2s shell of neon (K. Codling at al., Ref. 57, see also F. Wuilleumier and M.O. Krause, Ref. 13).

predictions by the Herman-Skillmann or Hartree-Fock single-particle model in the frozen structure approximation shows that this theory, which constituted the first advance over the previous hydrogenic theory, consistently overestimates the experimental cross sections by 10 to 40%. This is due to the fact that theory, in this formulation, calculated the sum of single and multiple processes for a given subshell (51), while the experiment selects the single-electron process only. At low photon energies additional EC interactions, namely EC within the p subshell and intershell correlation between s and p subshells, begin to influence the σ_{2s} (56). Figure 16 shows that the experimental data (13, 57) agree well with theoretical results (31, 56) that include these correlations, but poorly with a HF solution (30) that does not consider EC. The same EC effects are magnified in the σ_{ns} of the outer shells of Ar, Kr and Xe. In these elements, the s subshell can be shielded so effectively by the p subshell at a certain photon energy that σ_{ns} may vanish. Figure 17 compares the Ar 3s data (58,59) with the

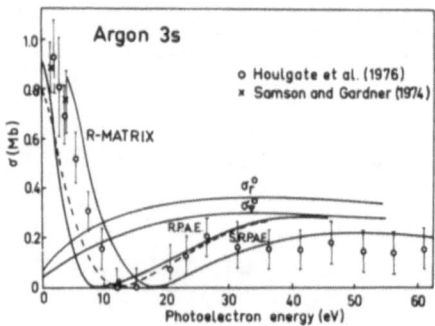

Fig. 17. Partial photoionization cross section of 3s shell of argon. (R.G. Houlgate et al., Ref. 58).

predictions of the RPAE-theory (60,61) and R-Matrix theory (31), both of which include the various EC effects. For the purpose of contrast, the HF result (62) is also shown in the length, σ_{ℓ}^{0}, and velocity, σ_{v}^{0}, forms.

Correlation also affects the asymmetry parameter β as demonstrated for 3p shell of Ar (17). The data on partial photoionization cross sections give evidence that electron correlation not necessarily associated with two-electron excitation can produce conspicuous effects. The dramatic variation of certain subshell cross sections is one of the strongest manifestations of EC in a single-electron property.

STATUS OF EXPERIMENTS

Studies performed to date on two-electron processes are summarized in Table II. The table shows clearly that while many key experiments have been performed there still remain gaps. This is true, although work has not been extended beyond investigations of closed-shell atoms. Experiments, in which photon energy and observation angle are the variables, are few. However, such experiments would be especially useful in assessing the contributions of the various EC effects, and ultimately establishing reliable systematics.

Only two investigations of the continuous photoelectron energy distribution have been made (10,45). To illustrate the general character of these

Table II. Status of Experiments on Multiple Electron Excitation of Free Atoms [a]

Shell	He	Ne	Ar	Kr	Xe	Hg	M[b]
1s	X̲ O̲	X̲ O̲	(X) (O)				
2s		X̲ O̲	X				
2p		X̲ O̲	X (O̲)				
3s			X O̲	(X)			
3p			X O̲	X			
3d			X	X			
4s			X O				
4p			X O	X	(X)		
4d				X	(X)		
4f					X		
5s				X O			
5p				X O			
outer[c]					X	X	

a) X indicates shake-up; O shake-off; X̲ or O̲ shake-up or shake-off obtained as a function of energy; and (X) or (O) incomplete data or analysis.
b) Zn; Cd; Ca; Sr; Ba; Eu; Yb; Pb.
c) Shells with binding energy less than 21.2 eV.

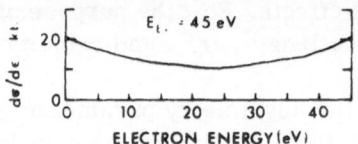

Fig. 18. Theoretical photoelectron spectrum of double ionization of neon liberating two L electron with total kinetic energy of 45eV. (T. N. Chang and R. T. Poe, Ref. 43).

spectra, Fig. 18 presents the theoretical curve (43) for double ionization in the L shell of neon at a low excess energy (Eq . 11). With increasing energy the minimum at half energy will become more pronounced pushing the spectral intensity to both ends of the spectrum.

Studies of the influence of electron correlation on single-electron pro-perties have just begun. Even so, these investigations have demonstrated that EC can cause conspicuous effects, such as the distortion of 4p photolines around Z=54 and the disappearance of partial photoionization cross sections at certain energies.

To date, work with open-shell atoms is non-existent. The only informa-tion in this area comes from solid-state systems and from atoms in which partially filled shells were created in two-electron processes.

ACKNOWLEDGEMENTS

Many colleagues provided me with their most recent and often still unpublished data. This generosity benefited lecturer and audience alike. Lecture and manuscript were prepared during the author's stay at the Laboratoire Curie, Paris and,under the Alexander-Von-Humboldt Stiftung's SeniorUS-Scientist Award Program, at the University of Freiburg. It is a pleasure to ackowledge the hospitality of Prof. J. P. Briand and Prof. W. Mehlhorn. This work was sponsered in part by the Energy Research and Development Administration under contract with the Union Carbide Corpo-ration.

REFERENCES

(1) K. Siegbahn, C. Nordling, A. Fahlman, R. Nordberg, K. Hamrin, J. Hedman, G. Johansson, T. Bergmark, S.E. Karlsson, I. Lindgren and B. Lindberg, " ESCA-Atomic, Molecular, and Solid State Structure Studied by Means of Electron Spectroscopy ", Nova Acta Regiae Soc. Upsal. (4) 20, Almqvist and Wiksells Boktryckeri AB, Uppsala (1967).

(2) K. Siegbahn, C. Nordling, G. Johansson, J. Hedman, P.F. Heden, K. Hamrin, U. Gelius, T. Bergmark, L.O. Werme, R. Manne and Y. Baer, " ESCA Applied to Free Molecules ", North-Holland Publ., Amsterdam (1969).

(3) K. Siegbahn, J. Electr. Spectr. 5, 4 (1974).

(4) K.D. Sevier, " Low Energy Electron Spectrometry " Wiley (Inter-science), New York (1972).

(5) M.O. Krause, " Electron Spectrometry " in Atomic Inner-Shell Pro-cesses, Vol. II, Academic Press, New York (1975).

(6) M.E. Rudd and J.H. Macek, Case Stud. At. Phys. 3, 47 (1972).

(7) T.A. Carlson, " Photoelectron and Auger Spectroscopy ", Plenum Press, New York, 1975.

(8) T.A. Carlson, in " Invited Lectures and Progress Reports of the VIII Internat. Conf. on the Physics of Electronic and Atomic Collisions " eds. B.C. Cobić and M.V. Kurepa, Inst. of Phys., Beograd, Yugosla-via, 1973.

(9) D.P. Spears, H.J. Fischbeck and T.A. Carlson, Phys. Rev. A9, 1603 (1974).

(10) M.O. Krause, T.A. Carlson and R.D. Dismukes, Phys. Rev. 170, 37 (1968).

(11) M.O. Krause, J. Phys. (Paris) 32, C4-67 (1971).

(12) M.O. Krause and F. Wuilleumier, J. Phys. B5, L143 (1972).

(13) F. Wuilleumier and M.O. Krause, Phys. Rev. A10, 242 (1974).

(14) U. Gelius, J. Electr. Spectr. 5, 985 (1974).

(15) T.A. Carlson, M.O. Krause and W.E. Moddeman, J. Phys. (Paris) 32, C4-76 (1971).

(16) S.P. Kowalczyk, L. Ley, R.L. Martin, F.R. McFeeley and D.A. Shirley, Faraday Disc. 60 (1975).

(17) R.G. Houlgate, J.B. West, K. Codling and G.V. Marr, J. Phys. B7, L470 (1974).

(18) J. Berkowitz, J.L. Dehmer, Y.K. Kim and J.P. Desclaux, J. Chem. Phys. 61, 2556 (1974).

(19) S. Süzer and D.A. Shirley, J. Chem. Phys. 61, 2481 (1974).

(20) See the following citations and the lectures or seminars given at this School especially those by T.A. Carlson and J.A.R. Samson.

(21) M.O. Krause, Adv. X-ray Anal. 16, 74 (1973) ; Phys. Fenn. 9, S1, 281 (1974).

(22) J.A.R. Samson and J.L. Gardner, Phys. Rev. Lett. 31, 1327 (1973).

(23) See D. Dill, this volume.

(24) J.A.R. Samson and A.F. Starace, J. Phys. B8, 1806 (1975).

(25) V. Schmidt, Phys. Lett. 45A, 63 (1973).

(26) A. Niehaus and M.W. Ruf, Z. Physik 252, 84 (1972).

(27) H.P. Kelly, Phys. Rev. A11, 556 (1975).

(28) T.D. Thomas and R.W. Shaw, Jr., J. Electr. Spectr. 7, 1081 (1974).

(29) M.O. Krause, T.A. Carlson and W.E. Moddeman, J. Phys. (Paris) 32, C4-67 (1971).

(30) D.J. Kennedy and S.T. Manson, Phys. Rev. A5, 227 (1972).

(31) P.G. Burke and K.T. Taylor, J. Phys. B16, 2620 (1975).

(32) F. Wuilleumier, Adv. X-ray Anal. 16, 63 (1973).

(33) E.J. McGuire, Sandia Lab. Report. SAND-75-0443 (Oct. 1975).

(34) S. Lundquist and G. Wendin, J. Electr. Spectr. 5, 513 (1974).

(35) S. Süzer, S.T. Lee and D.A. Shirley, (to be publ., 1976).

(36) H. Hotop (private communication).

(37) B. Brehm and K. Höfler, Int. J. Mass Spectr. and Ion Phys. 18, 338 (1975) ; H. Hotop and D. Mahr, J. Phys. B8, L301 (1975).

(38) S. Süzer, M.S. Banna and D.A. Shirley, J. Chem. Phys. 63, 3473 (1975).

(39) These designations can be considered equivalent, although $n\ell$, $\varepsilon\ell$ is the most general description, and the term shakeup could, but must not necessarily be restricted to cases where the sudden approximation (shakeoff theory) is valid. A third term, configuration interaction satellites, is also used to designate another variant of simultaneous excitation and ionization. An analogous consideration applies to shakeoff and $\varepsilon\ell$, $\varepsilon'\ell'$ nomenclature. Note that, here as elsewhere, " shakeoff" is often used as a general term that included " shakeup " ; similarly, " excitation " is often meant to include " ionization ".

(40) S. Goudsmit and L. Gropper, Phys. Rev. 38, 225 (1931).

(41) K. Codling, R.P. Madden and D.L. Ederer, Phys. Rev. 155, 26 (1967).

(42) G.K. Wertheim and A. Rosencwaig, Phys. Rev. Lett. 26, 1179 (1971).

(43) T.N. Chang and R.T. Poe, Phys. Rev. A12, 1432 (1975) ; T.N. Chang, T. Ishihara and R.T. Poe, Phys. Lett. 27, 839 (1971).

(44) M.O. Krause in " Proceedings of the 15th Annual Conf. on Mass Spectr. and Applied Topics ", ASTM- Conf. Denver (1967), p167.

(45) T.A. Carlson, Phys. Rev. 156, 142 (1967).

(46) Higher multipoles start to become discernible in the angular distribution at about 1.5 keV ; M.O. Krause, Phys. Rev. 177, 151 (1969).

(47) R.L. Martin and D.A. Shirley, Phys. Rev. A (to be publ. 1976).

(48) T. Åberg, this volume, lecture I.

(49) T.A. Carlson and C.W. Nestor Jr., Phys. Rev. A8, 2887 (1973) ; T.A. Carlson, C.W. Nestor Jr., T.C. Tucker and F.B. Malik, Phys.

Rev. 169, 27 (1968).

(50) R. Manne and T. Åberg, Chem. Phys. Lett. 7, 282 (1970).

(51) C.S. Fadley, J. Electr. Spectr. 5, 895 (1974).

(52) V. Schmidt, N. Sandner, H. Kuntzemüller, P. Dhez, F. Wuilleumier and E. Källne, Phys. Rev. A (to be publ. 1976) ; and Refs. therein.

(53) F.W. Byron and C.J. Joachain, Phys. Rev. 164, 1 (1967).

(54) V.L. Jacobs, Phys. Rev. A3, 289 (1971) ; V.L. Jacobs and P.G. Burke, J. Phys. B5, L67 (1972).

(55) J.W. Cooper and R.E. Lavilla, Phys. Rev. Lett. 25, 1745 (1972).

(56) M. Ya Amusia, V.K. Ivanov, N.A. Cherepkov and L.V. Chernysheva, Phys. Lett. 40A, 361 (1972).

(57) K. Codling, R.G. Houlgate, J.B. West and P.R. Woodruff, J. Phys. B, Lett. (to be publ. 1976).

(58) R.G. Houlgate, J.B. West, K. Codling and G.V. Marr, J. Electr. Spectr. (to be publ. 1976).

(59) J.A.R. Samson and J.L. Gardner, Phys. Rev. Lett. 33, 671 (1974).

(60) M. Ya Amusia, N.A. Cherepkov and L.V. Chernysheva, Zh. Eksp. Teor. Fiz. 60, 160 (1971) [Sov. Phys. JETP 33, 90 (1971)].

(61) C.D. Lin, Phys. Rev. A9, 181 (1974).

(62) J.W. Cooper and S.T. Manson, Phys. Rev. 177, 157 (1969).

THEORETICAL ASPECTS OF ELECTRON CORRELATIONS IN ELECTRON COLLISIONS[*]

Mitio Inokuti

Argonne National Laboratory

9700 S. Cass Avenue, Argonne, Illinois 60439, U.S.A.

ABSTRACT

Electron-correlation effects manifest themselves in diverse facets of electron collisions with an atom or molecule. A fast incident electron acts largely as an external agent, and its inelastic collisions probe the structure, including correlations within the target (in both the initial state and the final state), nearly in the same way as photoionization processes do. In inelastic collisions of a slow electron, the central object of study is the correlated motion of the incident electron and an electron excited out of the target core. This elementary observation is illustrated in the lecture by many examples and is elaborated by remarks on some current theoretical methods.

1. INTRODUCTION

1.1. Electrons Versus Photons. .How are They Different?

Electron collisions are much richer than photoionization in the variety of phenomena and data. On one hand, the variety of phenomena shows up as soon as one examines electron collisions at different kinetic energies. On the other hand, the variety of data

---

[*] Work performed under the auspices of the U.S. Energy Research and Development Administration.

is exemplified by terms such as "total," "differential," and "multiply differential," which indicate alternative levels of detail with which one looks at electron collisions. In compensation, the electron-collision data are usually of lower resolution.

The fundamental difference of electron collision from photoabsorption is the nonspecificity of excitation; an electron usually delivers to an atom or molecule only a fraction of its kinetic energy, while a photon (in the energy range of our interest) becomes completely absorbed in a single event. In other words, the energy delivered by each electron collision must be determined through observation (either of the scattered electron or of the target state after the collision).

1.2. Fast Electrons and Slow Electrons

It is useful to classify electron collisions into two kinds. The criterion of the classification is whether an incident electron is fast or slow compared to the internal motion of atomic electrons that pertain to a phenomenon under consideration. Electrons of a few-keV energy, for example, are fast with respect to all discrete excitations and to the majority of ionizing collisions with He, but are not fast with respect to the K-shell ionization of Ar. Phrased in sharper terms, the crucial distinction is whether the incident electron stays in the course of interactions as an external agent separate from the atomic electrons or it becomes incorporated into the target at least for a period of time. (This lecture mainly treats inelastic collisions, and touches upon elastic collisions only occasionally.)

2. FAST ELECTRONS

2.1. Elements of the Bethe Theory

Following Bethe [1,2], we treat the action of a fast electron as a weak perturbation, i.e., within the first Born approximation. Numerous results thus obtained are powerful as a framework of general understanding, as fully discussed in Refs. 3–5.

Bethe's initial point is that the momentum transfer $\hbar \vec{K}$ from the electron to an atom or molecule is a   key   variable. Suppose that an electron of kinetic energy T (assumed nonrelativistic, i.e., T < 10 keV) excites an atomic state n at excitation energy E_n measured from the ground state 0. The differential cross section is given as

$$d\sigma_n = \frac{4\pi a_0^2}{T/R} \cdot \frac{f_n(K)}{E_n/R} \cdot d[\ln(Ka_0)^2] \, , \tag{1}$$

where $a_0 = 0.529 \times 10^{-8}$ cm is the Bohr unit of length and $R = 13.6$ eV is the Rydberg unit of energy. The quantity $f_n(K)$ is the generalized oscillator strength (GOS) defined by

$$f_n(K) = (E_n/R)(Ka_0)^{-2} \left| (n \, | \sum_j \exp(i\vec{K} \cdot \vec{r_j}) | 0) \right|^2 \, , \tag{2}$$

where $\vec{r_j}$ is the position of the jth atomic electron and $(n| \quad |0)$ represents a matrix element between atomic states n and 0.

When the atomic final state belongs to a continuum, it is appropriate to consider the doubly differential cross section $(d/dE)(d\sigma)$ in place of $d\sigma_n$, the excitation energy E now being a continuous variable. Also, one defines the GOS density $df(K,E)/dE$ per unit range of E through replacing the index n by d/dE in Eqs. (1) and (2). Sometimes it is necessary to consider a finer breakdown of the cross section and the GOS, specified, e.g., by the energy and direction of an ejected electron resulting from fixed E.

Thus, the analysis of fast collisions reduces to the study of an atomic matrix element, just as the analysis of the photoeffect does. The similarity to the photoeffect is not only formal but deep-rooted in the physics. When the incident electron is scattered by a small angle, the momentum transfer $\hbar K$ is small. Intuitively, such a collision may be characterized by a large impact parameter[*] b. The force then exerted by the incident electron is uniform over the spatial extent of the atom, in the same way as implied by the dipole approximation for photoabsorption. Yet the same force is sharply peaked in time owing to the high velocity of the incident electron, and therefore its Fourier components are widely and smoothly distributed in frequency ω. Each of the Fourier components excites the atomic state at $E = \hbar\omega$. Consequently, electron collisions at small K are similar to irradiation with a photon flux having a wide and calculable spectrum.

---

[*]The variables K and b are roughly complementary to each other in the sense of Fourier analysis so that most collisions satisfy the relation $Kb \approx 1$.

To see this connection mathematically, one may expand the exponential in Eq. (2) and write the small-K behavior of $f_n(K)$ as

$$f_n(K) = f_n + (Ka_0)^2 f_n^{(1)} + \tfrac{1}{2} (Ka_0)^4 f_n^{(2)} + \dots, \tag{3}$$

where

$$f_n = (E_n/R) \left| (n| \sum_j x_j |0) \right|^2 \qquad [x_j = (\vec{K} \cdot \vec{r_j})/K] \tag{4}$$

is the dipole oscillator strength for the transition $0 \to n$, and $f_n^{(1)}$ and $f_n^{(2)}$ are combinations of certain matrix elements (p. 312 of Ref. 3). Thus, in principle, we expect to see in electron collisions at small K <u>whatever correlation effects we see in photo-absorption</u>.

There are some differences between electron collisions and photoabsorption, however. First of all, the limit $K \to 0$ is never precisely realized in electron collisions. The smallest K value for fixed E and T occurs at zero scattering angle, and is given by

$$(K_{min}a_0)^2 \cong E^2/(4RT) \tag{5}$$

where $E/T \ll 1$. The value of $(K_{min}a_0)^2$ is decisive for consideration of the non-dipole contributions expressed by the second and higher terms in Eq. (3). Second, by studying the K-dependence of $f_n(K)$, one can learn much about non-dipole excitations, which are often quite sensitive to electron correlations. For example, the $f^{(1)}$ value for the $1^1S \to 2^1S$ excitation of He is 0.13 according to a Hartree-Fock calculation, and is 0.084 according to a calculation including correlation effects [6].

Large-angle scattering is characterized by large K, i.e., by small b. If $\hbar K$ greatly exceeds the average momentum of atomic electrons, the role of electron binding should be inappreciable and the cross section should approximate the Rutherford cross section multiplied by Z, the number of atomic electrons. In other words, we should get $Z/(Ka_0)^2$ in place of $(R/E_n)f_n(K)$ in Eq. (1) and the GOS is nonvanishing only when $E \cong (\hbar K)^2/2m$. Equation (2) endorses this expectation. For large K, $\exp(i\vec{K} \cdot \vec{r_j})$ is strongly oscillatory in the configuration space. The matrix element therefore must vanish <u>unless</u> the product of the wavefunctions for initial and final states behave nearly as $\exp(-i\vec{K} \cdot \vec{r})$, i.e., <u>unless</u> one of the atomic electrons receives the entire momentum transfer $\hbar K$.

To put this argument more precisely, one can show that
(p. 336 of Ref. 3)

$$\frac{df(K,E)}{dE} \cong \frac{E/R}{(Ka_0)^2} \sum_j \left\langle \delta\left(\frac{(\hbar K)^2}{2m} + \frac{\hbar \vec{K} \cdot \vec{p_j}}{m} - E\right)\right\rangle , \quad (6)$$

where $\vec{p_j}$ is the momentum of the jth atomic electron, $\langle \ldots \rangle$ denotes
the expectation value in the initial atomic state, and $(\hbar K)^2/2m$ is
taken as much greater than the binding energy B of the atomic electrons. It is also possible to show that df(K,E)/dE differs significantly from zero in the range

$$\left|(Ka_0)^2 - E/R\right| < \xi (E/B)^{\frac{1}{2}} , \quad (7)$$

where ξ is a constant of the order of magnitude of unity. We name
the peaking of df(K,E)/dE in the region (7) as the Bethe ridge and
call this class of collisions quasi-free-electron scattering [7,8], a
subject of many recent measurements [9–12]. These measurements
primarily probe the momentum distribution in the target, and are pertinent to electron correlations insofar as they influence the momentum
distribution. Also, these measurements usually analyze ejected
electrons in coincidence with the monitoring of scattered electrons,
and therefore indicate a shell or subshell from which ejected electrons originate, giving a shellwise breakdown of df(K,E)/dE. Resulting data are often interpreted in a way similar to photoelectron
data by use of notions such as electron shake-off, even though
the mechanism of excitation greatly differs from photoeffect.

An important result concerns the T-dependence of the excitation
cross section σ_n, i.e., the integral of Eq. (1) over all possible K.
One can show in general that

$$\sigma_n = 4\pi a_0^2 (R/T) \left[(f_n R/E_n) \ln(4c_n T/R) + O(E_n/T)\right] , \quad (8)$$

where c_n is a constant derived from $f_n(K)$ (p. 325 of Ref. 3). As first
pointed out by Fano [13], a plot of $T\sigma_n$ versus ln T should behave
as a straight line for $T \gg E_n$, and its slope is determined by f_n/E_n.
The Fano plot enables one to extract f_n from electron-collision data.

For electrons with $T > 10^4$ eV, Eqs. (1) and (8) require modifications due to relativistic kinematics. (See p. 303 and p. 328 of
Ref. 3.)

2.2. Electron Correlations and GOS

For He discrete excitations, extensive calculations [6, 14−18] show the importance of electron correlations in obtaining accurate GOS. Here, decisive effects come from correlations in the ground state, especially from angular correlations expressed by the admixture of p^2 configurations [19].

As to discrete excitations in other atoms, recent calculations including correlation effects concern Mg, Ca, Sr, and Ba [20], Mg and Ca [21], Be and isoelectronic ions [22], Al [23], and Be, B, C, N, and O [24]. Although it is difficult to summarize the great volume of data, one can recognize a few points to be learned from these calculations. First, inclusion of correlations, especially angular correlations, normally allows the atom to contract and thus makes GOS significantly smaller than Hartree-Fock results at low K. The change of GOS at high K is often opposite. Second, all the calculations point to the importance of balanced treatment of the initial and final states; simultaneous improvement of the two states is necessary for good results. Third, sometimes introduction of electron correlations changes the character of GOS say from optically allowed to almost forbidden or vice versa. A most dramatic example [24] is the $2s^2 2p^3\ ^4S \rightarrow 2s2p^4\ ^4P$ transition in N. Fourth, one sometimes sees marked influence of correlations upon GOS near its minima, either as a function of excitation energy or as a function of K (p. 319 of Ref. 3). The $3s^2 3p \rightarrow 3s^2 nd$ transitions in Al [23] exemplify this point. Finally, for heavy atoms such as Ba [20] or Hg and Au [25], both relativistic effects and correlation effects are important.

The GOS for transitions to continua is more difficult to evaluate, except in the neighborhood of the Bethe ridge. The bulk of the GOS at small and moderate K lies in low-E regions, for which final-state electron interactions with the remaining ion core are in general complicated. Provided the ground-state wavefunction is reasonably known, measurements of GOS may be considered as a probe of these interactions. Especially the extensive data from coincidence measurements on scattered and ejected electrons [26,27] undoubtedly contain a great deal of physics which has hardly been unraveled by theory.

Examples of the continuum GOS calculations, including correlations, are not many. For He continua, calculations by Jacobs [28] and by Robb et al. [29,30] appear to be most definitive. These calculations use accurate wavefunctions obtained by close-coupling or R-matrix methods for the final state e + He$^+$, in which the outgoing electron is slow and therefore has ample time to be influenced by intricate interactions with the He$^+$ core, either in its ground state or in its excited state. Here we begin to see the intimate connection of fast-electron collisions with slow-electron collisions. Doubly excited states of He, degenerate with the single-ionization continuum, cause the well-known Beutler-Fano profile in the GOS, whose shape depends upon K as well as upon the angle of ejection resulting from autoionization [28, 31, 32]. Double ionization of He results in final states in which two electrons are slowly moving out of the He^{++} ion. The GOS for the double ionization should contain a great deal of physics on two-electron correlations, which have been investigated only experimentally [33].

Amusia and co-workers [34 – 37] have studied effects of electron correlations on the continuum GOS of other rare gas atoms by use of the random-phase approximation with exchange. The correlation effects are most dramatically seen in the 3p → εd transition in Ar, in the optically forbidden transition p → εp, and in transitions of the inner s electrons into the εp continuum.

3. SLOW ELECTRONS

3.1. Basic Observations

The dynamics of a slow-electron collision with an atom A is essentially concerned with excited states of the negative ion A$^-$, because the incoming slow electron becomes mechanically indistinguishable from outer atomic electrons during the collision. The dramatic manifestation of this observation is seen in resonances found for electron scattering by many atoms and molecules [38, 39]. The slow e-A collision is therefore closely related to the final states [40–42] of the photoabsorption by A$^-$. Likewise, the slow e-A$^+$ collision is closely related to the photoabsorption of A, and the slow e-A^{++} collision to the photoabsorption of A$^+$. Notice that the correspondence of the electron collision with the photoabsorption involves here a unit difference of the electron number in the initial target system.

Another principal distinction is that the final states in photoabsorption are usually limited by dipole selection rules to a few angular momenta, whereas in electron collisions there is only a loose limitation of angular momenta due to centrifugal forces. For example, the photoabsorption of H$^-$ leads to ^{1}P final states, but slow-electron collisions with H involve both singlet and triplet states with orbital characters S, P, D, etc., unless the electron energy is extremely small.

Moreover, the dynamics of slow _inelastic_ collisions

$$e + A \rightarrow e + A^* \quad \text{or} \quad 2e + A^+$$

is completely governed by correlations of two electrons that have energies much higher than the other electrons in the A^+ core. In this sense, electron correlations play an essential role in slow-electron inelastic collisions.

3.2. Electron Correlations in the Ground State

For illustration of several standard methods of treating electron correlations in general, let us consider briefly the ground state (1^1S) of two-electron systems such as He and H^-. As the starting point, let us consider the Hartree-Fock function

$$\psi = u(r_1)u(r_2) \tag{9}$$

where $\vec{r_1}$ and $\vec{r_2}$ are electron coordinates and $u(r)$ is an s orbital. (Throughout this section, spin variables are suppressed.) With the optimal $u(r)$, the function (9) gives 98.6% of the correct total energy for He, but gives many other properties unsatisfactorily.

The first kind of correlations, i.e., radial correlations, is described by the Eckart function

$$\psi = u(r_1)v(r_2) + v(r_1)u(r_2) \quad , \tag{10}$$

where $u(r)$ and $v(r)$ are different s orbitals. The importance of the radial correlations is easiest to see in H^-; function (10) succeeds in giving the electron affinity reasonably, whereas function (9) fails in this respect. Yet the mean square radius $\langle (\vec{r_1} + \vec{r_2})^2 \rangle$ comes out too large (by 10% for He and by 50% for H^-).

The second kind of correlation is angular, and one way of expressing this is to write

$$\psi = u(r_1)u(r_2) + \lambda [w_x(\vec{r_1})w_x(\vec{r_2}) + w_y(\vec{r_1})w_y(\vec{r_2})$$
$$+ w_z(\vec{r_1})w_z(\vec{r_2})] \quad , \tag{11}$$

where $w_x(\vec{r})$ is a p_x orbital and λ is a variational parameter. This function (11) exemplifies a general method of configuration inter-action (CI), and may be denoted by abbreviations $s^2 + \lambda p^2$. Notice that the terms in the square brackets of Eq. (11) may be put in the form $(\vec{r_1} \cdot \vec{r_2}) \times$ (function of r_1) $\times$ (function of r_2), and therefore function (11) depends explicitly on the angle between $\vec{r_1}$ and $\vec{r_2}$. With this function we start obtaining nonvanishing values of $\langle \vec{r_1} \cdot \vec{r_2} \rangle$ and $\langle \vec{p_1} \cdot \vec{p_2} \rangle$. Further, one can generalize Eq. (11) into a form such as $s^2 + \lambda p^2 + \mu s' + \nu d^2 + \ldots$, where λ, μ, ν, $\ldots$ are parameters. Note that s', i.e., a different s orbital, has been in-cluded; likewise, many different p and d orbitals may be included. (The above form shows configurations in the order of decreasing im-portance in He.)

Another way of generalizing Eq. (11) is to write

$$\psi = \sum_k \varphi_k(r_1,r_2) \, P_k(\cos \theta_{12}) \, , \tag{12}$$

where P_k is the Legendre polynomial and $\cos \theta_{12} = (\vec{r_1} \cdot \vec{r_2})/r_1 r_2$. If we recall that

$$2 r_1 r_2 \cos \theta_{12} = r_1^2 + r_2^2 - r_{12}^2 \, , \tag{13}$$

where $r_{12} = |\vec{r_1} - \vec{r_2}|$, we see that Eq. (12) is too restrictive; it contains only even powers of r_{12}, the crucial variable of the Coulomb potential $1/r_{12}$ between the electrons. Hylleraas' pioneer-ing work demonstrates the merit of introducing odd powers of r_{12} and using a new set of variables $s = r_1 + r_2$, $t = r_1 - r_2$, $u = r_{12}$ in

$$\psi = e^{-\kappa s} \sum_{\alpha,\beta,\gamma} c(\alpha,\beta,\gamma) \, s^\alpha \, t^{2\beta} \, u^\gamma \, , \tag{14}$$

where κ and $c(\alpha,\beta,\gamma)$ are parameters, and α, β, γ are non-negative integers [43]. A further generalized form [44]

$$\psi = e^{-\kappa s} \sum_{\alpha,\beta,\gamma} c(\alpha,\beta,\gamma) \, s^{\alpha-\gamma} \, t^{2\beta} \, u^{\gamma - 2\beta} \tag{15}$$

has been used also. The Hylleraas-type treatment has been exhaust-ively extended by Pekeris and co-workers [45–47].

Despite its great numerical success in variational calcula-tions, the functional form (14) or (15) cannot be the exact solution [48]. Fock [49] showed that the exact solution has the form

$$\psi = e^{-\kappa s} \sum_{n=1,\frac{3}{2},2,\dots} R^{2(n-1)} \sum_{p=0}^{[n-1]} g_{np}(\alpha,\theta_{12})(\ln R)^p ,$$

(16)

where $[n-1]$ represents the integral part of $n-1$,

$$R = (r_1^2 + r_2^2)^{\frac{1}{2}} ,$$

(17)

and

$$\alpha = \tan^{-1}(r_2/r_1) .$$

(18)

Fock's coordinates (R,α,θ_{12}) are a special case of the hyperspherical coordinates, which are appropriate for describing correlated motions in many-body systems in general. Indeed, these hypershperical coordinates have been used in a general discussion of chemical reactions [50], in a formulation of atomic-structure calculations (Ref. 51 and references therein), and in other contexts to be taken up in the next section. Furthermore, numerical calculations specifically designed to simulate the Fock expansion, Eq. (16), have proved successful for the He ground state [47,52].

3.3. Theories of Slow-Electron Collisions

In this section I shall review some electron-collision theories from our special point of view, namely, the description of electron correlations. My discussion will be illustrative rather than exhaustive; it will concentrate on certain topics that I have chosen from numerous possibilities, and deliberately leave out some subjects that are taken up by other lecturers, e.g., the random-phase approximation and the many-body perturbation theory.

Suppose that an electron of momentum $\hbar\vec{k}_0$ is incident on a neutral atom A in its ground state $\phi_0(\vec{r}_1,\dots,\vec{r}_N)$. After the collision one finds a scattered electron with momentum $\hbar\vec{k}_n$ and A in an eigenstate $\phi_n(\vec{r}_1,\dots,\vec{r}_N)$ at excitation energy E_n, where by energy conservation we have

$$(k_n a_0)^2 = (k_0 a_0)^2 - E_n/R .$$

(19)

The problem is to find the wavefunction of the A^- system under the boundary condition

$$\psi(\vec{r}_1,\dots,\vec{r}_N,\vec{r}_{N+1}) \rightarrow \phi_0(\vec{r}_1,\dots,\vec{r}_N)\exp(i\vec{k}_0\cdot\vec{r}_{N+1})$$
$$+ \sum_n \phi_n(\vec{r}_1,\dots,\vec{r}_N)A_n(\hat{k}_n)r_{N+1}^{-1}\exp(ik_n r_{N+1}) ,$$

(20)

as the scattered-electron position $\vec{r}_{N+1}$ tends to infinity. (For brevity, spin variables are suppressed.) The function $A_n(\hat{k}_n)$ is to be determined by calculation, and is called the scattering amplitude, from which all cross sections may be evaluated. The summation runs over all n for which $(k_n a_0)^2 > 0$. (Formalistically speaking, the summation includes continua to allow for ionization, but a precise form of the boundary condition for ionization is a complicated issue.) For collisions with an ion, say A^+, Eq. (20) must be modified to account for the asymptotic Coulomb force. This modification is tedious to describe, but is straightforward. This completes the statement of the problem.

By far the best known method of solution is the close-coupling method [53–56]. For the e + A collision, one writes the wavefunction as

$$\psi(\vec{r}_1, \ldots, \vec{r}_{N+1}) = \sum_{n=0}^{n_t} \phi_n(\vec{r}_1, \ldots, \vec{r}_N) F_n(\vec{r}_{N+1}) , \tag{21}$$

where $F_n(\vec{r})$ is a continuum orbital and the antisymmetrization is implied. Each term with fixed n represents a channel and the label n is called the channel index. The summation includes open channels ($k_n^2 > 0$) as well as closed channels ($k_n^2 < 0$). By treating Eq. (21) as a trial function in the Kohn variational principle, one obtains a set of simultaneous integro-differential equations for $F_n(\vec{r})$, similar in form to Hartree-Fock equations. Equivalently, one may take projections of the Schrödinger equation of the A^- system

$$(H - E) \psi = 0 \tag{22}$$

onto ϕ_n to arrive at the same result.

A basic argument for expansion (21) is that the totality of ϕ_n forms a complete set of functions and therefore any ψ should be expressible as Eq. (21), provided sufficiently many channels are included. This argument is indeed correct, and apparently is substantiated by considerable success of the method in many examples (as seen in Refs. 54–56). But it is unrealistic to say that one could always include as many channels as one wishes to obtain a good description of ψ. To appreciate this observation, we just have to recall the general structure of atomic energy levels. For any neutral atom, the energy levels become more and more dense as the first ionization energy I is approached, and then form continua above I, as forcefully discussed in Ref. 57.

The effect of the channel truncation is often discussed by use of Feshbach projection operators [58]. If P projects any operand onto included channels $(0 \leqq n \leqq n_t)$ and Q onto excluded channels $(n_t < n)$, the Schrödinger equation (22) may be rewritten as

$$P(H - E)(P + Q)\psi = 0 , \tag{23}$$
$$Q(H - E)(P + Q)\psi = 0 . \tag{24}$$

Solving Eq. (24) for $Q\psi$ as

$$Q\psi = - Q[Q(H - E)Q]^{-1} QHP\psi \tag{25}$$

and putting this into Eq. (23), we obtain

$$P\{ H - PHQ [Q(H - E)Q]^{-1} QHP - E\} P\psi = 0 . \tag{26}$$

The truncated close-coupling equation

$$P(H - E)P\psi = 0 \tag{27}$$

differs from the exact equation (26) in lacking the so-called optical-potential term

$$V_{opt} = - PHQ[Q(H - E)Q]^{-1} QHP . \tag{28}$$

It can be shown that at energies below the lowest excitation threshold not included in the expansion, V_{opt} is real and negative; in other words, Eq. (27) contains a fictitious repulsive potential owing to the truncation. At higher energies, V_{opt} becomes complex to account for loss of electron flux to excluded channels.

One way of diminishing the truncation effect is to use in Eq. (21) a set of suitably chosen functions χ_n instead of atomic eigen-functions ϕ_n. The functions χ_n, called pseudo-states, are in general superpositions of ϕ_n including highly excited states and con-· tinuum states, chosen, for example, to give the long-range asymptotic behavior of V_{opt} [59,60]. There is no general principle for the choice of the pseudo-states, apart from numerical efficiency.

Let us look at Eq. (21) again and compare its structure with various ground-state wavefunctions discussed in Sec. 3.2. Equation (21) treats the incident electron quite differently from the atomic electrons; $F_n(\vec{r}_{N+1})$ are determined numerically by solving the close-coupling equations over the entire spatial region, while $\phi_n(\vec{r}_n, \ldots \vec{r}_N)$ are simply presumed known in advance and are non-vanishing in a limited spatial domain. This sharp distinction between the incident electron and the atomic electrons originates from the boundary condition, Eq. (20), and Eq. (21), indeed, is a reasonable approximation for very low energies and for large r_{N+1}, i.e., in the beginning and toward the end of the collision process. But when the incident electron is close to the middle of the atom, its motion should be highly correlated with the atomic electrons, especially with the one that becomes excited. This situation is formally similar to the ground-state correlation, which one may describe, e.g., by simultaneously optimizing the two orbitals u(r) and v(r) in Eq. (10); an optimization of say u(r) under fixed v(r) would lead to a poor result.

Equation (21) specializes into

$$\psi(\vec{r}_1, \vec{r}_2) = \sum_n [\phi_n(\vec{r}_1) F_n(\vec{r}_2) \pm \phi_n(\vec{r}_2) F_n(\vec{r}_1)] \tag{29}$$

for two-electron systems such as $e - H$ or $e - He^+$, where the plus sign applies to a singlet state and the minus sign to a triplet state. We may add to Eq. (29) Hylleraas-type terms [Eq. (14) for the total S state]; then the trial function becomes capable of describing short-range correlations [61]. Alternatively, we may add to Eq. (29) CI-type terms [Eq. (11) for the total S state], which decay at large distances as bound-state orbitals [62−64]. An extreme idea in this direction is to abandon the close-coupling expansion altogether and to supplement Eq. (14) merely by terms necessary to satisfy the boundary condition for scattering. Thus, Schwartz [65] succeeded in obtaining the most accurate result for [1]S and [3]S $e - H$ states, although his calculation is limited to energies below the first excitation threshold of H. Similar work on [1]P and [3]P states has been performed also [66].

The use of Fock's coordinates for continuum states and doubly excited states suggests itself from the foregoing discussion. The great merit of Fock's coordinates is that they treat the two electrons in the $e - H$ or $e - He^+$ system on the same footing. Furthermore, the limit $R \to \infty$ includes alternative final-state possibilities: (a) $r_1 \to \infty$, $r_2 \to$ finite, (b) $r_1 \to$ finite, $r_2 \to \infty$, or (c) $r_1 \to \infty$, $r_2 \to \infty$. Case (a) or (b) corresponds to excitation and case (c) to ionization. Therefore, if one works out a reasonable theory in terms of Fock's coordinates, results should pertain to both excitation and ionization, and therefore to the competition between the two processes. From this point of view, it is not at all surprising [57] to see Wannier [67] and Rau [68] develop the theory of the threshold ionization by use of Fock's coordinates. Analyses of electron correlations in doubly excited states of H⁻ and He have recently been a subject of intense study [69−75]. When theory in this direction is fully developed, we shall have an adequate basis set in which a wavefunction $\psi(\vec{r}_1, \ldots, \vec{r}_N, \vec{r}_{N+1})$ for any e-A scattering state can be expanded more efficiently with full account of correlations between the incident electron and the excited atomic electron.

Finally, we must discuss a different line of development, i.e., the R-matrix theory [76-80]. We start with the crucial recognition that the close-coupling expansion, Eq. (21), is suitable only for large r_{N+1} and that, once the incident electron approaches the close neighborhood of the atomic electrons, all the electrons should be treated similarly, and in nearly the same way as in a bound state. (Rigorously, r_{N+1} here really means the greatest of the variables $r_1, \ldots, r_{N+1}$, in view of the implied antisymmetrization of the wave-function.) Therefore, it is logically suitable to separate the configuration space into two regions, i.e., an external region ($r_{N+1} > a$) and an internal region ($r_{N+1} < a$), a being a suitably chosen radius, and to treat the two regions differently. Because the basic forces in atomic interactions are Coulombic and thus of long range, one may wonder if a reasonable choice of the boundary radius a is possible at all. Indeed, the R-matrix theory was first developed by Wigner and co-workers [81-83] with specific reference to nuclear reactions for which basic forces are short-ranged.

As long as we consider electron collisions with an atom initially in the ground state (or in a low-lying state), the atomic eigenfunction is exponentially declining at large distances. Therefore, if we take $a \cong 10\ a_0$, practically all the ground-state charge distribution is enclosed in the sphere $r < a$, and atomic transitions always occur within it. Moreover, the exchange interactions between the incident electron and the atomic electrons, which give rise to complicated terms in the close-coupling equations, are non-vanishing only within the sphere.

The foregoing discussion has been made restrictive deliberately for the sake of definiteness; in particular, we have talked about the spherical boundary. In fact, the central idea of the R-matrix theory is to define an internal region somehow, i.e., in any shape appropriate for a specific problem under consideration, and then to treat the internal electron motion in as much detail as necessary. In other words, the boundary may be defined by $r = a(\hat{r})$, any smooth and positive-valued function of the radial unit vector $\hat{r}$ [79]. From this general point of view, how to treat the internal motion is a matter of technical consideration. Any of those methods for bound-state problems (discussed in Sec. 3.2) may be adapted to this end; for instance, the Hylleraas method may be suitable for one situation, and the CI method to another situation. Certainly, numerical calculations [76-80] have so far used a spherical boundary only and have treated the internal motion by means of the CI method or its variants, but it is important to appreciate the possibility of more flexible treatments in the general spirit of the R-matrix theory, e.g., for molecular photoionization or electron-molecule scattering.

Let us return to the restricted formulation again for definiteness, and see in some detail how the R-matrix theory is carried out in practice.

In the internal region $r < a$, the total wavefunction may be expressed as

$$\psi_k = \sum_{ij} c_{ijk} \, \Phi_i \, v_j + \sum_j d_{jk} \, \tilde{\phi}_j \, , \tag{30}$$

where Φ_i (the channel functions) are products of an atomic eigenstate ϕ_i and the spin angular part of the incident electron wavefunction, and $\tilde{\phi}_j$ are $(N+1)$-electron bound-state wavefunctions that describe any resonance explicitly. (Here again, the antisymmetrization of ψ_k is implied.) The radial functions $v_j(r)$ are characteristic of the R-matrix theory, and form a basis set over $0 \leqq r \leqq a$ with the boundary condition

$$v_j(0) = 0 \, , \tag{31}$$

$$\frac{a}{v_j(r)} \, \frac{dv_j(r)}{dr} \bigg|_{r=a} = b \, , \tag{32}$$

where b may depend upon the channel index and other parameters.

The discrete diagonalization of the Hamiltonian

$$(\psi_k, \, H\psi_{k'}) = E_k \, \delta_{kk'} \tag{33}$$

determines the expansion coefficients c_{ijk} and d_{jk}.

In the so-called eigenchannel method [79,80], E_k is preassigned to be the energy E of interest, and b come out as eigenvalues of the problem. Then ψ_k is nothing but the wavefunction ψ_E at energy E.

In the treatment by Burke and co-workers [76-78], b is fixed (usually at zero) and E_k are eigenvalues. Then ψ_E must be approximated as

$$\psi_E = \sum_k A_{Ek} \psi_k .$$
(34)

The R matrix is defined as[*]

$$R_{ij}(E) = \sum_k \frac{\gamma_{ki} \gamma_{kj}}{E_k - E} ,$$
(35)

where

$$\gamma_{ki} = \sum_j c_{ijk} v_j(a) .$$
(36)

The R matrix connects the radial wavefunctions $v_i(a)$ at the surface to the external radial wavefunctions $u_j(r)$ as

$$v_i(a) = \sum_j R_{ij}(E) \left[a \frac{du_j}{dr} - b u_j \right]_{r=a} .$$
(37)

In the external region $r > a$, one must solve a set of equations for u_j, which may be uncoupled or may be coupled only by multipole interactions expressed in terms of local potentials. However, it is necessary to obtain two independent solutions for each channel, say $s_j(r)$ and $c_j(r)$ which asymptotically behave as a sine and cosine function, respectively. An appropriate linear combination

$$u_j(r) = s_j(r) + \sum_i c_i(r) K_{ij}$$
(38)

then must be determined consistent with Eq. (37). The coefficients K_{ij} thus fixed form the K matrix, from which all cross sections are derived according to standard formulas. The resulting K matrix should come out insensitive to the boundary radius a, if the method is to be successful.

Results of many calculations [78] show the great power of the R-matrix method, especially for excitations of atoms to lower excited states by slow electrons. Currently unsolved problems are ionization and transitions involving very high discrete states. The connection of an internal solution with the external region is complicated for ionization where two electrons go out. Some merger of the

---

[*] Equation (35) will be useful in practice, only if the summation over k converges sufficiently rapidly. Although it is not obvious to see that it is indeed the case, a powerful technique (i.e., the so-called Buttle correction) has been worked out to ensure the rapid convergence of Eq. (35). See Ref. 78 for details.

R-matrix theory with the use of Fock's coordinates might be a suitable line of approach.

This observation leads to a personal outlook for future work. Suppose we attempt to solve the motion of two electrons in an internal region $R < a(\alpha, \theta_{12})$ of the Fock space spanned by R, α, θ_{12} for S states (more generally, by a few more variables as well). Because the potential is a function of the three or more variables, the problem is, so to speak, isomorphic to the scattering of a single particle in a nonspherical potential, e.g., to the electron scattering in a static molecular field; indeed, the problem is even more similar to a chemical reaction $A + BC \rightarrow AB + C$ on a single adiabatic-potential surface. In all these examples, the potential is essentially of two, three, or higher dimension in such a way that it disallows any exact separation of variables in the Schrödinger equation. Consequently, progress in any one of those superficially unrelated fields can be a rich food for growth in another of those fields. In this respect, recent developments of scattering theory for nonspherical potentials, e.g., in Refs. 84 and 85, deserve full attention.

ACKNOWLEDGEMENTS

I want to thank Professor U. Fano for valuable discussions throughout the preparation of the present text, and also Dr. J. L. Dehmer and Dr. C. M. Lee for critical reading of an earlier manuscript.

REFERENCES

1. H. Bethe, Ann. Phys. (Leipzig) 5, 325 (1930).
2. H. Bethe, in Handbuch der Physik, Ed. H. Geiger and K. Scheel (Springer, Berlin, 1933), Vol. 24/1, p. 273.
3. M. Inokuti, Rev. Mod. Phys. 43, 297 (1971).
4. M. Inokuti, in Invited Papers and Progress Reports, VIIth International Conference on the Physics of Electronic and Atomic Collisions, Amsterdam, 1971, Ed. T. R. Govers and F. J. de Heer (North-Holland, Amsterdam, 1972), p. 327.
5. E. N. Lassettre and A. Skerbele, in Methods of Experimental Physics, Vol. 3, Molecular Physics, Part B, Ed. D. Williams (Academic Press, New York, 1974), Chapter 7.2, p. 868.

6. Y.-K. Kim and M. Inokuti, Phys. Rev. 175, 176 (1968).
7. U. Amaldi, Jr. and C. C. degli Atti, Nuovo Cim. 66A, 129 (1970).
8. V. G. Levin, V. G. Neudatchin, A. V. Pavlitchenkov, and Yu. F. Smirnov, J. Chem. Phys. 63, 1541 (1975).
9. R. Camilloni, A. G. Guidoni, R. Tiribelli, and G. Stefani, Phys. Rev. Letters 29, 618 (1972).
10. A. G. Guidoni, G. Missoni, R. Camilloni, and G. Stefani, in Proceedings of the International Symposium on Electron and Photon Interactions with Atoms in Honor of Ugo Fano, Stirling, Scotland, 16−19 July 1974. (Plenum Press, New York, 1975) p. 149.
11. E. Weigold, S. T. Hood, and I. E. McCarthy, Phys. Rev. A 11, 566 (1975).
12. A. Ugbabe, E. Weigold, and I. E. McCarthy, Phys. Rev. A 11, 576 (1975).
13. U. Fano, Phys. Rev. 95, 1198 (1954).
14. Y.-K. Kim and M. Inokuti, Phys. Rev. 184, 38 (1969).
15. K. L. Bell, D. J. Kennedy, and A. E. Kingston, J. Phys. B 1, 204 (1968).
16. K. L. Bell, D. J. Kennedy, and A. E. Kingston, J. Phys. B 2, 26 (1969).
17. J. van den Bos, Physica 42, 254 (1969).
18. R. Vanderpoorten, Physica 48, 254 (1970).
19. K. E. Banyard and G. J. Seddon, J. Phys. B 7, 429 (1974).
20. Y.-K. Kim and P. S. Bagus, Phys. Rev. A 8, 1739 (1973).
21. W. D. Robb, J. Phys. B 7, 1006 (1974).
22. K. E. Banyard and G. K. Taylor, Phys. Rev. A 10, 1019 (1974).
23. C. A. Wells and K. J. Miller, Phys. Rev. A 12, 17 (1975).
24. S. L. Davis and O. Sinanoğlu, J. Chem. Phys. 62, 3664 (1975).
25. J. P. Desclaux and Y.-K. Kim, J. Phys. B 8, 1177 (1975).
26. H. Ehrhardt, K. H. Hesselbacher, K. Jung, and K. Willmann, Case Studies in Atomic Collision Physics II. Ed. E. W. McDaniel and M. R. C. McDowell (North-Holland, Amsterdam, 1972) Chapter 3, p. 159.
27. M. J. van der Wiel, Lectures this Institute.
28. V. L. Jacobs, Phys. Rev. A 10, 499 (1974).
29. W. D. Robb, S. P. Rountree, and T. Burnett, Phys. Rev. A 11, 1193 (1975).
30. T. Burnett, S. P. Rountree, G. Doolen, and W. D. Robb, to appear in Phys. Rev. A.
31. V. V. Balashov, S. S. Lipovetskiĭ, and V. S. Senashenko, Zh. Eksp. Teor. Fiz. 63, 1622 (1972). [Engl. transl.: Soviet Phys. —JETP 36, 858 (1973).]

32. V. V. Balashov, S. S. Lipovetskii, and V. S. Senashenko, Opt. Spektrosk. 35, 11 (1973). [Engl. transl.:Opt. Spectrosc. 35, 6 (1973).]

33. M. J. van der Wiel, Phys. Letters 41A, 389 (1972).

34. M. Ya. Amus'ya, N. A. Cherepkov, and S. I. Sheftel, Zh. Eksp. Teor. Fiz. 58, 618 (1970). [Engl. transl.: Sov. Phys. — JETP 31, 332 (1970).]

35. M. Ya. Amusia, in Atomic Physics 2. Proceedings of the Second International Conference on Atomic Physics, Oxford, 1970, Ed. P. G. H. Sandars (Plenum Press, London, 1971), p. 249.

36. M. Ya. Amusia, S. I. Sheftel, N. A. Cherepkov, and L. V. Chernysheva, Phys. Letters 40A, 5 (1972).

37. M. Ya. Amusia, N. A. Cherepkov, R. K. Janev, S. I. Sheftel, and Dj. Živanović, J. Phys. B 6, 1028 (1973).

38. G. J. Schulz, Rev. Mod. Phys. 45, 378 (1973).

39. G. J. Schulz, Rev. Mod. Phys. 45, 423 (1973).

40. A. Kasdan and W. C. Lineberger, Phys. Rev. A 10, 1658 (1974).

41. D. L. Moores and D. W. Norcross, Phys. Rev. A 10, 1646 (1974).

42. C. M. Lee, Phys. Rev. A 11, 1692 (1975).

43. E. A. Hylleraas, Rev. Mod. Phys. 35, 421 (1963).

44. T. Kinoshita, Phys. Rev. 105, 1490 (1957).

45. C. L. Pekeris, Phys. Rev. 126, 1470 (1962).

46. Y. Accad, C. L. Pekeris, and B. Schiff, Phys. Rev. A 4, 516 (1971).

47. K. Frankowski and C. L. Pekeris, Phys. Rev. 146, 46 (1966).

48. J. H. Bartlett, Jr., J. J. Gibbons, Jr., and C. G. Dunn, Phys. Rev. 47, 679 (1935).

49. V. A. Fock. Izv. Akad. Nauk S.S.S.R., Ser. Fiz. 18, 161 (1954). [Engl. transl.: Kgl. Norske Videnskab. Selskabs Forh. 31, 138 (1958).]

50. F. T. Smith, Phys. Rev. 120, 1058 (1960).

51. D. L. Knirk, Phys. Rev. Letters 32, 651 (1974).

52. A. M. Ermolaev and G. B. Sochilin, Dokl. Akad. Nauk S.S.S.R. 155, 1050 (1964). [Engl. transl.: Soviet Phys. — Doklady 9, 292 (1964).]

53. P. G. Burke and K. Smith, Rev. Mod. Phys. 34, 458 (1962).

54. K. Smith, The Calculation of Atomic Collision Processes (Wiley-Interscience, New York, 1971).

55. P. G. Burke and M. J. Seaton, in Methods of Computational Physics 10, 1 (1971).

56. P. G. Burke, Comments Atom. Mol. Phys. 3, 31 (1972).

57. U. Fano, Comments Atom. Mol. Phys. $\underline{1}$, 159 (1970).

58. H. Feshbach, Ann. Phys. (N.Y.) $\underline{19}$, 287 (1962).

59. R. J. Damburg and S. Geltman, Phys. Rev. Letters $\underline{20}$, 485 (1968).

60. S. Geltman and P. G. Burke, J. Phys. B $\underline{3}$, 1062 (1970).

61. A. J. Taylor and P. G. Burke, Proc. Phys. Soc. (London) $\underline{92}$, 336 (1967).

62. R. K. Nesbet, Comput. Phys. Commun. $\underline{6}$, 275 (1973).

63. R. S. Oberoi and R. K. Nesbet, Phys. Rev. A $\underline{8}$, 2696 (1973).

64. L. D. Thomas, R. S. Oberoi, and R. K. Nesbet, Phys. Rev. A $\underline{10}$, 1605 (1974).

65. C. Schwartz, Phys. Rev. $\underline{124}$, 1468 (1961).

66. R. L. Armstead, Phys. Rev. $\underline{171}$, 91 (1968).

67. G. H. Wannier, Phys. Rev. $\underline{90}$, 817 (1953).

68. A. R. P. Rau, Phys. Rev. A $\underline{4}$, 207 (1971).

69. J. H. Macek, Phys. Rev. $\underline{160}$, 170 (1967).

70. J. H. Macek, J. Phys. B $\underline{1}$, 831 (1968).

71. C. D. Lin, Phys. Rev. A $\underline{10}$, 1986 (1974).

72. U. Fano and C. D. Lin, in Invited Papers and Progress Reports, VIIIth International Conference on the Physics of Electronic and Atomic Collisions, Beograd, 1973. Ed. B. C. Čobić and M. V. Kurepa (Institute of Physics, Beograd, 1974), p. 229.

73. U. Fano and C. D. Lin, in Atomic Physics 4. Proceedings of the IVth International Conference on Atomic Physics, Heidelberg, 1974, Ed. G. zu Putlitz, E. W. Weber, and A. Winnacker (Plenum Press, New York, 1975), p. 47.

74. U. Fano, in Invited Papers and Progress Reports, IXth International Conference on the Physics of Electronic and Atomic Collisions, Seattle, 1975, Ed. J. S. Risley and R. Geballe (to be published by University of Washington Press, Seattle).

75. U. Fano, Lectures at this Institute.

76. P. G. Burke, A. Hibbert, and W. D. Robb, J. Phys. B $\underline{4}$, 153 (1971).

77. P. G. Burke, Comments Atom. Mol. Phys. $\underline{4}$, 157 (1973).

78. P. G. Burke and W. D. Robb, The R-matrix theory of atomic processes, to appear in Advances in Atomic and Molecular Physics (1975).

79. U. Fano and C. M. Lee, Phys. Rev. Letters $\underline{31}$, 1573 (1973).

80. C. M. Lee, Phys. Rev. A $\underline{10}$, 584 (1974).

81. E. P. Wigner and L. Eisenbud, Phys. Rev. $\underline{72}$, 29 (1947).

82. A. M. Lane and R. G. Thomas, Rev. Mod. Phys. $\underline{30}$, 257 (1958).

83. G. Breit, in Handbuch der Physik, Ed. S. Flügge (Springer, Berlin, 1959), Vol. $\underline{41}$/1, p. 1.

84. Yu. M. Demkov and V. S. Rudakov, Zh. Ekstp. Toor. Fiz. $\underline{59}$,
       2035 (1970). [Engl. transl.: Sov. Phys. — JETP $\underline{32}$, 1103
       (1971).]
85. D. Dill and J. L. Dehmer, J. Chem. Phys. $\underline{61}$, 692 (1974).

HIGH-ENERGY ELECTRON EXPERIMENTS

I. TOTAL AND DIFFERENTIAL CROSS-SECTIONS FOR MULTIPLE IONIZATION

M.J. Van der Wiel

F.O.M.-Institute for Atomic and Molecular Physics

Amsterdam, The Netherlands

ABSTRACT

The role of electron correlation in multiple ionization has been studied extensively by means of high-energy electron experiments. Some aspects of this work are discussed. Firstly, the ratio of total cross-sections for single and multiple ionization is considered in relation with the corresponding ratio for photon impact. Secondly, differential measurements of multiple ionization of the noble gases, as obtained in electron-ion coincidence experiments, are reviewed and compared with recent photon impact results and calculations.

1. INTRODUCTION

A classification of electron collision phenomena in categories of high and low energy should be based on a comparison of the projectile velocity with the orbital velocities of those target electrons that are involved in the collision. The present lectures (I and II) deal with the high-energy category, for which the Bethe theory [1] provides an adequate framework. Since electron impact experiments do not normally provide the spectroscopic detail one is used to from photon impact experiments, it is appropriate to confine our attention to ionizing processes, which generally give rise to smooth ionization cross-sections or energy-loss continua.

In addition, the emphasis will be on multiple ionization, since it is in these events that the phenomenon of electron correlation manifests itself most strongly. The reason is, that the single-particle nature of the transition operator requires the target electrons to interact in some way in order to permit multiple

electron transitions. This chapter deals with the impact energy dependence of total cross-sections for multiple ionization and describes
differential measurements of the same process, performed with electron-ion coincidence machines.

2. MULTIPLE IONIZATION CROSS-SECTIONS

In photon impact work one commonly observes or predicts that
the double-to-single ionization ratio -or satellite intensity ratio-
as a function of photon energy is either a constant from very near
threshold on (see chapter by Åberg [2] on shake theory) or at least
tends to a constant value for high energies. For electron impact, it
has often been claimed in the literature [3,4,5] that the same ratio
would be attained when going to sufficiently high impact energy
(fig. 1a). It will be argued here that the correspondence between the
two cases does not have such a simple form. The reason is that in the
case of electrons an increase of the impact energy does not imply a
corresponding increase of the energy transfers. Still, the low energy
transfers remain the most abundant. In fact, the contribution of
energy transfers in the region between the two thresholds for single
and double ionization (see fig. 1b), which obviously contributes to
the single process only, is normally rather important and hardly ever
negligible. So even in the idealized shake-off case, in which at every
energy transfer above the double-ionization threshold single and
double events occur with a constant ratio (S), the total cross-sections cannot have that same ratio due to the additional contribution
to single ionization.

Quantitatively, we can use the Bethe expression for the total
cross-sections to obtain (for the non-relativistic region, i.e.
$E_o < 20$ keV):

$$\frac{\sigma_{2i}}{\sigma_i} = \frac{M_{2i}^2 \ln c_{2i} E_o}{M_i^2 \ln c_i E_o} \quad , \tag{1}$$

where M_{ni}^2 is the integral of $1/E \, df_{ni}/dE$ over the oscillator
strength for the n + continuum and c_{ni} a constant depending on the
generalized oscillator strength. For a pure shake-off case eq. (1)
can be rewritten as:

$$\frac{\sigma_{2i}}{\sigma_i} = \frac{S[\int_{E_{2i}}^{\infty} \frac{1}{E} \frac{df_i}{dE} dE] \ln c_i E_o}{[\int_{E_i}^{E_{2i}} \frac{1}{E} \frac{df_i}{dE} dE] \ln c_i' E_o + [\int_{E_{2i}}^{\infty} \frac{1}{E} \frac{df_i}{dE}] \ln c_i E_o} \tag{2}$$

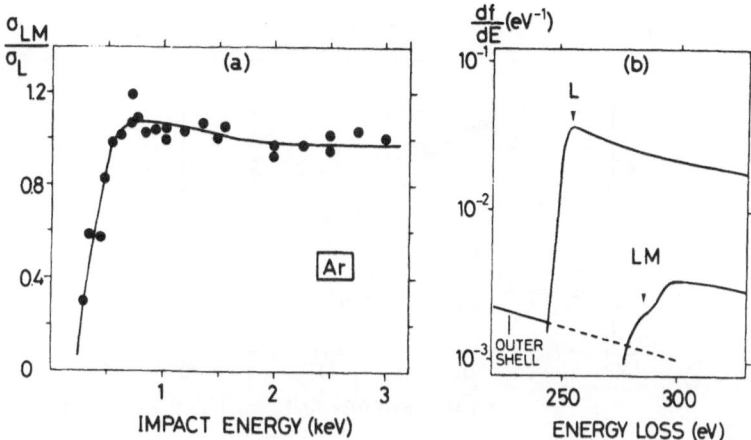

Fig. 1-a) Impact energy dependence of the relative cross section ratio for production of satellite (LM) and diagram (L) Auger lines of Ar [3] .

   b) Oscillator strengths for L- and LM-ionization of Ar. From the electron-ion coincidence experiment of ref.[6] , but renormalised to the total L-photoabsorption [7] . The contribution of the region between 250 and 300 eV to M_L^2 (see text) amounts to roughly one-half of the total M_L^2.

It is clear from eq.(2) that the cross-section ratio approaches the value S only if the first term in the denominator is negligible. This is evidently not the case for e.g. L-ionization of Ar (fig. 1b), where the weighted integral of df/dE in the region between E_L and E_{LM} amounts to approximately one-half the total M_L^2. One should therefore expect the experimental ratio of fig. 1a to decrease with energy and approach the value M_{LM}^2/M_L^2.

   Even in the case of Ne K-ionization, the region between the K- and KL- threshold [8] still contributes 15% to M_K^2; one should therefore be cautious to interpret the apparent constancy of the measured ratio [3] as the "asymptotic" shake-off ratio. In fact, experimental evidence for changes in X-ray satellite intensity -an analogous case- with electron impact energy over a very wide range (20 keV to 2 MeV) was reported recently [9] .

   For double and single ionization in one atomic shell one generally does not observe a constant ratio of oscillator strengths (see section 3). In such cases the total cross section ratio in the Bethe region decreases monotonically and is expected to approach a constant value M_{2i}^2/M_i^2 only at very high energies. Such a behaviour is demonstrated by e.g. He (fig. 2).

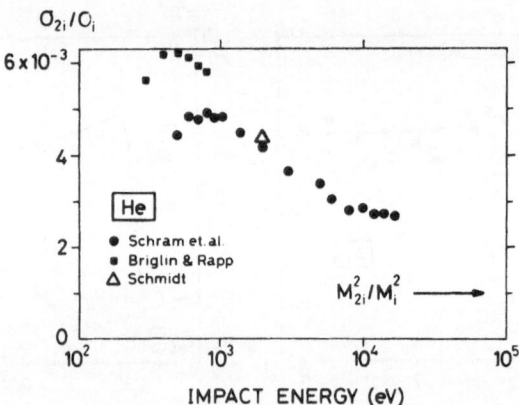

Fig. 2 - Impact energy dependence of total cross section ratio for double and single ionization of He [10-12] . The ratio M_{2i}^2/M_i^2 is a calculated value without relativistic correction.

3. ELECTRON-ION COINCIDENCE EXPERIMENTS

Much more detailed information on electron correlation is obtained when multiple transitions are observed differentially as a function of the energy loss of the incident electron. Two experiments of this type, in which scattered keV electrons are detected in coincidence with the ions, have been performed (see Table 1) by groups in Amsterdam [6] and in Leningrad [14] . The quantity measured in these experiments is the generalised oscillator strength $f(K)$, at one or a few values of the momentum transfer K, for formation of ions in a specific charge state.

The basis is again provided by the Bethe approximation [1] , which yields for the differential scattering intensity:

$$d^2\sigma(E,K) = \frac{2}{E} \frac{k_n}{k_o} \frac{1}{K^2} f(K) \qquad , \qquad (3)$$

where E is the energy transfer and k_o and k_n the magnitudes of the incident and scattered momenta. It is well-known, that for small K an expansion of $f(K)$ of the type:

$$f(K) = f(0) + K^2 f(1) + \tfrac{1}{2} K^4 f(2) + \ldots \ldots \ldots , \qquad (4)$$

opens the possibility of obtaining the optical oscillator strength $f(0)$. In order to do this, two methods have been applied extensively over the past ten years. Firstly, one can measure $f(K)$ over a range of K and extrapolate to K=0. This is a satisfactory method, applied e.g. by Lassettre and coworkers in energy loss experiments without coincidence [15] , but it is time-consuming in case one wishes to cover a wide and continuous range of energy losses. An alternative is to choose the experimental conditions such, that K is very small (i.e. high energy and small angle) and non-dipole terms in the expansion of $f(K)$ (eq. 4) can be neglected. Such a "dipole approximation" gene-rally leads to satisfactory oscillator strength spectra [6,16] , al-though an uncertainty remains with regard to its accuracy for special cases.

Recently, an efficient compromise between the two methods was shown to be succesful in removing the uncertainty inherent in the di-pole approximation [17] . The idea was to determine $f(K)$ from two energy loss spectra of 8 keV electrons at two scattering angles (0° and 1°), corresponding to values of K^2 equal to $K_{min}^2 = 2E/E_o$ and approximately 10^{-2} a.u., and make a linear extrapolation to K=0. This procedure leads to an accurate optical oscillator strength $f(0)$ and in addition yields an approximate value for $f(1)$, the derivative of $f(K)$ at K=0. An $f(1)$-spectrum for He is shown in fig. 3. One observes discrete structure of positive sign for the opticallly for-bidden states and negative sign for the allowed ones. The continuum has a shape which is more or less prescribed by the constraint of a sum rule:

$$\int_0^{\infty} f(1)\ dE = 0$$

The continuum spectrum passes through zero and has a positive region in order to make up for the normally negative discrete region. The 2s2p-resonance shows up clearly in this measurement, even with the poor resolution of 1 eV; possibly there is also in indication of the $2s^2$ resonance at 57 eV. The spectrum agrees quite well with that cal-culated by Bell and Kingston [19] , and is certainly of adequate accu-racy to enable an extrapolation of $f(K)$ to K=0 to be made. Note that in this case the dipole approximation, i.e. equating $f(K_{min})$ to $f(0)$, would have involved an error of less than 1%.

This section is concluded with a survey of the essential features of the two electron-ion coincidence experiments referred to above (table 1)

Table 1

group	refs.	E_o	angle	ion analysis	quantity measured
Amsterdam	6,16	8,10 keV	$0^{\circ},1^{\circ}$	time-of-flight	$f(K),f(o)$
Leningrad	14	3,4 keV	$0^{\circ}-10^{\circ}$	magnetic spectrometer	$f(K)$

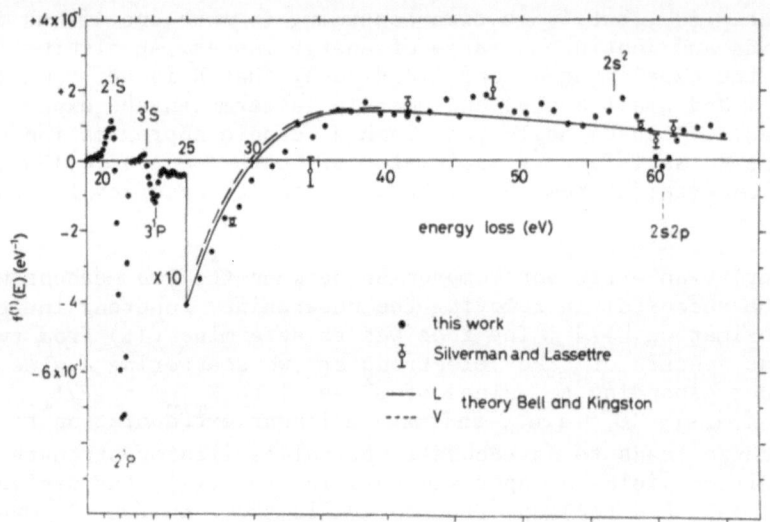

Fig. 3 - Derivative with respect to K^2 of the generalised oscillator strength for He, ●-8 keV impact energy [17] ; ○-500 eV impact energy [18] ; solid and dashed curves: "length" and "velocity" calculation [19].

4. CORRELATION IN THE NOBLE GASES

4.1. Intershell Correlation in Kr and Xe

    In a HF-approach the outer shells of Kr and Xe are predicted to have a smoothly decreasing single ionization oscillator strength in the region of the next inner d-subshell. However, for Xe evidence existed [20,21] for a rise in the single ion production, which is of a shape roughly similar to that of the d-cross section. Recently new and more detailed results have been obtained both theoretically and experimentally, which confirm this phenomenon of virtual d-excitations followed by excitation transfer to outer s- and p-electrons. Amusia et al. extended their calculations of the optical oscillator strengths for Kr^+ and Xe^+ [22] to include differential cross sections for keV electron scattering [23] . An analysis of this data for Xe^+ in terms of f(K) is shown in fig. 4, with inclusion of new experimental results.

    Firstly, the electron-ion coincidence method was used to re-measure these regions of interest with improved statistics, the conditions being: 8 keV impact energy and 0° scattering angle [24] . Secondly, West et al. obtained partial cross sections for photoionization of the outer s,p and d electrons in Xe. Addition of the s- and

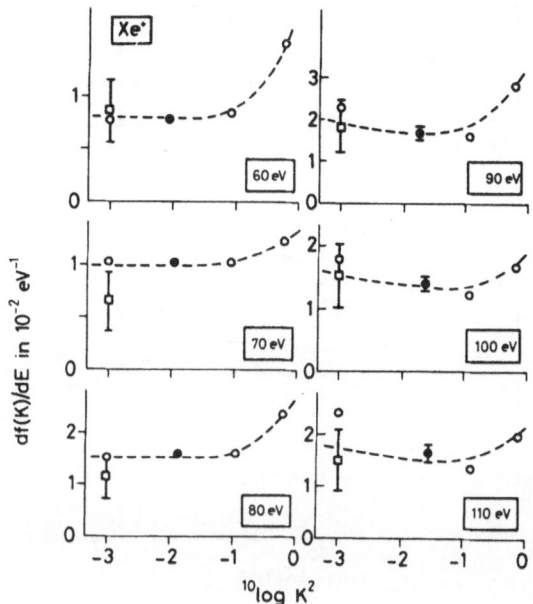

Fig. 4 - Generalized oscillator strengths for single ionization of Xe, at energies near the 4d-subshell threshold. ●-electron-ion coincidence experiment   (8 keV and K_{min}) [24]; □ - sum of 5s and 5p partial oscillator strengths   (photon impact) [25]; ○ - RPAE calculations [22,23]. Optical values have been plotted arbitrarily at a value of $K^2 = 10^{-3}$ a.u.

p-spectrum yields f(0)-values for Xe^+, which are plotted in fig. 4 arbitrarily at $\log K^2 = -3$. It may be concluded from fig. 4 that f(K) for these transitions shows the normal behaviour: rather flat for small K and rising towards the Bethe ridge at higher K. Near the "d-maximum" of the Xe^+ cross section at 90 eV, f(K) seems to pass through a shallow minimum. However, at all energy losses the electron impact measurements at K_{min} (for E_0 = 8 keV) appear to approach the optical limit quite closely and show better statistics than the photon impact results.

For Kr, the coincidence work [24] shows the intershell correlation to have a less dramatic influence (fig. 5). The RPA calculation by Amusia et al. [22] , however, appears to underestimate the effect and to put the maximum at a too high energy.

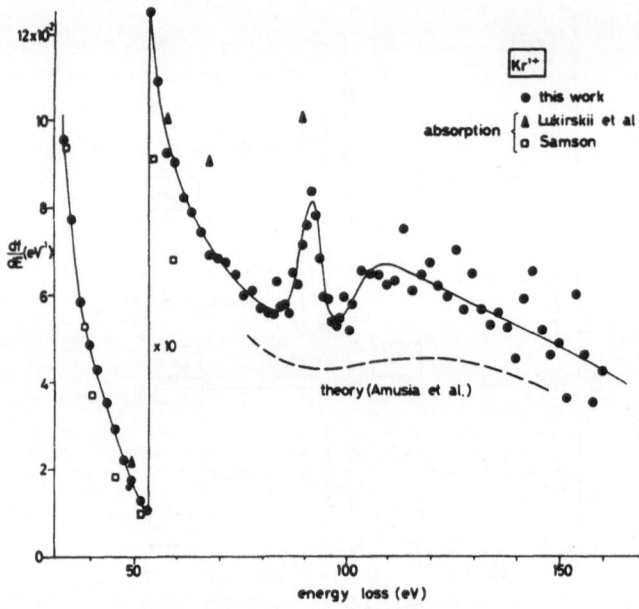

Fig. 5 - Oscillator strengths for single ionization of Kr, at energies near the 3d-subshell threshold. ● - generalised oscillator
strength for 8 keV impact energy and K_{min} [24] ; dashed curve: RPAE
calculation, [22] . The structure near 90 eV arises from discrete
excitation of 3d-electrons.

4.2. Outer Shell Multiple Ionization

Correlation effects on single transitions of the magnitude of
the d-maximum in Xe^+ are an exception. The observation of multiple
ionization, on the other hand, is in all cases a sensitive probe for
the phenomenon of correlation. The situation in literature with regard to multiple ionization in the outer shells of the noble gases
could be summarized as follows, until recently: For all gases, some
photon impact values of the double-to-single ionization ratio existed
[26] and rather extensive data from the electron-ion coincidence
method [6,21,27] . For He, Ne and Ar, large discrepancies up to a
factor of 1.5 existed, while for Kr and Xe the situation was quite
satisfactory. A detailed review is given elsewhere in this volume
[12] ; here we consider new results obtained with the coincidence experiment and compare them with recent synchroton data [12].

The largest amount of data seems to be available for the
Ne^{2+}/Ne^+ ratio. With reference to the chapter by Schmidt [12] , we
confine the discussion to two points. Firstly, the latest electron

impact data of the Amsterdam group [28] , taken in an improved
version of the coincidence machine, are lower than the previous ratios
by up to 25%. The reason for this discrepancy is not understood. The
improvement of the machine consisted of:
- replacement of the magnetic mass spectrometer by a wide time-of-
  flight tube with large area detector, in order to further reduce
  the problem of mass discrimination.
- reduction of the momentum transfer by going to zero scattering
  angle (instead of 8×10^{-3} rad)
Secondly, the present electron results are in excellent agreement with
the synchrotron data [12] , except for an apparent deviation at the
highest energies. This can probably be understood by considering the
ratio Ne^{2+}/Ne^{1+} as a function of K^2 (fig. 6) and including the extensive results of Gordeev et al. [14] . Figure 6 shows that is is
justified to compare ratios at K_{min} for 8 keV with optical values for
most energy losses except at 200 eV, where a slightly positive slope
with K^2 appears to exist, which tends to make the electron impact
ratio larger than the photon value.

Full agreement between the experiments on the synchrotron and
with the improved coincidence machine is reached for He and Ar [12].
In both cases Carlson's original data [26] appear to be somewhat too
low. For Ar, the situation is quite satisfactory now, while for Ne and
He there is clearly a need for more data at higher energies.

In conclusion, it should be remembered that a proper comparison
with theory should involve the double ionization oscillator strength
itself, rather than the ratios discussed above. Such an oscillator
strength spectrum is provided directly in the coincidence experiment
since the efficiency of the energy loss detector is constant over
the entire energy range. A similar measurement using photons adds a
serious complication to the experiment because of the required calibration of the photon flux monitor.

4.3. Multiple Auger Transitions

Auger transitions, in which the excess energy of the electron
filling the inner shell hole is shared by more than one ejected electron, were first observed in Auger spectra by Carlson and Krause [29]
as continua at the low-energy side of normal Auger peaks. An alternative way of observing such events [6] is to select an energy loss
of a fast projectile electron which is just sufficient to create an
inner shell hole (without additional outer shell ionization) and then
to look for formation of triply charged ions (fig. 7). For the case
of Ar-L ionization the two experiments metioned [29,6] conclude that
10-12% of the Auger events are double electron transitions. This probability is of the same order as that for shake-off accompanying the
initial creation of an L-hole (fig. 7). E.g. Ar-LM transitions are
found to occur in 15% of the cases. This latter number is in excellent

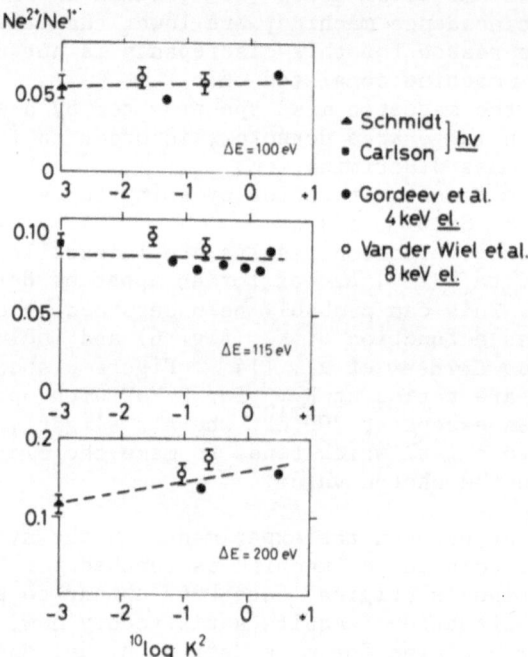

Fig. 6 - Ratio of double-to-single ionization in the n=2 shell of Ne, for various momentum transfers. ● - electron-ion coincidence experiment (8 keV) [28] ; ○ - electron-ion coincidence experiment (4 keV) [14] ; □ and △ - photon impact [12,26] .

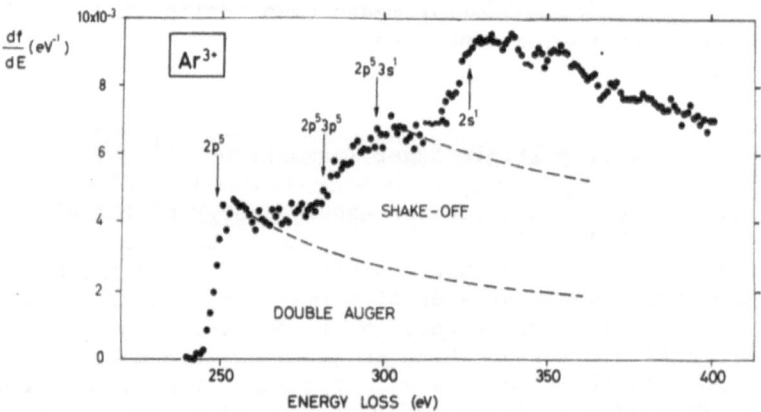

Fig. 7 - Oscillator strength for formation of Ar³⁺ through L-ionization [6] .

agreement with the calculated shake-off value.

In contrast to this, however, a dramatic difference exists between shake-off calculations [29] and measured probabilities for the double Auger transition. Using overlaps of the single L-hole state in Ar with of the triple M-hole state, a probability of 0.6% is calculated, which compares with the measured value of 10-12%. Evidently, we are dealing here with a case in which intra-shell correlation of the M-electrons has to account for the main fraction of the probability.

It is interesting to note that in the case discussed above one might expect to observe some kind of "post-collision" effect [30]. Consider an energy loss barely sufficient to remove an L-electron in Ar. The electron is then still so close to the atom when the Auger rearrangement occurs, that it might influence the decay. For instance ejection of two Auger electrons might produce such a change in the central field, that it causes the outgoing L-electron to be captured. Thus, a variation of the Ar^{3+}/Ar^{2+} ratio might be observed (see e.g. [6]); without any post-collision interaction this ratio would simply be constant with energy loss. Another effect could be a shift or broadening of Auger lines.

REFERENCES

1. M. Inokuti, Rev. Mod. Phys. 43, 297 (1971).
2. T. Åberg, this volume.
3. T.A. Carlson, W.E. Moddeman and M.O. Krause, Phys. Rev. A1, 1406 (1970).
4. M.H. Mittleman, Phys. Rev. Letters 16, 498 (1966).
5. B. Peart, D.S. Walton and K.T. Dolder, J. Phys. B. 4, 88 (1971)
6. M.J. van der Wiel and G. Wiebes, Physica 53, 225 (1971)
7. A.P. Lukirskii and T.M. Zimkina, Bull. Acad. Sci. USSR, Phys. Ser. (English Transl.) 27, 808 (1963).
8. F. Wuilleumier and M.O. Krause, Electron Spectroscopy, Edt. D.A. Shirley, North-Holland Publ. Co. (Amsterdam, 1972) p. 259.
9. N. Cue and W. Scholz, Phys. Rev. Letters 32, 1397 (1974).
10. D.D. Briglia and D. Rapp, Bull. Am. Phys. Soc. 11, 69 (1966).
11. B.L. Schram, A.J.H. Boerboom and J. Kistemaker, Physica 32, 185 (1966).
12. V. Schmidt, this volume.
13. F.W. Byron and C.J. Joachain, Phys. Rev. 164, 1 (1967).
14. Yu.S. Gordeev, S.G. Schemelinin and E.P. Andreyev, Abstracts of IX ICPEAC, Univ. of Wahington Press (1975), p.893.
15. E.N. Lassettre and co-workers, J. Chem. Phys. 40, 1208-1275 (1964)
16. M.J. van der Wiel, Physica, 49, 411 (1970).
17. C. Backx, R.R. Tol, G.R. Wight and M.J. van der Wiel, J. Phys. B. 8, 2050 (1975).
18. S.M. Silverman and E.N. Lassettre, J. Chem. Phys. 40, 1265 (1964).

19. K.L. Bell and A.E. Kingston, J. Phys. B. $\underline{8}$, L265 (1975).
20. R.B. Cairns, H. Harrison and R.I. Schoen, Phys. Rev. $\underline{183}$, 52 (1969).
21. Th.M. El-Sherbini and M.J. van der Wiel, Physica $\underline{62}$, 119 (1972).
22. M.Ya. Amusia, L.V. Chernysheva and V.K. Ivanov, Phys. Lett. $\underline{43A}$, 243 (1973).
23. M.Ya Amusia, N.B. Berezina and L.V. Chernysheva, Abstracts of IX ICPEAC, Univ. of Wash. Press (1975), p. 889.
24. M.J. van der Wiel and G.R. Wight, Phys. Lett $\underline{54A}$, 83 (1975).
25. J.B. West, P.R. Woodruff, K. Codling and R.G. Houlgate, J. Phys. B., to be published.
26. T.A. Carlson, Phys. Rev. $\underline{156}$, 142 (1967).
27. M.J. van der Wiel and G. Wiebes, Physica, $\underline{54}$, 411 (1971).
28. G.R. Wight and M.J. van der Wiel, J. Phys. B., to be published.
29. T.A. Carlson and M.O. Krause, Phys. Rev. Letters $\underline{17}$, 1079 (1966).
30. F.H. Read, Radiation Research (1975), to be published.

HIGH ENERGY ELECTRON EXPERIMENTS

II. BINARY AND DIPOLE (e,2e) METHODS

M.J. Van der Wiel

F.O.M. Institute for Atomic and Molecular Physics

Amsterdam, The Netherlands

ABSTRACT

After a general introduction on the variety of physical situations investigated in (e,2e) experiments, a more detailed discussion is given of two high-energy experiments, which operate in the limits of large and small momentum transfer, the "binary" and "dipole" experiments, respectively. In addition, a survey is given of evidence for multiple electron transitions in small molecules, obtained using the methods of (e,2e) and electron-ion coincidence (previous chapter).

1. INTRODUCTION

Electron impact ionization experiments, in which the two outgoing electrons are detected in coincidence after angular and energy analysis, are commonly referred to as (e,2e) experiments. Several experiments of this general type, but concerned with quite divers physical situations, have been reported over the past five years [1-9] A very qualitative survey of four such situations is given in Fig. 1.

Fig. 1 represents the "intermediate" and as such the most general case, the experiments by Ehrhardt et al. [1] , where intermediate refers to the parameters impact energy and momentum transfer. The scattered projectile (momentum $\vec{k}_s$) is observed at a fixed angle (5-15°) and coincidences with the second outgoing electron are recorded as a function of the angle. Two lobes, indicating the coincidence intensity, are usually found, one of which points more or less in the direction of the momentum transfer $\vec{K}(= \vec{k}_o - \vec{k}_s)$, while the second has the opposite direction. However, a considerable deviation away from the direction of $\vec{K}$ towards larger angles occurs, in particular for

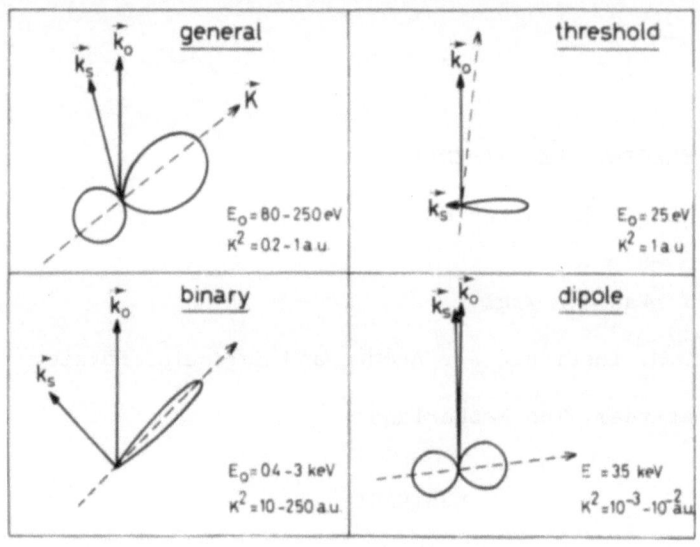

Fig. 1 - Survey of situations investigated in (e,2e) experiments. $\vec{k}_0$ and $\vec{k}_s$ are incident and scattered momenta; the momentum transfer ($\vec{K}$) direction is indicated by the dashed arrows. The solid curves represent coincidence intensity as a function of ejected electron angle. a-general case [1] ; b-threshold case [2] ; c-binary case [3-7] ; d-dipole case [8,9] .

lower impact energies. The obvious reason is the mutual repulsion of the two outgoing electrons, accompanied by transfer of some "recoil" momentum to the nucleus.

The remaining figs. 1b-d deal with limiting cases of this gene-ral situation. Fig. 1b represents the threshold case, i.e. impact energy of ~ 25 eV in He, as investigated by Cvejanovic and Read [2] in a very elegant fashion, using time-of-flight analysis of the ener-gies of the two outgoing electrons. With reference to the chapter of Dolder [10] , we just mention two points: the momentum transfer looses all relevance for the symmetry of the problem; instead the two electrons come out at approximately 180° with respect to each other. Figs. 1c and d, finally, refer to two high-energy cases, in which the momentum transfer does provide the axis of symmetry for the distri-bution of ejected electrons. For very high K, one approaches the si-tuation of "binary" scattering off a free electron, i.e. the ejection of an electron is confined to the direction of $\vec{K}$, with a narrow ope-ning angle about that direction due to the initial momentum distri-bution of the target electron. In contrast, under conditions of very low momentum transfer, we simulate the situation of "dipole" exci-tation or photoabsorption.

In summary, we saw two cases (fig. 1a and b) in which simple
targets like He were studied in order to gain more insight in scattering theory. On the other hand, for the two latter cases (figs.
1c and d) the scattering theory is well-known, at least in principle,
and the experiment is used to study complicated properties of a variety of targets. The remainder of this chapter deals with the "binary" and "dipole" cases.

2. BINARY (e,2e) METHOD

Experiments of this type have been reported by groups in Rome [3,4]
and in Adelaide [5-7] . In the analysis of the data the basic assumption is that the projectile interacts with only one target electron,
i.e. the binding forces in the target play no role during the collision except in that they define the initial target state. The presence of the other target electrons is commonly accounted for by considering distorsion of the incoming and outgoing waves (off-shell-
distorted wave impulse approximation [5]   ). The basic assumption
leads to a factorization of the transition matrix element T:

$$T = T_{ee} \cdot F(q) \tag{1}$$

where T_{ee} is the Mott scattering matrix element and the function F
of the atomic momentum q can be written as:

$$F(q) = \int e^{iqr_i} dr_i \ <\psi_{ion}(N-1)|\psi_{atom}(N)> \tag{2}$$

The overlap is taken over the coordinates common to the atom (N electrons) and the ion (N-1 electrons), such that a dependence on the
ejected electron coordinate r_i (or momentum q) remains. This is the
essential point of the method: it allows determination of such overlaps as a function of one atomic coordinate, or alternatively, it
produces the shape of atomic wavefunctions. In eq. (2) ψ(N-1) represents all eigenfunctions of the (N-1) system, i.e. multiple electron
transitions are possible through non-orthogonality of the two wavefunctions, or through configuration interaction in initial and (or)
final state.

The experimental arrangements employed so far have been either
of the "coplanar symmetric" or of the "non-coplanar symmetric" geometry. The first geometry is characterised by: incoming and outgoing
electrons (1 and 2) in one plane; $E_1 = E_2$ and $E_1 + E_2 = E_0 - E_{binding}$;
symmetric variation of the two angles around 45°. For the second
geometry we have: detection of one electron outside the scattering
plane; variation of its azimuth while $\theta_1 = \theta_2 = 45°$; same conditions
for the energies.

Having energy transfers nearly one-half the primary energy and
scattering angles near 45°, the magnitude of the momentum transfer is

sufficient to justify the use of the impulse approximation. For the coplanar geometry the factor T_{ee} in eq. (1) varies rapidly with E_o and θ and therefore makes it a suitable arrangement for studying the validity of the approximations. Once this is established, the non-coplanar set-up, in which T_{ee} is constant with ϕ, is the more convenient for a measurement of F(q) without kinematical corrections. A final remark concerns the direction of q measured with these methods. The coplanar set-up measures momenta parallel to the incident beam direction, whereas in the non-coplanar case one observes momenta in a plane perpendicular to $\vec{k}_o$. This could be an important asset of the method, in case one wishes to study oriented targets, as e.g. molecules adsorbed on a surface.

A representative result of the method is shown in fig. 2 for ionization of three states in the Ar M-shell [6] : $3p^{-1}$, $3s^{-1}$ and a band designated as $3p^{-2}nd$. It is interesting to note (fig. 2b) that the latter two have the same "s-wavefunction" behaviour. This is a direct demonstration of the known fact that the two-electron transitions $3p^{-2}nd$ derive their intensity from the $3s^{-1}$ hole state through configuration interaction. The form of the overlap in eq. (2) implies that F has the same dependence on q, if only in the final state configuration interaction plays a role. If, however, CI is also important in the ground state, a different angular behaviour for different eigenstates of the ion may be expected. One such case is shown in fig. 3, where the ratio of formation of He^+ in the n=2 and n=1 states is given as a function of q [7] . Firstly, the ratio is not constant with q and secondly, a simple Hartree-Fock calculation yields an incorrect behaviour. Only the use of a correlated ground-state wave function [7] leads to agreement with experiment.

One might ask about a possible connection of this ratio with the corresponding number measured in photoelectron spectroscopy. In general there is no direct relation, except with the photon ratio at asymptotic energy. There the momentum of the ejected electron requires the nucleus to take up an equally large amount of recoil momentum. For such an exchange it is necessary that the electron happened to be at r=0 - equivalent to large q - at the instant of absorption of the photon. For that reason the asymptotic photon ratio [11] has been plotted at 1/q = 0, and indeed the (e,2e) ratio shows a steep rise towards the asymptotic value with increasing q.

3. DIPOLE (e,2e) METHOD

Conditions of dominantly dipole excitation are selected when observing fast scattered electrons at an energy loss of only a small fraction of the incident energy and at small scattering angle (for a detailed discussion see the previous chapter). Once the (small) momentum transfer is fixed, the distribution of ejected electrons is the same as that of photoelectrons ejected by radiation with linear po-

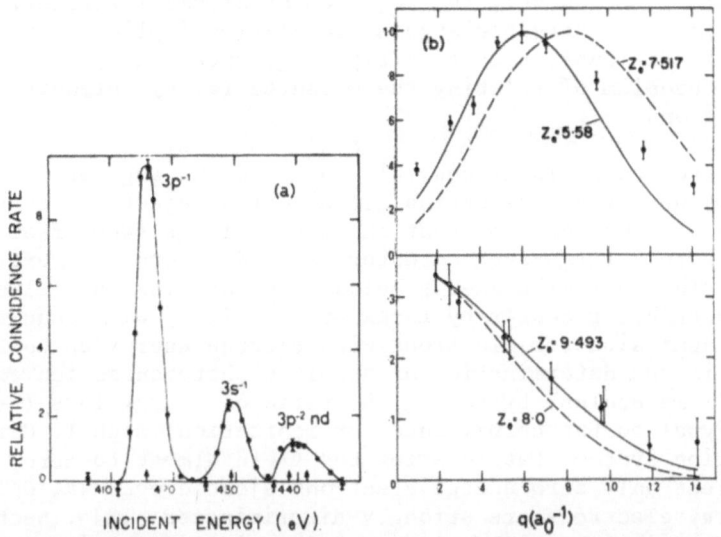

Fig. 2 - a) Binding energy scan of the Ar M-shell [6] , using
$E_0 - E_{binding} = E_1 + E_2$ ($E_1 = E_2 = 200$ eV);
        b) Electron momentum distributions for each of the peaks of
fig. a, and comparison with hydrogenic calculations using various
values of Z_{eff}.

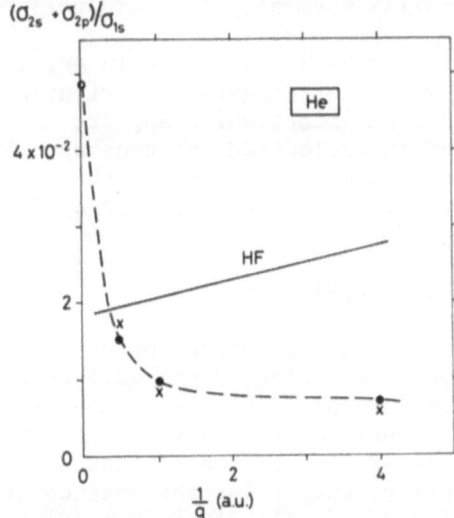

Fig. 3 - Ratio of formation of He$^+$ in the n=2 and n=1 state [7] . (q
is electron momentum in the ground state). Solid line: HF calculation;
crosses-calculation with correlated wavefunction; open circle-calcu-
lated asymptotic photon impact ratio [11] .

larization along K. However, the requirement of small scattering angle
at an experimentally feasible angular resolution implies that normally
the $\vec{K}$-direction is subject to a fairly large uncertainty. This com-
plicates the problem of relating the measured (e,2e) intensity to that
of photoelectrons.

Only in two cases is it possible to avoid the angular distribution problem, namely by collecting all electrons ejected over the entire solid angle. One case is that in which a large sweep field in
conjunction with a large-area detector is used in order to extract
all ejected electrons with energy below a certain maximum value. Such
a system, described recently by Backx et al. [12] , was found to have
complete transmission for electrons with ejected energy up to 20 eV.
It was used in the determination of absolute photoionization efficiencies of small molecules, by taking the ratio of energy loss (or absorption) signal to the coincidence (or ionization) signal. Using the
same extraction system, but lowering the field almost to zero, one is
able to collect only zero energy electrons ejected over 4π, while all
higher energy electrons are strongly discriminated [12] . Such a
scheme of detecting threshold electrons in coincidence with the energy
loss (fig. 4a) is the equivalent of a threshold photoelectron experiment. The interest of this work is that oscillator strengths are obtained for ejected electron energies for which it is evidently very
difficult to operate ordinary electron spectrometers. Moreover, the
method permits direct measurement of FC envelopes (ejected electron
energy constant) instead of a convolution of FC envelope and varying
electronic transition matrix element, as in ordinary photoelectron
work.

When the ejected electrons are analyzed in angle [8,9,13,14], we
must concern ourselves with their angular distribution. We then have
a regular dipole (e,2e) experiment, which appears to simulate rather
closely the situation of photoelectron spectroscopy. Knowing the distribution of $\vec{K}$ directions which contribute to the scattering in a
small cone about zero degree, it is possible to derive for the coincidence intensity [8,9] :

$$I_{coinc} = A_{kin} \frac{df}{dE} (1-c_\alpha(E)\beta) \tag{3}$$

where A_{kin} is a kinematical factor, df/dE the partial oscillator
strength for the observed transition, β the well-known photoelectron
angular anisotropy parameter and $c_\alpha(E)$ a function of E, which depends
on impact energy, opening angle of the forward scattering and ejection
angle α. Eq. (3) is quite similar to the corresponding expression for
the photoelectron intensity, except for the absence of A_{kin} and re-
placement of $c_\alpha(E)$ by $\frac{1}{2}(3/2\sin^2\alpha-1)$. Although $c_\alpha(E)$ is normally different from the corresponding optical expression for the same α ex-
cept for $E \to 0$, an analysis by Hamnett et al. [13] shows that $c_\alpha(E)$
is equal to zero at all E for α = 54.7°, i.e. the same "magic angle"
exists for the dipole (e,2e) case and for photon impact. This opens
up the way for measurements of partial oscillator strengths which are

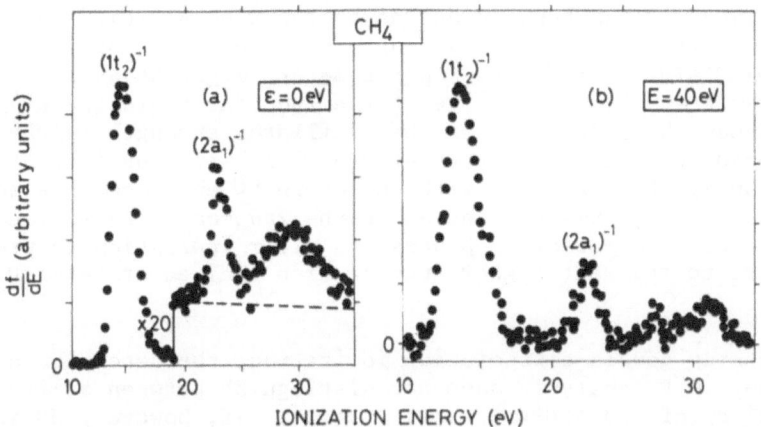

Fig. 4 - a) Threshold oscillator strengths for ionization of CH_4 [12] (scan of energy loss in coincidence with ejected electrons of zero energy, $\varepsilon = 0$)

b) Partial oscillator strengths for ionization of CH_4 [15] (scan of ejected electron energy at fixed energy loss of 40 eV).

independent of the angular anisotropy ([14] and fig. 4b for CH_4 [15])

4. TWO-ELECTRON TRANSITIONS IN MOLECULES

In contrast to the situation for atoms, hardly any calculations exist for molecules of oscillator strengths for two-electron transitions. In this section some experimental evidence for such transitions is described, with the aim of indicating their importance and of suggesting an area for further theoretical work. The evidence was obtained with both (e,2e) methods mentioned above and with the "dipole" electron-ion coincidence experiments (previous chapter). In the latter experiment, the ion extraction system is the same as that used for ejected electrons (with all voltages reversed) and therefore has complete transmission for energetic fragment ions with energy below 20 eV. Detection of the ions in coincidence with the scattered electron permits analysis of the mass based on the ion time-of-flight (TOF) [16,17] .

4.1. H_2

A two-electron excitation similar to the 2s2p-transition in He has long been expected to occur also in H_2. Since such a state is most likely to be repulsive, it intersects the FC zone over a wide energy range and would therefore be hard to detect on a large background of continuous absorption. Detection of the H^+-formation

channel, however, might be a more sensitive probe for such an autoionizing state. It has been suggested by Backx et al. [17] that a
feature of their H^+ oscillator strength spectrum (fig. 5a), obtained
with the high-transmission TOF spectrometer, might be ascribed to a
two-electron transition. The rise at the $2p\sigma_u$-threshold (28 eV) and
the subsequent sharp decrease of the oscillator strength in the region of 35-40 eV cannot be interpreted as a normal behaviour of the
$2p\sigma_u$-continuum, after folding with the known FC envelope. The additional intensity in this region, above the $2p\sigma_u$-contribution, could
well be due to a two-electron process. A clear indication is present
of the onset to the next higher two-electron excited state of H_2^+,
the $2p\pi_u$.

As regards double dissociative ionization, the detection scheme
used in the T.O.F. analysis does not distinguish between arrival of
a single H^+ or of two protons simultaneously. If, however, HD is used,
two ions H^+ and D^+ arising from decay of HD^{++} are not treated equally
but only the faster H^+ is detected. Thus, the ratio H^+/D^+, which for
single ionization is equal to unity (no isotope effect is expected
for dissociation via repulsive states), is caused to increase at
energies beyond the double ionization threshold (fig. 5b). This
fraction of increase, multiplied by the H^+ oscillator strength (fig.
5a), provides a first measurement of the double ionization oscillator
strength of H_2 [18].

4.2 N_2 and CO

For six ionic states of N_2 and of CO, including "two-electron"
states, branching ratios for energies up to 60 eV were obtained by
Hamnett et al. [19] using the dipole (e,2e) method. Their set of data
is fully consistent with the ionic fragmentation pattern as studied
in the electron-ion experiment [20]. It is found that in both molecules the dissociation channel is dominated by decay of "two-electron"
states. One example is the $C^2\Sigma$ state of CO^+, which is known to exhibit extensive configuration interaction (CI) with the $B^2\Sigma$ state. A
CI calculation [21] can be expected to predict the asymptotic ratio
of relative intensities of these two states; it is interesting that
the experimental result for this ratio (fig. 6) [20] indicates that
already at 40 eV the limiting value is approached. The high maximum
near threshold presumably represents the contribution of "direct"
collisions, i.e. energy transfer from the first outgoing electron to
excite another valence electron. Since a transfer of only 2 eV is
needed (the difference in ionization potentials of the $B^2\Sigma$ and $C^2\Sigma$
states), such a direct collision is a very probable process near
threshold.

4.3 CH_4

A first indication of the occurrence of two-electron transitions

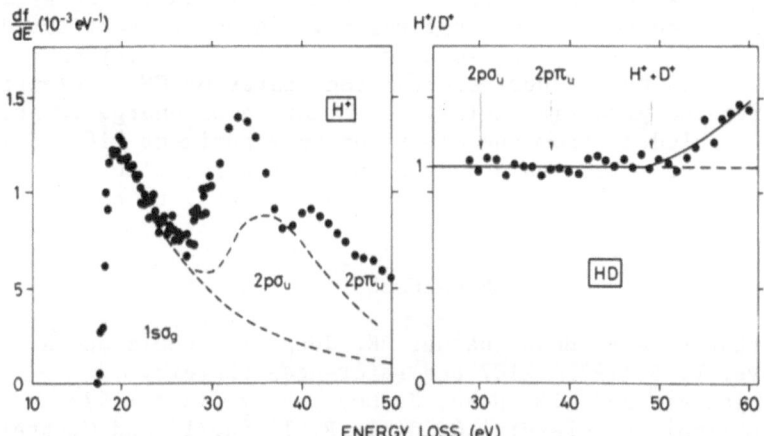

Fig. 5 - a) Oscillator strength for formation of H$^+$ from H$_2$ [17] (ion transmission = 100% for proton energies ≤ 20 eV)

b) Formation of H$^+$ and D$^+$ from HD. When ion pairs are formed in double ionization, only the faster ion (H$^+$) is detected.

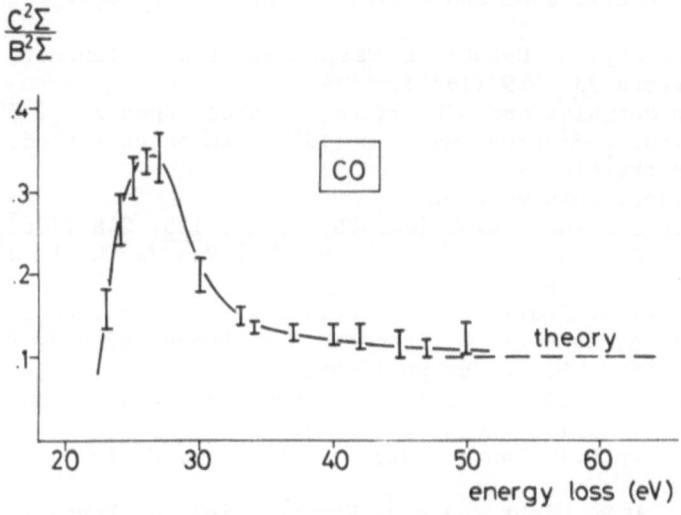

Fig. 6 - Oscillator strength ratio of two states of CO$^+$, which are connected through configuration interaction [20] . Asymptotic ratio (theory) : [21] .

in CH$_4$ was obtained by Hood et al. [22] in their (e,2e) measurements as an unresolved tail in the binding energy spectrum beyond the two one-electron orbitals. More recent work with the dipole (e,2e) methods [12,15] shows structure in this energy region, which appears to be adequately resolved in the spectrum of fig. 4b (scan of ejected elec-

tron energy at 40 eV energy loss). In fig. 4a (scan of energy loss at
0 eV ejected electron energy) the relative intensities of the two
bands seems quite different, such that the higher energy band shows
up only as a weak hump. These excited ion states of CH_4 are entirely
dissociative and give rise mainly to formation of energetic protons,
as can be concluded from the electron-ion experiment [16]. Similarly
as for the other molecules discussed here, the two-electron states
contribute quite significantly (up to 25% for CH_4) to the total ionization at higher energy losses.

REFERENCES

1. H. Ehrhardt, K.H. Hesselbacher, K. Jung, M. Schulz and K. Willmann,
    J. Phys. B. 5 (1972) 2107 and references therein.
2. S. Cvejanović and F.H. Read, J. Phys. B. 7, 1841 (1974).
3. R. Camilloni, A. Giardini Guidoni, R. Tiribelli and G. Stefani,
    Phys. Rev. Letters 29, 618 (1972).
4. A. Giardini Guidoni, G. Missoni, R. Camilloni and G. Stefani,
    Proceedings of Symposium on "Electron and Photon Interactions
    with Atoms", Stirling (1974).
5. S.T. Hood, I.E. McCarthy, P.J.O. Teubner and E. Weigold, Phys.
    Rev. A 8, 2494 (1973).
6. E. Weigold, S.T. Hood and P.J.O. Teubner, Phys. Rev. Letters 30,
    475 (1973).
7. I.E. McCarthy, A. Ugbabe, E. Weigold and P.J.O. Teubner, Phys.
    Rev. Letters 33, 459 (1974).
8. M.J. Van der Wiel and C.E. Brion, J. Elec. Spectr. 1, 309 (1973).
9. C.E. Brion, Radiation Research (1975), to be published, and re-
    ferences therein.
10. K.T. Dolder, this volume.
11. E.E. Salpeter and M.H. Zaidi, Phys. Rev. 125, 248 (1962).
12. C. Backx, G.R. Wight, R.R. Tol and M.J. Van der Wiel, J. Phys.
    B. 8 (1975), to be published.
13. A. Hamnett, W. Stoll and C.E. Brion, J. Phys. B., to be published.
14. W. Stoll, A. Hamnett, G. Branton, C.E. Brion and M.J. Van der
    Wiel, J. Phys. B., to be published.
15. M.J. Van der Wiel, W. Stoll, A. Hamnett and C.E. Brion, Chem.
    Phys. Lett. (1975), to be published.
16. C. Backx and M.J. Van der Wiel, J. Phys. B. 8 (1975), to be pu-
    blished.
17. C. Backx, G.R. Wight and M.J. Van der Wiel, J. Phys. B. 8 (1975),
    to be published.
18. G.R. Wight and M.J. Van der Wiel, to be published.
19. A. Hamnett, W. Stoll and C.E. Brion, J. Electr. Spectr. (1975),
    to be published.
20. G.R. Wight, M.J. Van der Wiel and C.E. Brion, J. Phys. B., to
    be published.
21. P. Bagus, private communication.
22. S.T. Hood, E. Weigold, I.E. McCarthy and P.J.O. Teubner, Nature
    (Physical Science) 245, 65 (1973).

SOME EXPERIMENTS ON COLLISIONS BETWEEN ELECTRONS AND ATOMS OR IONS WHICH INVESTIGATE CORRELATION EFFECTS

K.T. Dolder

Professor of Atomic Physics

The University, Newcastle Upon Tyne NE1 7RU, England

1. INTRODUCTION

During the last few years there has been a growing interest in the subtleties of interactions between charged projectiles and atomic systems and the effects of correlations have received special attention. This paper will describe various experiments which examine correlation effects in low energy collisions but, to simplify the discussion, we will concentrate on interactions between charged projectiles and two-electron targets. Our attention will therefore be primarily focussed on the He atom and the Li^+ and H^- ions. In our discussion of electron-ion collisions we will however review some recent experiments which show autoionization and inner shell ionization in some singly-charged ions of alkalis and alkaline earths.

Let us first consider two new experimental approaches which promise greatly to enhance our knowledge of the excitation and ionization of atoms by electron impact. These techniques are particularly interesting because they hold the promise of further development and much wider application.

2. MEASUREMENTS OF TRIPLE DIFFERENTIAL CROSS SECTIONS FOR IONIZATION

Consider the ionization of helium,

$$e + He \longrightarrow He^+ + e + e \tag{1}$$

In the last 40 years there have been several measurements of <u>total</u>

cross sections for this process (1) and comparison of the results suggests that the accuracy is little better than $\pm$ 10%. This is insufficient to discriminate between more sophisticated theories of ionization now being developed.

Recently a radically new experimental approach has been pioneered and reviewed by Ehrhardt and his colleagues (2). They detected in coincidence the two electrons which emerge from an ionizing event and defined a triple differential cross section;

$$\frac{d^3\sigma}{dEd\Omega_a d\Omega_b} = f(E_o, E_a, \theta_a, \theta_b, \varphi) \tag{2}$$

where $d\Omega_a$ and $d\Omega_b$ are the angular apertures of the two electron collectors, E_a and E_b represent the electron energies whilst θ_a and θ_b are the directions of their trajectories with respect to the incident beam. In these experiments the beams were coplanar so the azimuthal angle φ was zero.

This type of measurement is of great value for the following reasons,

(a) With the exception of the effects of spin, the triple differential cross section provides a complete description of the reaction kinetics.

(b) Triple differential cross sections can be compared directly with theory. To obtain a total cross section theoretical results must be integrated over all angles ; this causes the loss of much information which is often crucial in assessing the value of a particular theory.

(c) This approach is particularly valuable (as we shall see in section 4) in probing threshold ionization where correlation effects are especially strong.

The apparatus of Ehrhardt et al (2,3) is illustrated by Figure 1. Electrons from the cathode (K) are selected in energy ($\Delta E_o \approx 0.03eV$) by an electrostatic selector (S) before passing through an atomic beam of helium. Two rotatable collectors are used for the outgoing electrons and each incorporates an electrostatic selector to define E_a and E_b.

The severe difficulties encountered by these experiments arise mainly from the effects of stray electric and magnetic fields and the very low rate of coincident counts (typically 0.01 to 1.0 sec^{-1}). A particularly illuminating discussion of these problems and their remedies is included in the review by Ehrhardt et al (2).

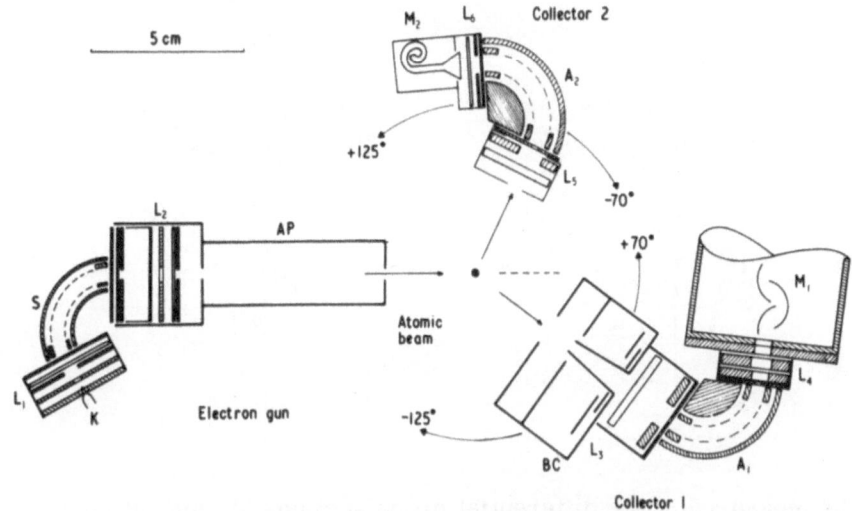

Figure 1. Plan view of apparatus of Ehrhardt et al. Energy resolved
electrons from the gun ionize a helium beam. The two outgoing
electrons are detected in coincidence by rotatable collectors
which incorporate energy resolution.

Figure 2 illustrates the value of these results. It presents some of
the triple differential cross sections measured by Ehrhardt et al (3) for
the ionization of helium by 256eV electrons. The radial positions of the
points are proportional to $d^3\sigma$ whilst the angular circumferential scales
denotes θ_b. The chart on the left refers to E_a=229eV, E_b=3eV and θ_a=6°
whilst that on the right shows E_a=182eV, E_b=50eV and θ_a=8°. Predictions
of a plane wave Born calculation are shown by the continuous curves in
both plots and the inadequacy of this theory is clearly revealed, especially
when $\theta_b \approx$ 120°. If we were to compare measured total cross sections
with Born theory at these energies we would find that the differences
would scarcely exceed the probable experimental error. The triple differential cross sections therefore provide a much more extensive and exacting
test of theory.

3. MEASUREMENTS OF TRIPLE DIFFERENTIAL CROSS SECTIONS
   FOR EXCITATION

Analogous experiments for excitation have been performed by Kleinpoppen and his colleagues. Consider the reaction,

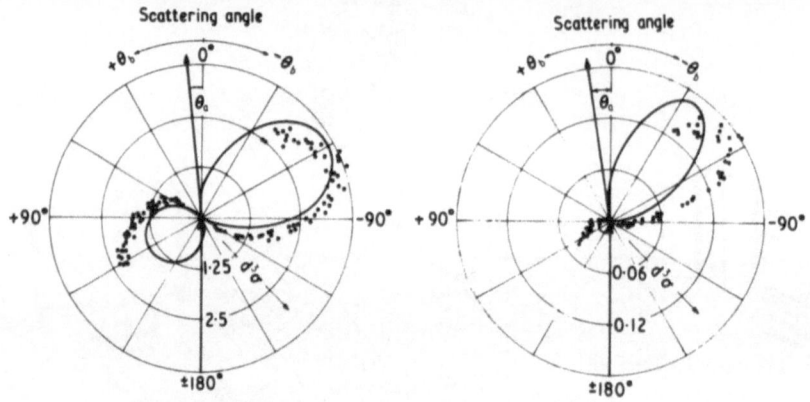

Figure 2. Measured triple differential cross sections for ionization of
helium compared with plane wave Born calculations (continuous
line).

$$e + He \, (\, 1^1S \,) \; \longrightarrow \; He \, (\, 2^1P \,) + e + h\nu \qquad\qquad (3)$$

Eminyan et al (4, 5) measured the rate of coincident detection of the
scattered electron and emitted photon. This type of experiment has the
same attractions as those just described but it has two further advantages
which deserve to be emphasized.

One not only obtains, in principle, a complete description of the
reaction kinetics (with the exception of spin) but, for the 1^1S - 2^1P
transition of helium it is possible, for the first time, to deduce relative
differential cross sections for the three degenerate magnetic sub-levels
of the upper state. Moreover, total cross sections for excitation are
notoriously difficult to measure accurately (the absolute determination
of very weak photon fluxes poses severe and subtle experimental problems).
Coincidence experiments provide such a wealth of detail that they ought to
discriminate effectively between rival theories of excitation without the
need for absolute calibration.

The theory of this experimental approach has been developed for He
(1^1S - 2^1P) by Macek and Jaecks (6). Neglecting spin, the excitation is
completely described at a given incident energy by three parameters, σ,

λ and χ. These are defined in terms of the excitation amplitudes a_0, a_1 and a_{-1} of the sub-levels by,

$$\sigma = 2 \left| a_1 \right|^2 + \left| a_0 \right|^2 \qquad (4)$$

$$\lambda = \frac{\left| a_0 \right|^2}{2 \left| a_1 \right|^2 + \left| a_0 \right|^2} \qquad (5)$$

whilst χ is the relative phase of a_1 and a_0. In the experiments the incident electrons and both detectors are coplanar whilst λ and χ are deduced from their known dependence on the coincidence count rate and the two detector angles.

The power of this technique is illustrated by Figure3 which compares measured and calculated values of λ for incident electron energies of 60 and 80eV and a range of electron scattering angles.

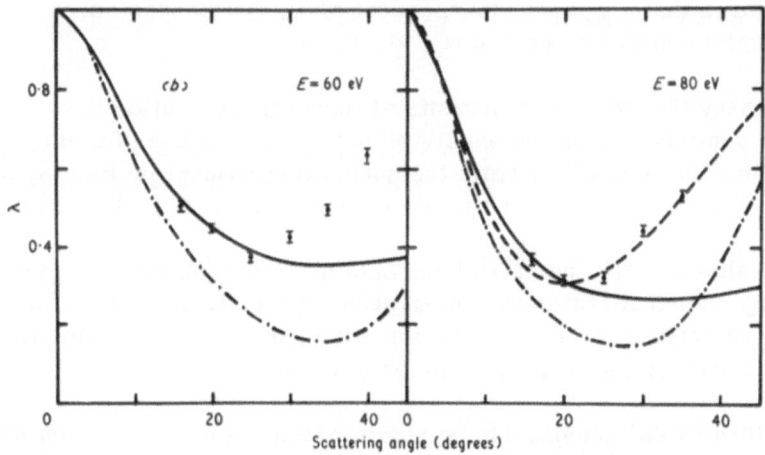

Figure 3. Measured and calculated values of the parameter for the excitation of He (2^1P) by 60 and 80eV electrons. The continuous, dashed and chain-dashed curves respectively represent results of Born's, Distorted Wave and Many Body Green's function calculations.

4. THRESHOLD IONIZATION BY ELECTRON IMPACT

Simple classical considerations are sufficient to demonstrate that threshold ionization is dominated by correlations between the scattered and ejected electrons. There must be strong radial and angular correlation between the two outgoing electrons because, at threshold, they are constrained to move in opposite directions with the same energy. If one electron were more energetic than the other it would draw ahead and so experience a weaker (partially shielded) ionic field. This electron would then escape, carrying with it the majority of the available energy. The second electron would therefore remain trapped in a highly-excited state so that ionization would not occur. This constraint provides the basis of Wannier's (7) classical law which predicts that, close to threshold, the ionization cross section (σ_i) varies as

$$\sigma_i \propto E^n \tag{6}$$

where E is the difference between the incident electron energy and the ionization energy and n = 1.127. It also predicts the distributions of the energies $P(E_a)$ and $P(E_b)$ of the two outgoing electrons and the angles $P(\theta_{ab})$ between them. Specifically, $P(\theta_{ab})$ is found to be a maximum when θ_{ab} = 180° and the width of this distribution is proportional to $E^{0.25}$. It also follows that $P_E(E_a)$ is independent of E_a and is constant over all possible values in the range $0 < E_a < E$.

The many theoretical treatments of threshold ionization have included quantum, semi-classical and wholly classical approaches. Similar results to those just outlined follow from the quantum treatment by Rau (8) whilst Fano (9, 10) pointed out that correlations are implicit in both formulations.

The value n=1.127 has also been obtained semi-classically by Peterkop (11) and by classical trajectory integrations (e.g. Gruyic (12)) although quantum treatments of Temkin and his colleagues (e.g. Temkin (13)) and Kang and Kerch (14) give values closer to unity.

The theoretical results can be summarised as follows. When studied, $P_E(E_a)$ has always been found to be uniform whilst $P(\theta_{ab})$ had a maximum value at 180° and width proportional to $E^{0.25}$. Various treatments gave values of n ranging from 1.0 to 1.5 but the energy range of this power law is not determined by any theory.

Experimental results tend to favour the value n=1.127. For example, Marchand et al (15) obtained n=1.16±0.03 for energies within 1eV of the threshold of helium whilst Cvejanovic and Read (16) devised experiments

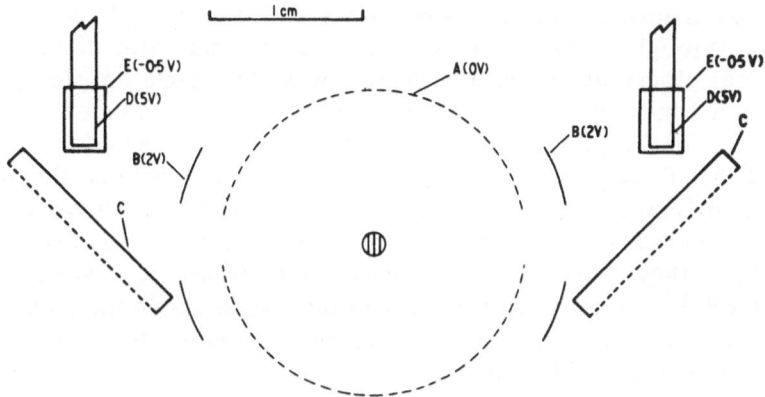

Figure 4. Apparatus used to investigate threshold ionization. An electron
beam (not shown) collided with a helium beam at the centre of a
cylindrical cage. Two outgoing electrons from ionization were
deflected by the energy selectors (C) and detected in coincidence.

to test the main predictions of Wannier's theory with an apparatus illustrated
by Figure 4. Energy resolved electrons ($\Delta E_0 \approx 0.03 eV$), moving parallel
to the axis of a cylindrical collision region, (i.e. perpendicular to the
plane of Figure 4) collided with a helium beam. The outgoing electrons
escaped through two diametrically opposite apertures ($\theta_{ab} = 180°\pm20°$)
and were deflected by novel electrostatic energy selectors which had low
resolution but high collection efficiency. This was essential because, near
threshold, the rate of ionization is inevitably small. Electrons transmitted
by the selectors were then detected in coincidence to ensure that they
emerged from the same event.

In a second experiment the apparatus was modified to make
$\theta_{ab} = 150\pm20°$.

The results were, within their limits of error, compatible with
Wannier's predictions. It was found that the width of P(θ_{ab}) varied as
$E^{0.19\pm0.07}$ and $P_E(E_a)$ was uniform (within 15%) for energies between
0.2 and 0.8eV above threshold. It was also established that P_θ (θ_{ab}) was
several times larger for θ_{ab}=180° than for θ_{ab}=150°.

An ingenious modification of the apparatus was used to determine n. Only one electron analyser was employed and it was tuned to have a high transmission efficiency only for very slow electrons. The effect of field penetration through the aperture in the cylindrical cage also discriminated against faster electrons so that the system was, in effect, an energy filter with a narrow bandwidth, ΔE, of about 15meV. If it is assumed (and the measurements of Cvejanovic and Read and various theories do not dispute this point) that $P_E(E_a)$ and $P_E(E_b)$ are uniform, it follows that the number of electrons detected will be proportional to $\sigma_i . \Delta E/E$, which varies as E^{n-1}. Thus, by measuring the transmitted electron current for various values of E_0, they obtained points illustrated by figure 5. A smooth curve varying as $E^{0.127}$ is seen to fit closely to the points above the ionization threshold but when the various sources of error were evaluated it was concluded that $n = 1.131 \pm 0.019$.

Below the ionization threshold peaks corresponding to Rydberg states of helium can be seen. These states have large radii and can be formed only if one electron has receded to a great distance before the second electron enters the bound state. Otherwise interactions between the two electrons would exchange energy and momentum and thus reduce the probability of forming these highly-excited states. In fact we see from the figure

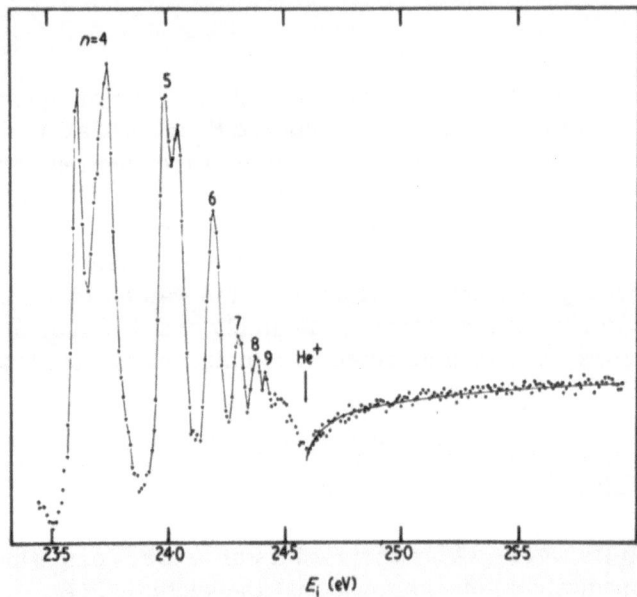

Figure 5. Yield of very slow electrons from helium plotted against incident electron energy. The continuous line above the ionization threshold varies as $E^{0.127}$.

that for principal quantum numbers greater than 9 the peaks merge and form a " cusp " which approximately mirrors the curve above the ionization energy. The width of this cusp indicates the energy range for which these correlations are appreciable.

If we intend to apply Wannier's law " close " to threshold we ought to specify, more exactly, the energy range over which it is valid. This problem has been investigated by Spence (17). We have already noted that correlation provides the basis of Wannier's law and that correlation is also manifested by the appearance of a " cusp " in the electron scattering cross section at energies just above the ionization threshold. The width of a cusp therefore indicates the energies for which correlations are strong and hence the range of the threshold ionization law.

Spence used a variation of the " trapped electron " technique to examine the total inelastic scattering of electrons by helium. He collected electrons which had excited or ionized helium and had lost <u>additional</u> energy less than an amount W, which was set in successive experiments at 0.07, 0.40, 1.00 and 2.00eV. Excitation functions for these values of W are illustrated by Figure 6. The region to the right of the line I-I (the ionization threshold)

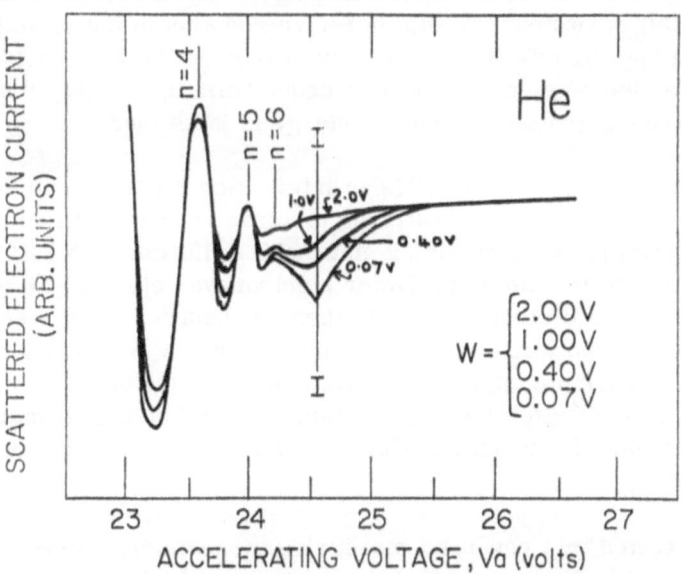

Figure 6. Yield of scattered electrons which have excited or ionized helium and have lost addition energy less than corresponds to the voltage W. Results for four values of W are illustrated.

refers to ionization whilst to the left of this line the curves show peaks due to the excitation of bound states. When W is small (0.07eV) the curve refers to scattered electrons which have ionized helium close to threshold and we see a pronounced cusp in the excitation function. This cusp disappears when W is increased beyond 1eV and so it follows that for helium Wannier's law is only valid for energies less than 2eV above threshold.

5. THRESHOLD EXCITATION OF AUTOIONIZING STATES

It can easily be seen that threshold excitation of autoionizing states is particularly sensitive to the effects of electron correlation. Consider an electron with energy E_0 which collides with helium. It may simultaneously excite both bound electrons, e.g.

$$e + He (1s^2)\ ^1S \longrightarrow e + He (2s^2)\ ^1S \tag{7}$$

provided that $E_0 > E_e$, where E_e represents the total excitation of both electrons. Energy can then be transferred between the excited electrons and so it might be expected that one of them would be ejected with an energy $(E_e - E_i)$, where E_i (25.4eV) is the ionization energy of helium. But, near threshold, it must be remembered that the projectile electron has given up almost all of its energy and therefore remains in the vicinity of the target for an appreciable time, partially screening the ionic field. Consequently the ejected electron, as it recedes from the target, " sees " a smaller attractive potential so that its energy is increased.

The magnitude of this " post collision interaction " (PCI) can easily be estimated. For example, if a scattered electron receded with an energy of 1eV from an atom in an autoionizing state with a lifetime of 5×10^{-15} s it would travel only 3 nm before the faster electron was ejected. Since the potential between the two electrons would then be about 0.5 V, the energy of the scattered electron would be appreciably enhanced. According to Hicks et al (18) this simple approach predicts an increase of $\Gamma/2\ [\ (E_0 - E_e)/I\ ]^{-1/2}$ in the energy of the scattered electron where Γ is the energetic width of the autoionizing state and I is the ionization energy of atomic hydrogen.

Hicks et al tested this result by measuring the numbers and energies of electrons ejected at 70° from autoionizing states of helium. Figure 7 illustrates their results. Each curve shows four peaks which correspond, from left to right, to the $(2s^2)\ ^1S$, $(2s2p)^3P$, $(2p^2)\ ^1D$ and $(2s2p)\ ^1P$ states and the numbers above each curve show the amounts $(E_0 - E_e)$ by

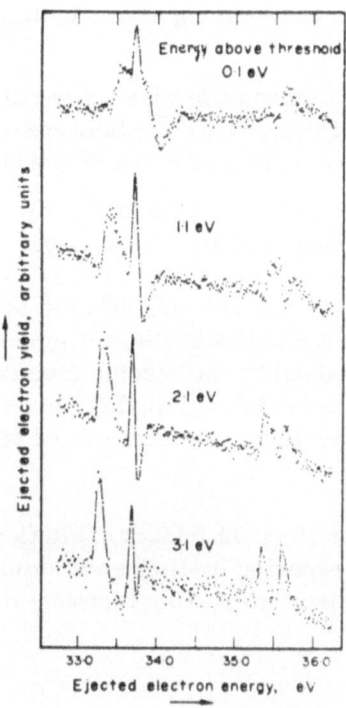

Figure 7. Electrons ejected from autoionizing states of helium when the incident electron energies were 0.1, 1.1, 2.1 and 3.1eV above the threshold. The peaks refer, from left to right, to ($2s^2$) ^{1}S, (2s2p) ^{3}P, ($2p^2$) ^{1}D, (2s2p) ^{1}P autoionizing states.

which the incident electron energy exceeded the autoionization threshold (this difference was held constant as E_o was varied). It is apparent that the energetic positions of the ^{1}S peaks show the expected increase as ($E_o - E_e$) is decreased whilst the positions of the ^{3}P peaks are almost unchanged. This is to be expected because the life of the ^{3}P state is an order of magnitude longer than for ^{1}S so the scattered electron has travelled distance from the helium atom when the energetic electron is ejected. The measured shifts are, as predicted, proportional to ($E_o - E_e$) $^{-1/2}$ but the quantitative agreement with theory is poor. A thorough theoretical treatment has not yet appeared but the following imperfections of the simple post collision interaction model were discussed by Hicks et al (18),

(a) Classical arguments have been applied even though quantum theory can easily be shown to be more appropriate.

(b) The scattered (slow) electron has been assumed to be at rest throughout the interaction.

(c) Stark shifting of the autoionizing states by the scattered electron has been ignored.

(d) No account has been taken of interference effects (which have been demonstrated experimentally to exist) between electrons ejected from neighbouring resonances or of overlaps between resonances.

(e) It has been assumed that the peaks in the measured curves correspond to the true energetic centres of the resonances.

A different experimental approach was adopted by Spence (private communication) who measured shifts in the energy lost by the scattered electrons rather than that gained by the ejected electrons. He used a " trapped electron " technique which verified the principal conclusions of Hicks et al and gave cross sections, as functions of energy, for the excitation of the ^{1}S and ^{3}P states.

Further observations of electron impact excitation of autoionizing states of helium have been reported by Hicks and Comer (19) in a paper which includes some extremely interesting technical details of the Manchester group's apparatus.

6. THE EXCITATION OF HIGH RYDBERG STATES

An electron which has autoionized helium near its threshold loses almost all of its energy and it then loses even more energy as the result of PCI. Clearly, if the energy lost by PCI is too large the electron cannot escape and so we are left, not with a He$^+$ ion, but with a highly-excited helium atom. High Rydberg states can , of course, also be populated by direct electron impact excitation and so we expect interference between these two channels in the excitation functions near autoionization threshold. This has been observed in electron energy loss spectra byKing et al (20) who obtained the excitation functions, illustrated by figure 8, for forward scattering. These results refer to states of helium with principal quantum numbers from 3 to 8. Structure in optical excitation functions which almost certainly has the same origin has been observed byHeideman et al (21,22). King et al discussed their results in terms of a " shake-down " model which is a quantum equivalent of the classical treatment of PCI discussed in sections 6 and 9.

7. DOUBLE IONIZATION OF TWO ELECTRON SYSTEMS

Consider the simplest examples of multiple ionization namely,

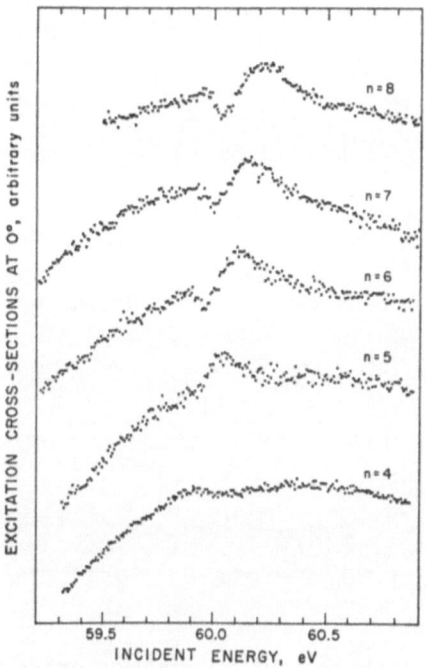

Figure 8. Structure in energy loss spectra for states of helium with principal quantum numbers for n=4 to 8. The abscissa denotes the
incident electron energy.

$$e + He \longrightarrow He^{++} + e + e + e \tag{8}$$

and the double ionization of Li^+ and H^-. We now have a situation in which
these are three outgoing electrons and correlation effects must be important
although they are specifically excluded by the most elementary theoretical
approach, namely the " sudden approximation " (SA) (e.g. Mittleman
(23)). This assumes that a projectile removes one electron from helium
without perturbing the second bound electron. The second electron was
originally in a stable state of He but must now choose between proceeding
to a bound state of He^+ or to the continuum. Neglecting interaction between
the bound electrons it follows that the cross section (σ_{++}) for double
ionization must be proportional to those for single ionization (σ_+).
Consequently σ_+/ σ_{++} should be independent of the projectile energy.

Figure 9 illustrates the inaccuracy of this prediction at high and low
energies by comparing it with experimental results for He (van der Wiel
et al (24)), Li^+ (Peart and Dolder (25))and H^- (Peart et al (26)), although,
at intermediate energies, the measurements for He and Li^+ are not greatly

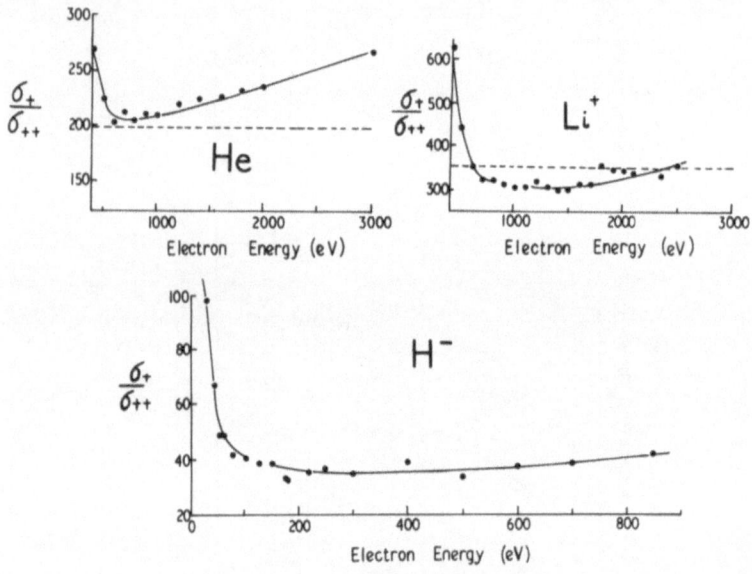

Figure 9. The ratio of single to double ionization cross sections for He,
Li$^+$ and H$^-$ plotted against electron energy. Predictions of the
" sudden approximation " for He and Li$^+$ are represented by
horizontal lines.

at variance with results of the SA which are represented in the figure by
horizontal lines.

Tweed (27, 28) has discussed the application of Born I and Born II
approximations to the double ionization of two-electron systems. Born II
includes successive interactions between the projectile and the two bound
electrons but Tweed considered it unsuitable for the present problem
because correct phases could not be assigned to these interactions. The
inclusion of exchange was similarly precluded. He therefore employed
the Born I approximation but difficulties arose in deciding the effective
ionic charge which interacted with the outgoing electrons. If it was assumed
that both ejected electrons had the same velocity they would each, at large
distances, experience the full ionic charge ; this is the " no shielding
approximation " (NSA). Alternatively one could assume that the electrons
have different velocities and calculate an effective protential by taking a
spherical average (SAA) over their momentum vectors.

Theory is very sensitive to the choice of effective potential. For He and Li$^+$ the NSA gave good agreement with experiment at high energies but was a few times too low near threshold, whilst results of the SAA were an order of magnitude too large.

The SAA led to difficulties when applied to H$^-$ because it implies that the faster ejected electron " sees " a zero (fully screened) ionic potential whilst the NSA gave results an order of magnitude too small. One expects H$^-$ to be especially sensitive to correlation potentials so the two ejected electrons might interfere strongly. This does seem to happen and Tweed found good agreement ($\approx$ 20%) between Born I and experiment for H- when an effective charge of 0.6 was assumed to act on each electron.

8. THE FORMATION OF SHORT-LIVED EXCITED NEGATIVE IONS

A celebrated experiment by Schulz (29) revealed a sharp resonance in the cross section for scattering of electrons by helium at energies around 19.3eV. This was interpreted as the formation of He$^-$ (1s2s^2) ^{2}S$_{1/2}$ with a life of order 10^{-13} s. Numerous other doubly-excited negative ions were subsequently discovered and studied in various laboratories (see review by Schulz (30)). More recently evidence of short-lived, triply-excited, doubly-negative ions was obtained by Walton et al (31) and Peart and Dolder (32). Their experiments employed inclined beams of electrons and H$^-$ ions to measure detachment cross sections for the reaction,

$$H^- + e \longrightarrow H + e + e \tag{9}$$

Inclined beams facilitate the study of reactions at low interaction energies with good energy resolution. Walton et al found a pronounced structure in the detachment function with a width $\approx$ 1eV at energies close to 14.2eV. Taylor and Thomas (33) then used a computational technique (the " stabilization method ") to demonstrate that Hartree-Fock wavefunctions with the configuration (2s^{2}2p) ^{2}P^0 are stable over a range of energies, about 1eV wide, centred on 14.8eV. This represents satisfactory agreement with the experiment because the observed resonance is broad and so its centre cannot be exactly located from the measurements. The calculation also suggested that a second resonance with a largely 2p^3 configuration might exist at a higher, unspecified, energy. A more refined experiment was therefore performed by Peart and Dolder (32) which revealed a second, sharper, resonance at 17.26eV and subsequently Thomas (34) calculated that a state which is 60% (2p)^3 and 20% (2s)^2(2p) can exist at 17.26eV. Both resonances are illustrated by figure 10.

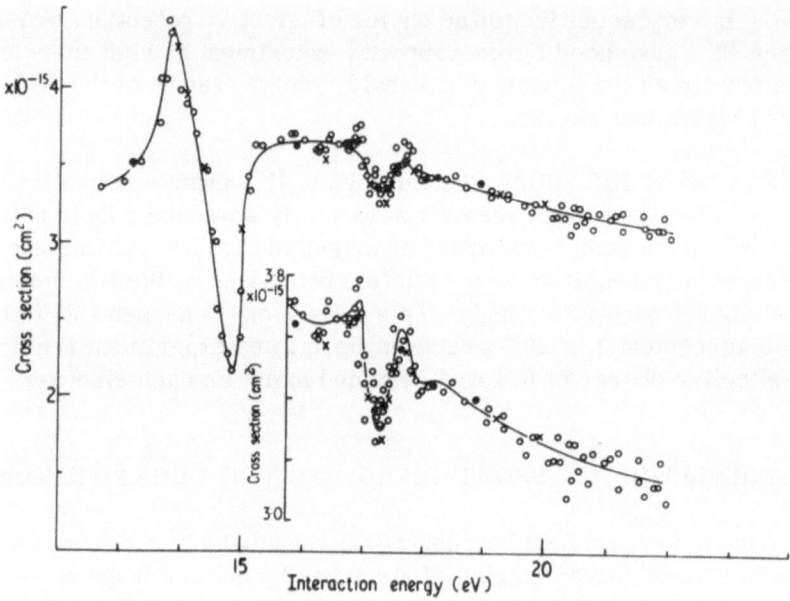

Figure 10. Structure in the electron impact detachment cross section of H⁻.

Mass spectrometric measurements seemed to suggest that stable doubly-negative ions of heavier elements might exist (Baumann et al (35)) but it proved impossible to confirm these results in other laboratories (e.g. Spence and Chupka, private communication). Consequently there is no convincing evidence for the existence of long-lived doubly-negative ions.

9. EXPERIMENTS IN WHICH POSITIVE IONS ARE USED TO PROBE ATOMIC SYSTEMS

Although we are primarily concerned with the effects of electron collisions there are some interesting related processes which arise when positive ions are used as projectiles. Ions, by virtue of their large mass, carry sufficient energy to produce inelastic reactions, even when their velocity is low. Consequently, an ion may cause a reaction and then remain in the vicinity of the target long enough to interact further with the atomic electrons.

In section 5 we discussed post collision interactions which follow threshold autoionization by electron impact. Essential features of this model were first proposed by Barker and Berry (36) who observed decreases in energy of the ejected electrons when helium was autoionized by He^+ with 1 to 4keV energy. In this case the positive potential of the ion attracted the ejected electron and so reduced its energy.

Since H^- is so susceptible to correlation effects it seemed worthwhile to study interactions between H^+ and H^- ions. Two reactions have, so far, been observed,

$$H^+ + H^- \nearrow \quad H + H \qquad\qquad\qquad\qquad (10)$$
$$\searrow \quad H^+ + H^+ + e \qquad\qquad\quad (11)$$

Peart et al (unpublished) recently re-measured cross sections for reaction (10) and they are now engaged on measurements of reaction (11). The question arises whether the electron lost by the H^- ion will be captured by the proton (reaction 10) or ejected (reaction 11). Preliminary results suggest that the former reaction is very strongly favoured unless the proton energy exceeds a few keV.

10. COLLISIONS BETWEEN ELECTRONS AND POSITIVE IONS

During the last 14 years techniques have been developed to study collisions between electrons and ions and these have been reviewed, for example by Dolder (37). Ions are of practical interest because they are common constituents of most laboratory and stellar plasmas whilst, at a more fundamental level, comparisons between the properties of ions and their isoelectronic atoms are sometimes very revealing.

Autoionization and inner shell ionization seems to be particularly important in many positive ions. Peart and Dolder (38) have discussed their measurements for the ionization of Li^+, Na^+, K^+, Rb^+, Cs^+, Mg^+, Ca^+, Sr^+ and Ba^+ by electron impact, e.g.

$$e + Ba^+ \longrightarrow Ba^{++} + e + e \qquad\qquad\qquad (12)$$

In the case of alkali ions they found strong evidence of inner shell ionization only for Rb^+ and Cs^+ whilst there were very large contributions from autoionization for all of the alkaline earth ions except Mg^+, as can be seen from figure 11. Interpretation of these results is based on Hansen's (39,40) Hartree-Fock calculations. For autoionization he argues that the dominant inner shell excitation is of the type $np \longrightarrow nd$ where n=3, 4 and 5 for Ca^+, Sr^+ and Ba^+, respectively. By analogy one would need a $2p \longrightarrow 2d$ transition to produce similar autoionization in Mg^+ but, since 2d states cannot exist,

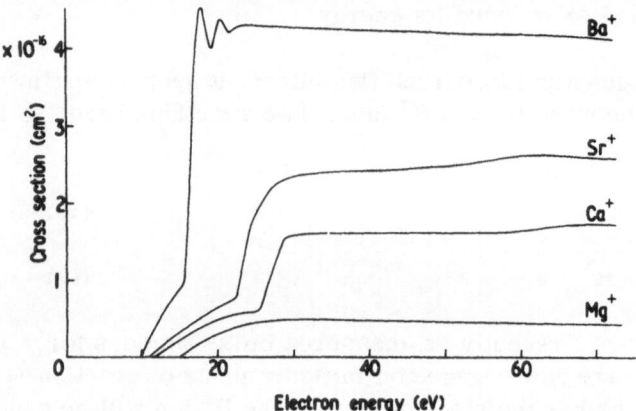

Figure 11. Measured electron impact ionization cross sections for Mg^+,
        Ca^+, Sr^+ and Ba^+ ions.

we can understand the marked difference between Mg^+ and other ions
illustrated by Figure 11.

Other topics in electron ion collisions which are currently of interest
include the very careful measurements by Dunn's group at JILA of resonant
excitation of ions which are important in astrophysics and aeronomy (e.g.
Ca^+, Ba^+ and N_2^+). These experiments are restricted to ions which emit
visible resonant radiation and they are characterised by very careful absolute
calibration of the optical detector (e.g. Crandall et al (41)). A perplexing
problem is the unexplained differences between theoretical and experimental
results for the electron impact excitation of He^+ (1S - 2S) discussed by
Seaton (42) whilst, for molecular ions, Peart and Dolder (43) have described
a series of measurements of cross sections for the various inelastic
channels in electron H_2^+ interactions.

Acknowledgements

The author is indebted to Drs. F. Read and D. Spence for providing
experimental results in advance of publication.

References

(1) H.S.W. Massey and E.H.S. Burhop, Electronic and Ionic Impact Phenomena, Vol.1, eds. W. Marshall and D.H. Wilkinson, Oxford University Press, 1969.

(2) H. Ehrhardt, K.H. Hesselbacher, K. Jung & K. Willmann, Case Studies in Atomic Collision Physics 2, 161 (1972).

(3) H. Ehrhardt, K.H. Hesselbacher, K. Jung, M. Schulz & K. Willmann, J. Phys. B 5, 2107 (1972).

(4) M. Eminyan, K.B. MacAdam, J. Slevin & H. Kleinpoppen, Phys. Rev. Lett. 31, 576 (1972).

(5) M. Eminyan, K.B. MacAdam, J. Slevin & H. Kleinpoppen, J. Phys. B 7, 1519 (1974).

(6) J. Macek & D.H. Jaecs , Phys. Rev. A4, 2288 (1971).

(7) G.H. Wannier, Phys. Rev. 90, 817 (1953).

(8) A.R.P. Rau, Phys. Rev. A4, 207 (1971).

(9) U. Fano, Comments in Atom. and Molec. Phys. 1, 159 (1970).

(10) U. Fano, J. Phys. B 8, (1975) in press.

(11) R. Peterkop, J. Phys. B 4, 513 (1971).

(12) P. Grujic, J. Phys. B 5, L137 (1972).

(13) A. Temkin, J. Phys. B 7, L450 (1974).

(14) I.J. Kang & R.L. Kerch, Phys. Lett. 31A, 172 (1970).

(15) P. Marchand, C. Paquet & P. Marmet, Phys. Rev. 180, 123 (1969).

(16) S. Cvejanovic & F.H. Read, J. Phys. B 7, 1841 (1974).

(17) D. Spence, Phys. Rev. A11, 1539 (1975).

(18) P.J. Hicks, S. Cvejanovic, J. Comer, F.H. Read & J.M. Sharp, Vacuum 24, 573 (1974).

(19) P.J. Hicks & J. Comer, J. Phys. B 8, 1866 (1975).

(20) G.C. King, F.H. Read & R.C. Bradford, J. Phys. B 8, 2210 (1975).

(21) H.G.M. Heideman, G. Nienhuis and T. Van Ittersum, J. Phys. B 7, L493 (1974).

(22) G. Nienhuis and H.G.M. Heideman, J. Phys. B 8, 2225 (1975).

(23) M.H. Mittleman, Phys. Rev. Lett. 16, 498 & 779 (1966).

(24) M.J. Van der Wiel , Th. M. El-Sherbini & L. Vriens, Physica 42, 411 (1969).

(25) B. Peart & K. Dolder, J. Phys. B 2, 1169 (1969).

(26) B. Peart, D.S. Walton & K. Dolder, J. Phys. B 4, 88 (1971).

(27) R.J. Tweed, J. Phys. B 6, 259 (1973).

(28) R.J. Tweed, J. Phys. B 6, 270 (1973).

(29) G.J. Schulz, Phys. Rev. Lett. 10,104 (1963).

(30) G.J. Schulz, Rev. Mod. Phys. 45, 378 (1973).

(31) D.S. Walton, B. Peart & K. Dolder, J. Phys. B 4, 1343 (1971).

(32) B. Peart & K. Dolder, J. Phys. B 6, 1497 (1973).

(33) H.S. Taylor & L.D. Thomas, Phys. Rev. Lett. 28, 1091 (1972).

(34) L.D. Thomas, J. Phys. B $\underline{7}$, L97 (1974).

(35) H. Baumann, E. Heinicke, H.J. Kaiser & K. Bethge, Nucl. Inst. & Meth. $\underline{95}$, 389 (1971).

(36) R.B. Barker & H.W. Berry, Phys. Rev. Lett. $\underline{151}$, 14 (1966).

(37) K. Dolder, Case Studies in Atomic Collision Physics $\underline{1}$, eds. E.W. McDaniel & M.R.C. McDowell) N. Holland, Amsterdam, 1969.

(38) B. Peart & K. Dolder, J. Phys. B $\underline{8}$, 56 (1975).

(39) J.E. Hansen, J. Phys. B $\underline{7}$, 1902 (1974).

(40) J.E. Hansen, A.W. Fliflet & H.P. Kelly, J. Phys. B $\underline{8}$, L127 (1975).

(41) D.H. Crandall, P.O. Taylor & G.H. Dunn, Phys. Rev. $\underline{10}$A, 141 (1974).

(42) M.J. Seaton, Adv. in Atomic & Mol. Phys. (1975) in press.

(43) B. Peart & K. Dolder, J. Phys. B $\underline{8}$, 1570 (1975).

General Review

The topics on electron helium collisions have been reviewed by F.H. Read as an invited paper at IX ICPEAC (Seattle, 1975) and also in " Radiation Research", Proceedings of a symposium at London, Ontario, December 5-7, 1974. Electron ion collisions have been reviewed by K. Dodler & B. Peart, Reports on Progress in Physics (in preparation).

THE QUASI-MOLECULAR MODEL IN HEAVY PARTICLE COLLISIONS

M. BARAT

Laboratoire des Collisions Atomiques, Université Paris-

Sud, Bât. 220 - 91405 ORSAY

In atom-atom collisions, heavy incident particles can strongly perturb the electronic edifice and are therefore not a probe as "clean" as are photons and electrons, to study electron correlations in an isolated system (atom, ion, molecule). On the other hand, after the "hurricane" caused by such violent collisions, the electronic clouds are left in very "exotic" states which are not commonly encountered with lighter perturbers : multiple excited and/or ionized states will occur. Furthermore, when the electronic clouds strongly interpenetrate the excitation process itself involve the interaction of several electrons of each partner. It is therefore no need to stress the importance of *electron correlation* in such processes.

I POINT LIKE PROJECTILES

If we consider the simplest case of a *point like* projectile, I mean a particle without any electronic structure such as a proton or an α-particle, the excitation mechanism is actually very similar to that produced by electron impact, at least for very high velocities[1]. The excitation (or ionization) process can be understood as *direct* coulomb interaction between the bare projectile (charge Z_1) and the target electron involved in the excitation process. Using the notations defined in fig. 1. The interaction is given by

$$Z_1 \frac{e^2}{|\vec{R}-\vec{r}|}$$

229

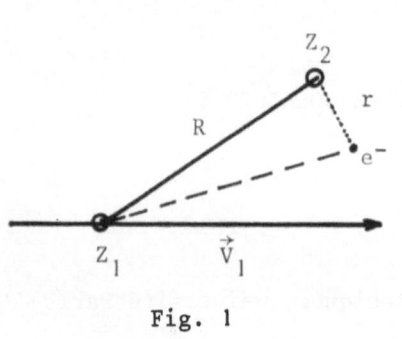

Fig. 1

The transition probability is then expressed in a first order perturbation theory as the squared matrix element of this interaction taken between the initial and final states. As far as this process is a small perturbation, of the system, the cross section can be calculated using the Born approximation or more specific the plane wave Born approximation (*PWBA*).[2] It is also possible to take advantage of the validity of the classical picture descri-bing the heavy particle *trajectory* and thus, use a semi-classical approximation (*SCA*).[3] In this method, the electrons are treated quantally whereas the nuclei can be considered following a classical trajectory. A purely classical treatment has also been worked out[4] and leads to essentially the same results. (Binary Encounter Method, *BEA*).

A nice feature of the BEA and PWBA theories, is the possibility for scaling the cross sections, allowing for a universal expression to be used for all targets :

$$\frac{n^2\sigma}{z_1^2} = f(E/\lambda U)$$

where

λ is the projectile/electron mass ratio
E is the energy of the incident particle.
n is the binding energy of the electron in the target
and σ is the total cross section for a given excitation.

The theoretical method described above apply for outer shell as well as inner shell excitations. For instance figures 2(a) and 2(b) give examples showing the results for K shell excitation by proton impact. In fig. 2(a) the experimental data obtained for a large variety of targets are seen to fit reasonably well these universal curves. Data shown in fig. 2(b) are obtained from Auger electron measurements [6]. It is seen that the scattering of the data points about the universal curve is considerably reduced as compared with the data of fig. 2(a) which were deduced from X-Ray measurements. This might reflect the problems dealing with X-Ray measurements due to the uncertainties of the fluorescence yield.

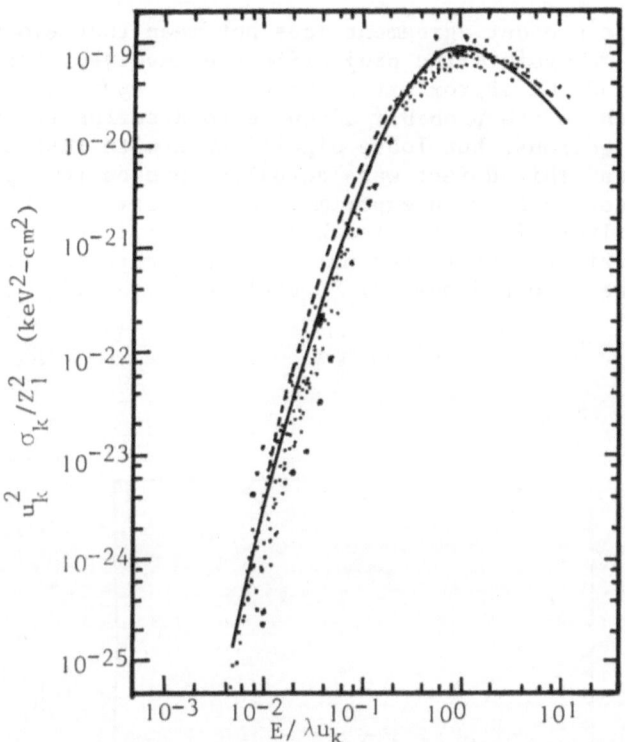

Fig. 2(a) Scaled cross section for proton impact K shell excitation
of many targets from Be to Au[5]

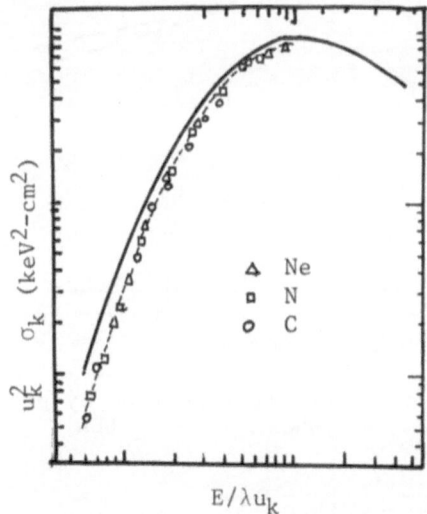

Fig. 2(b) Scaled cross sections for K shell ionisation of Ne, N, C,
vs proton energy E[6]. BEA theory[4]

However, the present agreement does not mean that everything is solved as far as point-like projectiles are involved. For instance, PWBA and SCA approximations made use of *hydrogenic* wave functions which are probably adequate to describe the very inner K shell electrons, but loose significance with respect to L shell electrons. This defect was actually invoked to explain the disagreement observed between experiment and theory for the L shell excitation of Silver (fig. 3a). Yet hydrogenic wave functions were found accurate enough to account for the excitation of the deeper (and hence more hydrogenic) L shell of gold (fig. 3b)

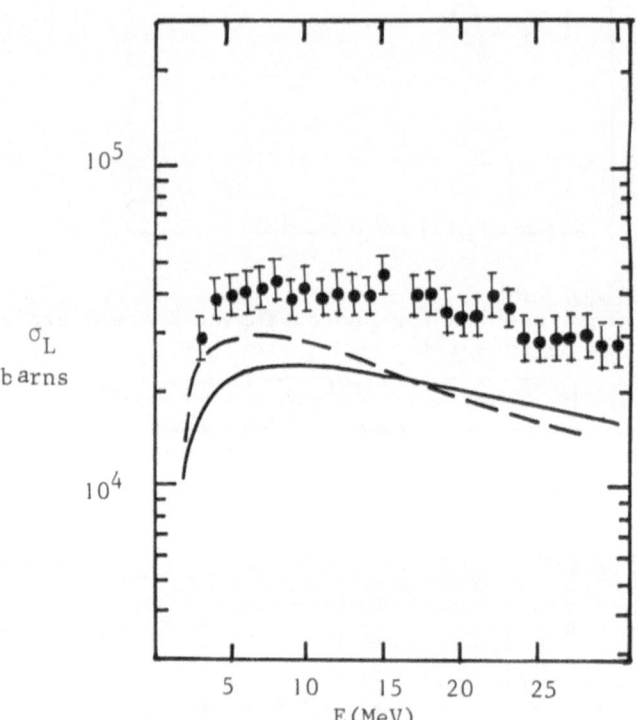

Fig. 3(a) L shell excitation cross section by proton impact for Silver(7). Theory :   ———PWBA,   ---- BEA

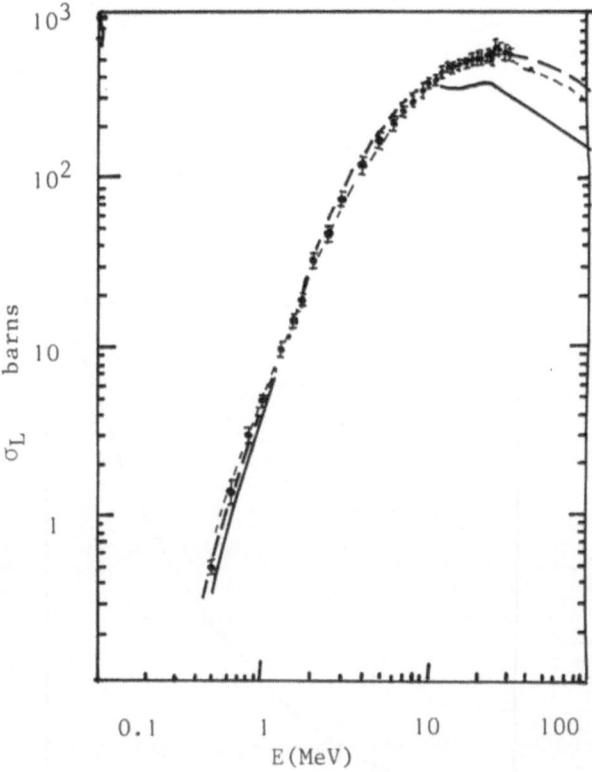

Fig. 3(b) L Shell excitation cross section by proton impact for Gold[8]. Theory : ———— SCA,·········· PWBA,— — — —   BEA

Since the Z_1^2 dependence of the cross-section is a typical signature of the approximations involved in the PWBA and BEA theories, a departure from this law will yield information on the importance of higher order terms neglected in these theories. This has been well demonstrated using completely stripped heavy projectiles which create a much stronger coulomb field. Data obtained with Z_1 values as large as 9 (fig. 4)[9], seems to indicate a Z_1^3-type contribution[10]. If this is true, one can expect different cross sections for two projectiles of same masses but having opposite charge. This would lead for instance to different "stopping power" for π^+ and π^- particles.

I do not want to discuss this subject further since improvements of the theory requires either refinements of the scattering theory (a topic which is outside the scope of this lecture), or improvement of the atomic wave function to be used, which is not specific of heavy particle impact problems.

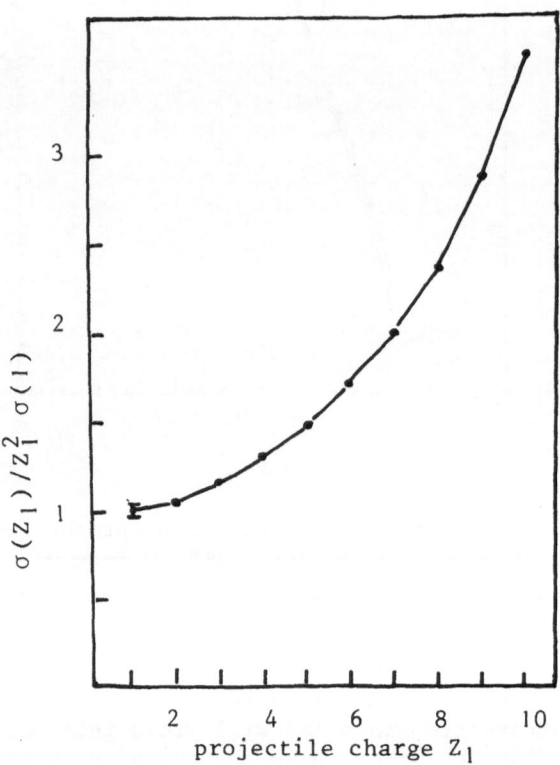

Fig. 4. Reduced cross sections for Ar K shell excitation[9] as a function of the Z_1 charge of the projectile

II MOLECULAR ASPECT OF HEAVY PARTICLE COLLISIONS

The next question to be raised is : what happens when the projectile is no longer "structureless" i.e surrounded by an electron cloud ? Are the BEA, PWBA, SCA theories still working well ?

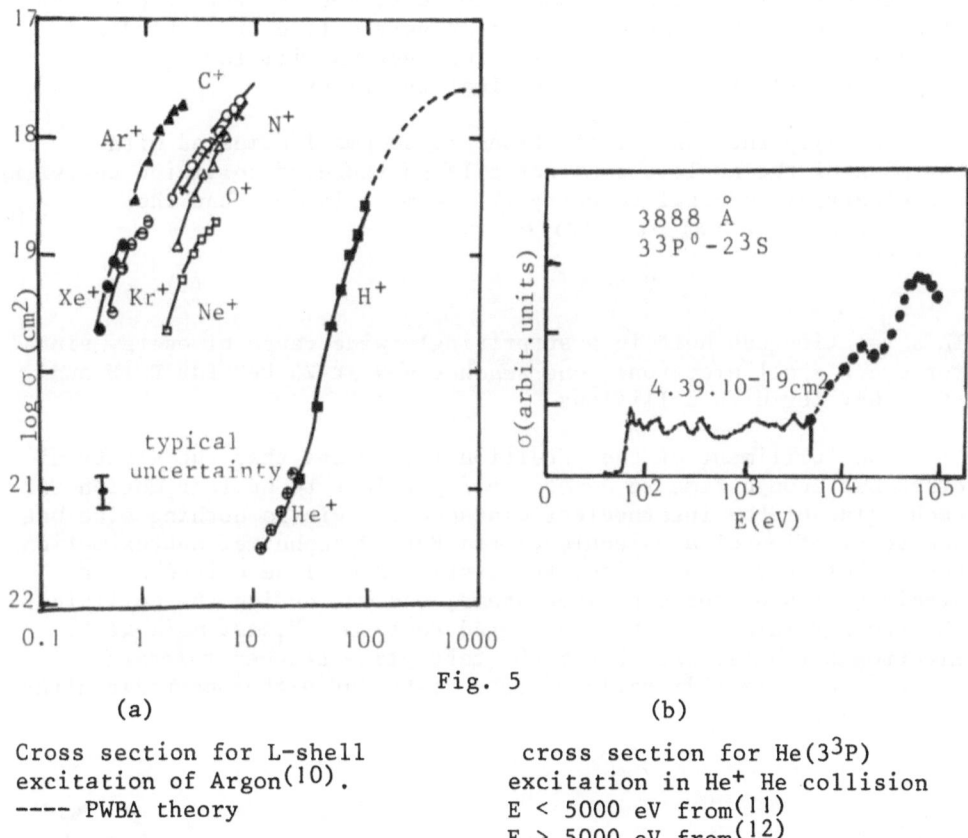

Fig. 5

(a)

(b)

Cross section for L-shell
excitation of Argon[10].
---- PWBA theory

cross section for He(3^3P)
excitation in He$^+$ He collision
E < 5000 eV from[11]
E > 5000 eV from[12].

The answer is obviously *no* as indeed illustrated in fig. 5(a) and (b) which display data for two different types of processes. The cross sections obtained for carbon K shell X Rays (fig. 5(a)) show that the data corresponding to structureless projectiles (H$^+$, He^{++}) agree well with the PWBA curve, whereas the results for heavier ions are completely "off" this line and lie 3,4 or 5 orders of magnitude higher. As a second example, the excitation cross sections of He(3^3P) atoms by Helium ions, in a wide range of the projectile energies (fig. 5(b)) clearly demonstrate the

complete failure of Born approximation to account for the large
cross section obtained at low energy (below a few keV).

It is clear that these very large discrepancies indicate an
excitation mechanism which is different from *direct* coulomb excitation. During hard collisions, the electron clouds of both particles strongly interpenetrate, and interaction between the electrons
of both particles, actively participate to the excitation processes.
Although this *many body* problem seems very difficult to handle
a specific property of *such* colliding systems will fortunately
help us to simplify their theoretical treatment.

Actually the mass of electrons is so small compared with
the mass of the nuclei, that for a large range of collision energies,
the electronic orbital velocity (v) is much larger than the
relative nuclear velocity (V) :

$$V \ll v \qquad\qquad\qquad (1)$$

This relation can hold in a surprisingly wide range of energy since
for e.g. K shell electrons, one reaches v=V at 25 keV for $H^+ + H$ and
at 45 GeV for $U^+ + U$ collisions.

The fulfilment of the condition (1), means that the electronic
cloud has enough time to *adiabatically* adjust to nuclear motion for
each value of the internuclear distance R. This is nothing else but
the description of a molecule in the Born–Oppenheimer approximation.
Hence, before the collision, the system is well described by the
atomic states of the separated atoms, whereas during the collision
the two particles will be considered to form a *"quasi molecule"*.
Electronic transitions that might take place between *molecular
states,* could finally yield excited states of either particle after
the collision (fig. 6).

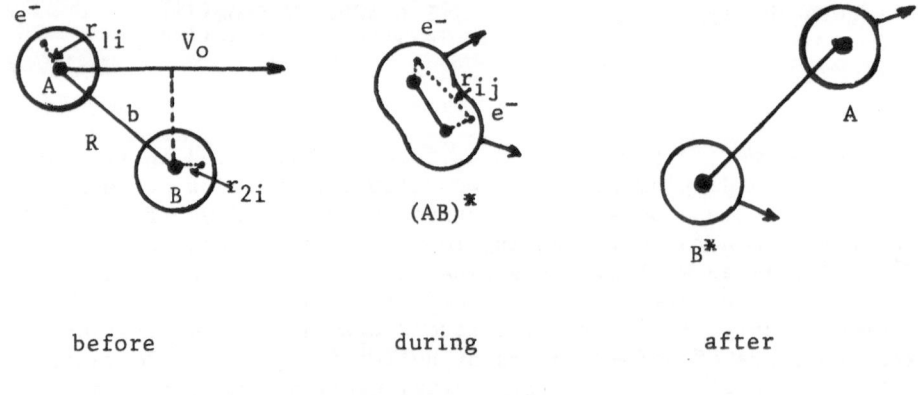

before   during   after

Fig. 6

Quasi-molecular formalism for the treatment of atom-atom collisions[13]

In the spirit of the above considerations, we can try to separate the *nuclear* motion (R) from the electronic motion (r) as it is done in the Born-Oppenheimer approximation.

Let $\psi(\vec{R},\vec{r})$ be the wave function of a system governed by the total Hamiltonian :

$$\mathcal{H} = -\frac{1}{2\mu}\nabla_R^2 + \frac{Z_1 Z_2}{R} + \sum_i(-\nabla_i^2 - \frac{Z_1}{r_{1i}} - \frac{Z_2}{r_{2i}}) + \sum_{j>i}\frac{1}{r_{ij}} \qquad (2)$$

the coordinates are given in fig. 6 (atomic units are used throughout this section) μ is the reduced mass of the nuclei. $\mathcal{H}$ can be written as the sum of the electronic hamiltonian $H_{e\ell}$ and the nuclear kinetic energy operator $T_R = \frac{-1}{2\mu}\nabla_R^2$:

$$\mathcal{H} = T_R + He\ell \qquad (3)$$

The total wave function ψ can be expanded over a complete orthonormal basis set of real "electronic wave functions" $\phi(R,\vec{r})$ which depend on R only parametrically :

$$\psi(\vec{R},\vec{r}) = \sum_i F_i(\vec{R})\ \phi_i(R,\vec{r}) \qquad (4)$$

The functions $F_i(\vec{R})$ describe the nuclear motion. The Schrödinger equation can then be written as :

$$\mathcal{H}\psi = (T_R + He\ell)\sum_i F_i(\vec{R})\ \phi_i(R,\vec{r}) = E\sum_i F_i(\vec{R})\ \phi_i(R,\vec{r}) \qquad (5)$$

Multiplying (5) on left by $<\phi_j|$ and integrating over the electronic coordinates $\vec{r}$, one gets an infinite set of coupled equations :

$$<\phi_j|T_R|\ F_j\phi_j> + (<\phi_j|He\ell|\ \phi_j> - E)\ F_j =$$

$$-\sum_{i\neq j}(<\phi_j|T_R|F_i\ \phi_i> + <\phi_j|He\ell|\ \phi_i>F_i(R) \qquad (6)$$

We have put the diagonal terms in the left hand side of the equation. We can also write (6) as :

$$\left(T_R + < \phi_j | He\ell | \phi_j > \underbrace{\frac{1}{2\mu} < \phi_j | \nabla_R^2 | \phi_j > - E}_{} \right) F_j$$

$$E_j(R)$$

$$= - \sum_{i \neq j} \left(\underset{\substack{\text{potential} \\ \text{coupling}}}{< \phi_j | He\ell | \phi_i >} - \underbrace{\frac{1}{2\mu} (<\phi_j | \vec{\nabla}_R | \phi_i > \vec{\nabla}_R - < \phi_j | \nabla_R^2 | \phi_i >)}_{\text{dynamic coupling}} \right) F_i(\vec{R})$$

$$(7)$$

In these coupled equations, the *left* hand side represents the scattering by a potential $E_j(R)$, and the right hand side represents the coupling between the jth channel with the other i channels.

Dynamic coupling terms

The labeling "dynamic" coupling becomes more obvious if the collision velocity is explicitly introduced. This is readily achieved in a time-dependant treatment. Using the semi-classical approximation, it can be shown[14] that the dynamic coupling terms which are expressed by the second term of the right hand side of eq.(7) imply the time derivative of the "electronic" wave function :

$$< \phi_j | \frac{d}{dt} | \phi_i > \qquad (8)$$

since $\dfrac{d}{dt} = \dfrac{\partial}{\partial R} \dfrac{dR}{dt} + \dfrac{\partial}{\partial \theta} \cdot \dfrac{d\theta}{dt}$

$\dfrac{d}{dt} = v_R \dfrac{\partial}{\partial R} + \dot{\theta} \dfrac{\partial}{\partial \theta}$

or $\dfrac{d}{dt} = v_R \dfrac{\partial}{\partial R} + \dot{\theta} i L_y$

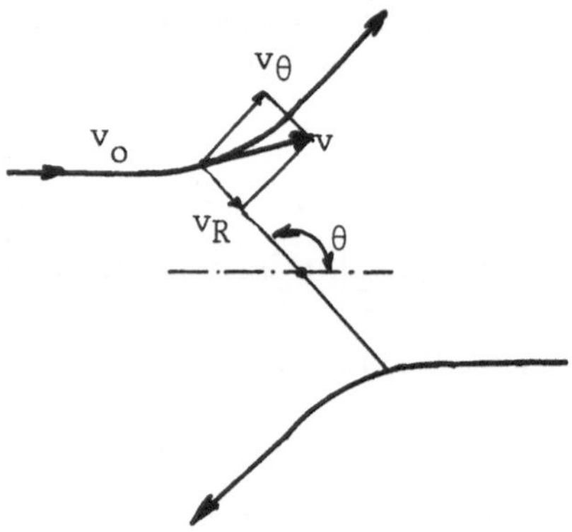

where $L_y = \dfrac{1}{i} \dfrac{\partial}{\partial \theta}$ is the component of the total electronic angular momentum operator along an axis perpendicular to the collision plane. The matrix element in (8) can then be expressed as :

Fig. 7.

$$v_R < \phi_i | \frac{\partial}{\partial R} | \phi_j > + \dot{\theta} i < \phi_i | L_y | \phi_j > \qquad (9)$$

(i) the first term called *radial* coupling induces transitions between molecular states having the same projection of the total electronic angular momentum along the internuclear axis (Λ symmetry), as for example two $^1\Sigma$ states.

(ii) the second term called *rotational* (or Corriolis) coupling, satisfies the selection rule $\Delta\Lambda = \pm 1$, as for instance a Σ and a Π state.

Choice of the Molecular Basis

Adiabatic representation. Up to now, no particular choice of the electronic wave functions ϕ_i has been made. We can first choose the ϕ_i's to be the eigen functions of the electronic Hamiltonian as it is done in the Born-Oppenheimer approximation :

$$\text{Hel } \phi_i^a = \varepsilon_i^a(R)\phi_i^a, \quad H_{ij} = \varepsilon_i \, \delta_{ij} \tag{10}$$

The $\varepsilon_i^a(R)$ are then the *adiabatic* potential energy curves of the quasi-molecule. With Eq.(10), the set of coupled equation (7) becomes :

$$\left[T_R - \left(E - \varepsilon_j^a(R) + \frac{1}{2\mu} < \phi_j^a | \nabla_R^2 | \phi_j^a > \right) \right] F_j(R) =$$

$$\sum_{i \neq j} \frac{1}{2\mu} \left(< \phi_j^a | \nabla_R^2 | \phi_i^a > + < \phi_j^a | \vec{\nabla}_R | \phi_i^a > \vec{\nabla}_R \right) F_i(R) \tag{11}$$

In the adiabatic representation the electronic transitions are thus induced by the "dynamical" terms which involve the derivatives ∇_R of the molecular wave function.

Diabatic representation[13]. We can alternatively look for electronic wave functions ϕ^d for which the $<\phi_i | \nabla_R | \phi_j >$ terms are very small so that they can be neglected. Of course these new wave functions are no longer eigenfunctions of Hel, and the coupling term responsible for the inelastic transitions is now given by the potential coupling $H_{ij} = < \phi_i^d | \text{Hel} | \phi_j^d >$ off-diagonal terms, since (7) reduces to :

$$\left[T_R - \left(E - \varepsilon_j^d(R) \right) \right] F_j(R) = -\sum_{i \neq j} < \phi_j^d | \text{Hel} | \phi_i^d > F_i(R) \tag{12}$$

where $\varepsilon_i^d(R) = < \phi_i^d | \text{Hel} | \phi_i^d >$

At this point, the distinction between *adiabatic* and *diabatic* representations remains purely formal. Yet it is necessary to answer the following questions :

What is the most appropriate basis for treating an atom-atom collision problem ? - What is the underlying physical meaning hidden
behind the words diabatic and adiabatic ? I would like to make
it clear within a few examples :

Example 1.- Let us consider first a two state problem (fig. 8)

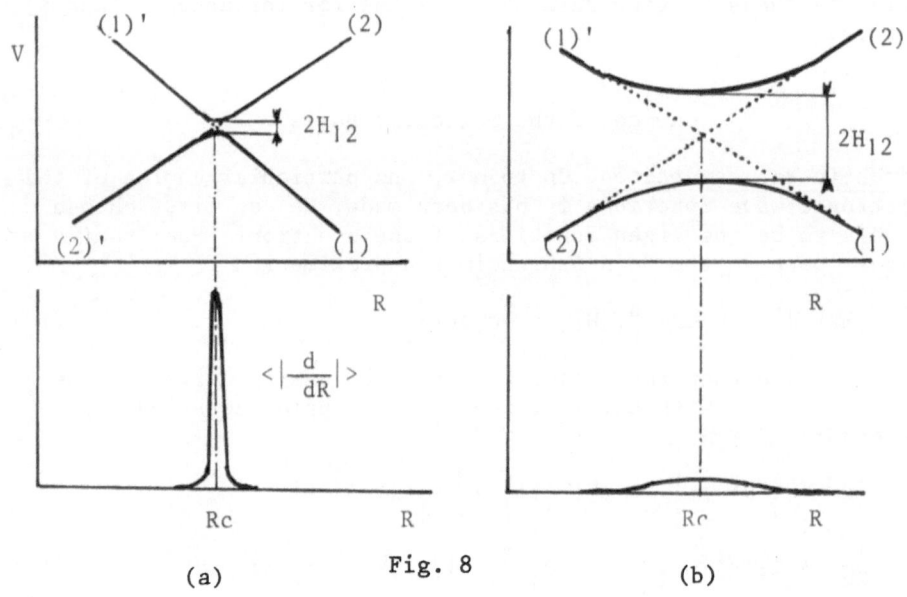

(a)   Fig. 8   (b)

where two potential energy curves (1) and (2) for states of the
same symmetry (say $^2\Sigma^+$) tend to cross at R = Rc. Due to the Wigner
von-Neuman[15] non crossing rule, this crossing is avoided in the
adiabatic representation (full line) and in case (a) of fig. 8,
this avoidance gives rise to a very sharp bend in the potential
curves near Rc. Hence, the $<|\frac{\partial}{\partial R}|>$ matrix element which is very
weak outside the crossing region, becomes *important* near R_C. This
causes the transition probability (1) → (1)' to be strong between
the *adiabatic* curves, and the probability of remaining on the
adiabatic curve (1) → (2)' to be small[13]. A more realistic view
is to consider new potential curves from which the sharp bendings
are removed ("smooth" dashed curves in fig. 8a). These curves
should consequently correspond to states for which the $<|\frac{\partial}{\partial R}|>$
couplings are nearly zero. This is exactly what we expect from
diabatic states. Hence with these new "diabatic" potential curves,
the transition (1) (1)' → (2) (2)' is small and is caused by the
small off-diagonal term H_{12} of the electronic Hamiltonian. In this
case the diabatic representation is much more appropriate. On the
other hand, if the crossing is strongly avoided (case b in fig. 8),
the bend of the adiabatic curves is strongly reduced. In contrast,
the potential coupling H_{12} which amounts to half of the energy sepa-

ration between the adiabatic curves[16] becomes important. In case
(b) the transition probability (1) → (1)' is *weak* and consequently
the adiabatic representation would be a better choice. It thus
appears that in each case, we must choose the representation which
minimizes the coupling terms.

However, since the $<|\frac{\partial}{\partial R}|>$ coupling is weighted by v_R (formula
9), the comparison between cases (a) and (b) stands only for
comparable velocities. Actually, it is always possible to find a
collision velocity small enough for which case (a) behaves adiabatically and high enough to get a diabatic behaviour in case (b)

Example 2.- Another typical example is given by the elastic scattering in He$^+$-He collisions[17].
The initial A $^2\Sigma_g^+$ potential
curve (fig.9) representing
the incoming channel is
crossed by a complete series
of excited states of same
symmetry which have only a
very *weak* interaction with
it . In the adiabatic representation the infinite series
of crossings are hardly
avoided. Hence the calculation of the elastic scattering cross section requires
the consideration of the
infinite set of strongly
coupled A $^2\Sigma^+$, n$^2\Sigma^+$ states
in the adiabatic represen-
tation, whereas only *one*
state A'$^2\Sigma^+$ is needed in
the diabatic representation !

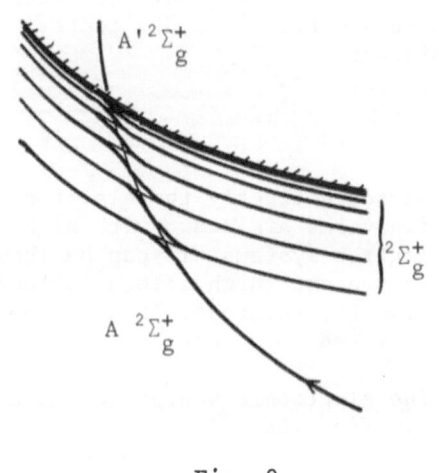

Fig. 9

Example 3.- Let us, consider a complete orthonormal basis of real
function $|\phi^d>$. If according to Smith[13] one defines the diabatic
states such that $<\phi_i^d|\frac{\partial}{\partial R}|\phi_j^d> = 0$ holds for all states and for all
values of R, this implies that $\frac{\partial}{\partial R}|\phi_j^d> = 0$ which means that the
$|\phi_j^d>$'s are no longer R dependent. In fact, they are nothing else
but the *atomic* wave functions[18]. Such "drastic" diabatic states
are appropriate to deal with very high collision velocities, such
that, during the very short time of the encounter, the electronic
clouds have no time to adjust adiabatically and they can be considered as frozen in their *atomic* states. Thus it seems more realistic to define the diabatic states as those for which :

$$v_R < \phi_i \left| \frac{\partial}{\partial R} \right| \phi_j > \ll H_{ij}(R)$$

in the transition region*.

---

*In practice it is not always possible to find a *unique* basis which minimizes the couplings in the whole range of R values. In particular, adiabatic states are often required to reproduce the *correct* atomic behaviour at large internuclear distances. For this purpose, basis changes can be worked out to provide the "best" states in each region of internuclear distances[19].

III THE H$^+$ H COLLISION : A ONE ELECTRON PROBLEM

The H$^+$-H system provides the simplest case of heavy particle collisions. The electronic hamiltonien describes the motion of a single electron in a two center coulomb field. By using spheroïdal coordinates

$$\xi = \frac{r_1 + r_2}{2R} , \qquad \zeta = \frac{r_1 - r_2}{2R} ,$$

one can solve exactly the electronic problem so that potential and coupling terms are known with high accuracy[20]. Furthermore in one-electron systems, it can be shown that an additional symmetry [21] is present which allows for potential energy curves of the same symmetry to cross. Thus in the H$^+$ H system, *diabatic* curves are generated.

The electronic energy curves of the H$_2^+$ molecule are schematically shown in fig. 10a.

Let us consider the case where the system is initially in the ground state. As the collision proceeds, the evolution of the potential energy is described by the $1s\sigma_g$ curve of fig. 10a. For an internuclear distance smaller than R_0, the system has an equal chance to follow the $1s\sigma_g$ curve as the $2p\sigma_u$ curve. On the way out, the probability amplitudes resulting from the scattering by each of the two curves will coherently interfere, thus leading to an oscillatory behaviour of the elastic differential cross section (see fig. 10c). However the $2p\sigma_u$ state is degenerate with the $2p\pi_u$ state at small internuclear distances. Consequently electronic transitions from $2p\sigma_u$ to $2p\pi_u$ can be induced by rotational coupling near R = 0[24]. This mechanism gives rise to *selective* excitation of H(2p) since the higher levels (n ⩾ 3) are not strongly coupled with any of the two incident states. Such a collision mechanism is well confirmed by the experiment[22]. Before the results are shown, I shall describe briefly the experimental setup used.

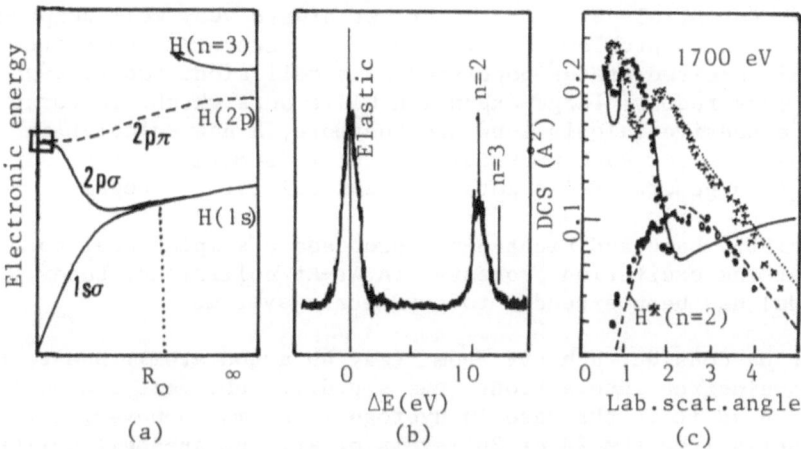

Fig. 10(a) Schematic potential energy curves of H_2^+. The nuclear repulsion $\frac{Z_1 Z_2}{R}$ is omitted to allow for the curves to be drawn to the "united" atom limit R = 0. (b) energy loss spectrum showing the dominance of H(n = 2) excitation[22]. (c) differential cross sections (—— elastic, ········ exchange into the ground state, ----
H(n = 2) excitation). The theoretical curves are from ref.(23).

A mass-selected H^+ beam is crossed by a thermal H beam issued from a radio-frequency discharge. The H^+ ions scattered at a given angle are detected after having passed an electrostatic energy analyzer. The energy losses are characteristic of the various atomic levels excited by the collision. In the H^+ H case, the spectrum of fig. (10b) shows that only the n = 2 levels are excited. This is in full agreement with the theoretical predictions. If the number of particles having suffered a given energy loss, is determined as a function of the scattering angle, the corresponding differential cross section (DCS) can be deduced. Figure 10c shows the experimental DCS for H(n = 2) excitation and as for elastic scattering, the agreement with theory[23] is very good.

IV THE MANY ELECTRON PROBLEM : THE "ELECTRON PROMOTION MODEL"

As soon as more than one electron are involved in the collision system, the inter-electronic interaction $1/r_{ij}$ no longer allows for a separation of the electronic hamiltonian in prolate shperoïdal coordinates. Therefore the Schrödinger equation cannot be solved exactly. Yet, potential curves and coupling terms can, of course, be determined approximatively by using e.g. Hartree Fock or configuration interaction methods or even effective hamiltonian (pseudo

potential) methods[25]. These methods which essentially lead to
adiabatic potentials are at present not always very well adapted
to the collision problem. For instance, to accurately describe
the highly excited states populated in a collision, *"ab initio"*
calculations require large expansion basis sets which, in turn,
lead to expensive calculations. Furthermore, a new calculation
has to be done for each particular case; this makes it very difficult to find general rules governing the collision process.

In 1965, Fano and Lichten[26] proposed a simple model to
interpret the excitation processes in Ar-Ar collisions. Later,
this model has been extended to asymmetric systems[27].

Let us consider a heavy atom (say an argon atom). Due to the
electron-electron interactions, the s,p,d... sublevels are no longer
degenerate as it is the case in hydrogenic atoms. However, for
inner shells like the 2s or 2p levels of Ar, the subshell splitting
is relatively small compared to the energy difference between the
shells. In other words, this means that the energy of the inner
shell electrons is dominated by the Coulomb field of the nuclei
almost completely screened by inner most shells. In this case the
effect of incomplete shielding as well as that of the innershell
penetration by the outer shell electrons remains rather weak. This
dominance of the Coulomb field allows us to describe the quasi-
molecule by using a one-electron two center Coulomb model (or H_2^+
like model). Gerstein and Krivshenkov[28] have solved the problem
of one electron in the field of two Coulomb centers using spheroïdal
coordinates. They have derived the rules to connect the levels of the
separate atoms ($R \to \infty$) with those of the "united" atom ($R = 0$).
Using their formula, Barat and Lichten[27] have suggested the
following rules to construct diabatic correlation diagrams:

(i) The number of nodes n_r in the radial part of the orbitals is
conserved for all R. (n_r being given by $n_r = n-\ell-1$ where n and ℓ
are the principal and azimuthal atomic quantum numbers respectively).

(ii) the projection λ of ℓ on the internuclear axis is a "good"
molecular quantum number and σ, π, δ... MOs are generated.

(iii) In symmetric systems the u, g symmetry is conserved.

The atomic orbital levels of the "united" atom (R=0) and those
of the "separate" atoms (R=∞) are connected together using these rules.
Fig. 11 shows a typical exemple of such a correlation diagram. It is
seen that, as in a two center Coulomb problem, two MOs *having the
same spatial symmetry do cross*. We will call such MOs: *diabatic MOs*.
These crossings should of course be avoided if the whole actual
electrostatic interaction was considered.

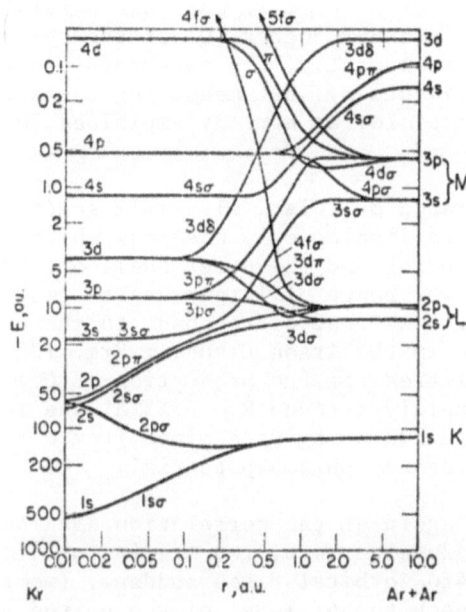

Fig. 11 Ar$_2$ Diabatic MO correlation diagram (from ref. 26)

L shell excitation of Argon

We can illustrate the use of correlation diagrams for the interpretation of inner shell excitation with the historical Ar$^+$-Ar example. The pionneering work of both the Leningrad Group and of the Connecticut Group[29] consisted in measuring the number scattered particles in coïncidence with the recoil particles (Fig. 12a).

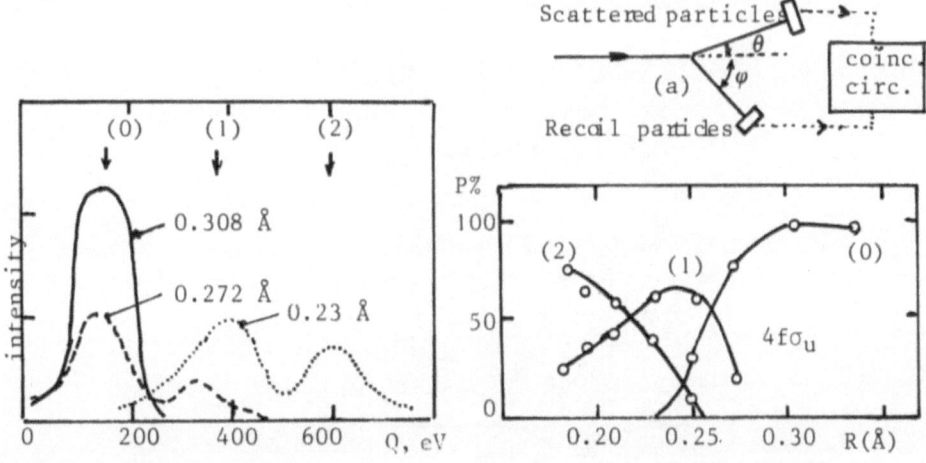

Fig.12 (b)Q measurements and (c) relative probability for (0),(1) and (2) L shell vacancy creation in Ar$^+$ Ar collisions[41].

From the knowledge of θ and φ it was then possible to deduce the
variation of the internal energy (Q) of the system, i.e the excitation energy. More recently, the same results were obtained[29] with
more simple experiments by just measuring the energy loss undergone
by a scattered particle, as already explained for the H+ H experiment.

For a wide range of relatively small scattering angles, the
energy loss spectra display only one peak which corresponds to both
the elastic scattering and the outer shell excitation processes.
However in a narrower range of larger angles, two additional peaks
are found   (fig. 12b) that correspond to the creation of *one*
and *two* vacancies in the Argon L shell. Fig. 12c shows that the
relative probabilities for the production of 0,1 and 2 L-shell
vacancies vary rapidly arround R ≃ 0.25 Å. The relation between the
scattering angle and R being determined from classical scattering
by a suitable screened Coulomb potential.

Let us look again at the correlation diagram of fig.11.
Among the twelve 2p electrons, two are in the $4f\sigma_u$ MO. Around
R ≃ 0.25 Å, the $4f\sigma_u$ orbital rises suddenly (we say that it is
"promoted") to reach the 4f level of the united atom. This promotion
causes a set of crossings between the $4f\sigma_u$ MO and several excited
(and empty) orbitals. At these crossings *one* or *two* electron
transitions can occur, thus leaving one or two vacancies in the 2p
shell after the collision.

Fastrup *et al.*[29] have carried out similar experiments on
slightly asymmetric systems. Their data show that the *L shell
excitation* takes place almost *exclusively in the lower Z* partner.

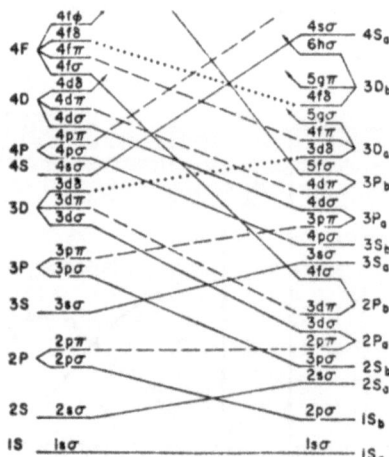

Fig. 13. Diabatic MO correlation diagram for a slightly asymmetric
system (from ref. 27)

This can easily be understood when looking at the correlation diagram of Fig. 13. The promoted 4fσ MO is correlated to the 2p orbital
of lower Z atom, one or two vacancies are easily produced in the
4fσ orbital at the crossings with empty orbitals whereas creation
of vacancies in the 2p level of the higher Z atom requires a more
complicated mechanism.

Excitation of K shell electrons in Ne-Ne collisions can also
be explained in the same framework[29]. Again from Fig. 11 one sees
that one or two vacancies can be formed in the $2p\sigma_u$ orbital by the
$2p\sigma_u \to 2p\pi_u$ electron transition induced by *rotational coupling*
near R = 0. This mechanism supposes of course that one or two
vacancies are initially present in the outer 2p shell of Neon. This
means that the probability of K shell excitation will strongly
depend on the outer shell configuration. Indeed, it has been shown
that the cross section for K shell excitation for the Ne^{++} Ne
collision is about twice as large as that of the Ne^+ Ne collisions[29].

It is impossible in two lectures to review the whole field of
inner shell excitation which has so rapidly expanded in the few
past years. However I must say some words about some of the most
striking developments :

(i) Vacancy sharing

Consider again the K shell excitation in slightly asymmetric
systems. We have shown that the creation of a 1s vacancy preferentially takes place in the higher Z partner of the collisions
($1s_B$ in Fig. 14). However the spectra of ejected electrons (30) as
well as X ray measurements[31] demonstrate that in addition to
the creation of vacancies in $1s_B$ (fig. 14) excitations of $1s_A$
electrons also occur with a much smaller probability. This has
been understood with the help of a *two step* mechanism which accounts
successfully for the experimental results. In the first step,
a $1s_B$ vacancy is produced by a $2p\sigma-2p\pi$ electron transition. Then,
part of these vacancies are transfered to $1s_A$ by a semi localised
transition (fig. 14) between the $2p\sigma$ and $1s\sigma$ MOs The latter charge

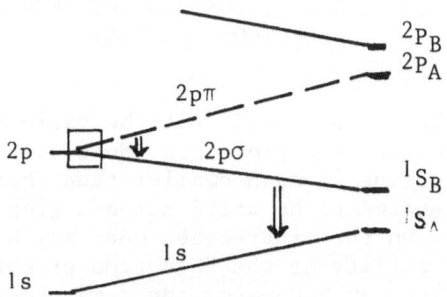

Fig. 14. Production of a ^1S_A vacancy through two step
processes (K shell vacancy sharing)

transfer mechanism initially proposed by Demkov[32] was successfully
applied to inner shell vacancy sharing by Meyerhoff[31].

(ii) Molecular X-Rays

Up to now, we have only considered that the decay of an excited
state takes place after the collision partners have separated. The
X-rays emmitted are then characteristic of the atom. However it is
also possible for the decay to takes place during the collision.
Suppose that an argon atom with an *initial* vacancy in the 2p shell
collides with an other argon atom in the ground state. This vacancy will "follow" (fig. 11) the 2pπ orbital correlated with the 2p
level of krypton and the 2pπ vacancy could be filled by an outer
electron with simultaneous emission of an X-ray photon. Due to the
very short collision time, this "quasimolecular" X-ray emission
would be extremely weak. However this effect can be relatively
enhanced by the decrease of the vacancy life time during the collision. For instance, the lifetime of a L shell vacancy in argon
which amounts to about 4.10^{-15} sec. is reduced to 5.10^{-16} sec. close
to the krypton "united" atom.

Obviously, since the decay takes place in a wide range of
internuclear distances, the corresponding X-ray spectrum will exhibit
a broad distribution reflecting the energy variation of the 2pπ MO.
Such "MO-X-rays" have been observed for the first time by Saris *et
al.*[33]. Since this discovery, "MO-X-rays" have been observed
and studied extensively for many colliding systems[34,35]. It is
noteworthy that the inner vacancy can also decay by an *Auger process
(or autoionization)*. In the case of a "quasimolecular" decay, this
gives rise to a continuous distribution of ejected electrons. It has
been shown that this mechanism is responsible for ionization processes in outer shell[40] excitation. In this case, the extremely small
fluorescence yield makes it very difficult to observe the "MO ultraviolet light".

IV OUTER SHELL EXCITATION. TWO-ELECTRON TRANSITION BY
ELECTRON CORRELATION

The molecular model would still be valid for the interpretation
of outer shell excitation processes. However, since the velocity of
outer shell electrons is much smaller than that of the inner electrons,
the model is expected to be valid for energies below a few keV or
few tens of keV. On the other hand, one may have serious doubt
about the applicability of the "electron promotion model" since the
subshell splitting may become of the same order of magnitude as the
shell energy, which indicates an important departure from the pure
coulomb case. At this point, we can ask the following questions :

(i) Do H_2^+ like MOs still represent good *diabatic* MOs ?
(ii) Are every features in the collision processes, accounted for
in the framework of the independant particle model ? Or is the use of
a many electron picture required ?

At present, no simple answer may be given to question (i).
However we can still consider that the "H_2^+ like MO correlation
diagram" represents a zero order approximation. Let's take an
example. During an Argon-Argon collision, among the twelve 3p-
electrons, two of them will "follow" the promoted $5f\sigma_u$ orbital
(fig. 11) and will be promoted. At small internuclear distance, these
two electrons become two (equivalent) *Rydberg* electrons in the quasi
coulombic field of an Ar_2^{+} core. If we just neglect the electron-
electron correlation in the $5f\sigma_u$ MO, it is clear that the one-
electron H_2^+ like correlation diagram should also apply in this
region of internuclear distances. As seen from figure 11, one thus
expects one or two electron transitions at the $5f\sigma_u$-$4p\pi_{u_0}$or $5f\sigma_u$-
$5p\sigma u$ crossings in a very narrow region around R $\simeq$ 0.75 A. This
view is nicely confirmed by the experimental results[36] which
demonstrate the similar behaviour of the inner $4f\sigma_u$ MO and the outer
$5f\sigma_u$ MO (compare fig. 15 and fig. 12).

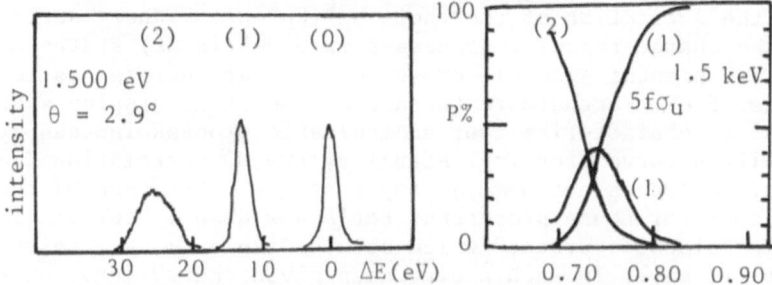

Fig.15 (a) energy loss spectrum showing the excitation
of (0) (1) and (2) outer 3p electrons in Ar Ar collision[36]. (b) relative probability for (0) (1) and (2)
electrons excitation.

Let's then consider the case of a very asymmetric system, as the
He-Ar collision, for instance. The diabatic correlation diagram
constructed with the correlation rules of section III (fig. 16)
displays a crossing between two *valence* MOs. From this diagram, the
selective excitation of the Helium atom is expected. This is in
complete contradiction with the experimental findings[37] which show
an exclusive excitation of the argon atom. This clearly gives evidence for a strong avoidance of the $3d\sigma$-$2p\sigma$ crossing between two
valence shell MOs (full line in fig. 16). Such an "adiabatic"
behaviour of *valence* MOs seems to be a general rule in very asymmetric systems. However, in cases where a quasi-u-g symmetry is present, two *valence* MOs can still retain the H_2^+ like behaviour and
display a diabatic crossing. In Mg^+ Ne collisions (fig. 17), for

instance, the 4fσ –3sσ crossing (corresponding to the 4fσ$_u$–3sσ$_g$
crossing in the parent symmetric Ne-Ne system) is diabatic leading
to one or two electron excitation in the Neon atom[37].

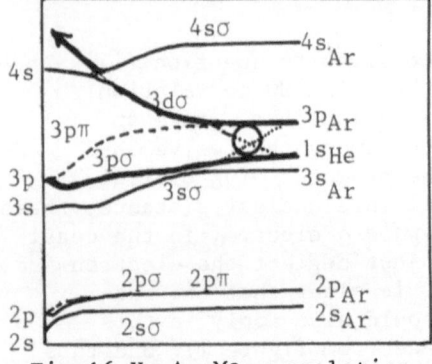

Fig.16 He Ar MO correlation
diagram

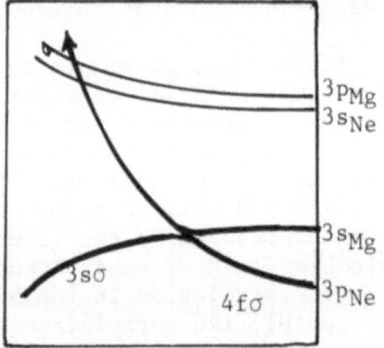

Fig.17 Mg Ne⁺ MO correlation
diagram

In order to answer question (ii), let's discuss the He⁺ Ne
collision. The experimental results[38] show that the Ne*(2p⁵ 4s)
and Ne**(2p⁴ 4s²) induced by one and two electron transitions at the
3dσ–4sσ crossing (DI in fig. 18) are only important at keV energies.
However, the excitation of the whole Ne*(2p⁵ nℓ) Rydberg series as
well as the charge transfer processes into He*(1s nℓ) states are also
observed and present sizeable cross sections at energies as low as
a few tens of eV. Furthermore these cross sections display a behaviour that is characteristic of an inelastic process induced by a
well localized curve crossing. Actually, the MO correlation diagram (fig. 18a) does not display any crossing outer than DI that
could account for these processes. Let's now draw a "correlation
diagram for diabatic states" which should display *many electron*
processes (if any). In such a diagram, a given energy curve corresponds to a single configuration state which is built up by filling

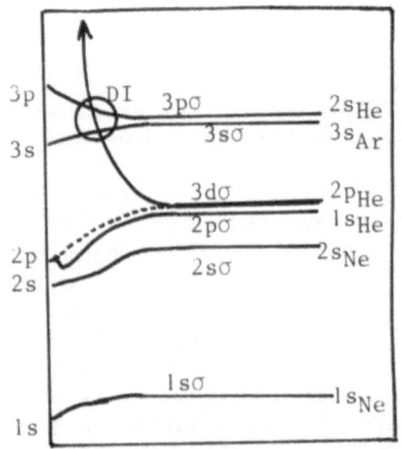

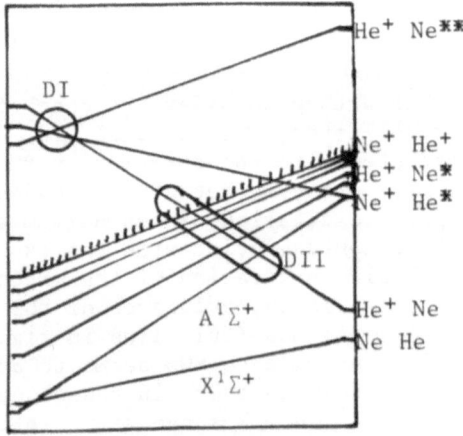

Fig. 18. He Ne⁺ correlation diagram (a) MOs, (b) states.

the appropriate MOs of fig. 18a with the right number of electrons according to Pauli principle (fig. 18b). It is seen that, in addition to the triple state crossing (DI) which reflects the DI MO crossing, this diagram displays an infinite set of DII crossings (called diabatic II)[36] subsequent to the promotion of the incident $B^2\Sigma^+$ state. The new crossings *do not* reflect any MO crossing . In addition it has been shown by Sidis and Lefebvre Brion[39] that, at the DII crossings the $< |\frac{d}{dR}| >$ interaction between the states involved vanishes. Thus these states are diabatic in the sense defined by Smith[13] and are called "Quasi-diabatic". The only interaction inducing the transition at each of the B-C crossings is due to *configuration interaction* $<3d\sigma^2 | 1/r_{12} | 2p\sigma\ n\ell\sigma >$. In cases of isolated crossings, the collision problem is easy to treat and experimental results could thus bring a *direct* estimate of the electron-electron correlation.

It is noteworthy that Diabatic II processes are absent in the parent neutral He Ne collision. Actually the different behaviour of the neutral and ionic systems is understandable by noting that the initial $1s_{He}$ vacancy which is specific of the ionic system, becomes an inner vacancy in the quasi-molecule (fig. 19b). The incident $B^2\Sigma^+$ state then becomes a "core excited" state in the quasi-molecule and hence crosses the excited states that have a ground state core. At shorter internuclear distances the $B^2\Sigma^+$ state even crosses the continuum and becomes a molecular autoionizing state since it is correlated to the $3p^53d^2$ autoionizing state of the Mg^+ united atom state. The decay of this autoionizing state *in* the quasi-molecule gives rise to an Auger ionization process (we say quasi-molecular ionization). The measurement of the distribution of the electron energy yields information[40] on the electron $< 3d\sigma^2 | 1/r_{12} | 2p\sigma\ \phi >$ interaction. (ϕ being a continuum one electron wave-function).

ACKNOWLEDGMENT.- I would like to thank Drs.V. Sidis and N. Stolterfoht for their help in preparing this paper.

REFERENCES

1. N.F. Mott and H.S.W. Massey, The Theory of Atomic Collisions Clarendon Press (1963).
2. E. Merzbacher and H.W. Lewis, Handbuch der Physik 34, 166 (1958)
3. J. Band and J.M. Hansteen, Kl . Danske Videnskab Jelskab Mat-Fys Medd 31, n°13 (1959)
4. J.D. Garcia, Phys. Rev. A1, 280 (1970), A1, 402 (1970)
5. J.D. Garcia, R.J. Fortner and T.M. Kavanagh, Rev. Mod. Phys. 45, 111 (1973)
6. N. Stolterfoht and D. Schneider, Phys. Rev. A11, 721 (1975)
7. G.A. Bissinger, S.M. Shafroth and A.W. Waltner, Phys. Rev. A5, 2046 (1972)
8. S.M. Shafroth, G.A. Bissinger and A.W. Waltner, Phys. Rev. A7,566

(1973)

9. J.R. McDonald, L.M. Winters, M.D. Brown, L.D. Ellsworth,
   T. Chiao and E.W. Pettus, Phys. Rev. Let. $\underline{30}$, 251 (1973)

10. R.C. Der, T.M. Kavanagh, J.M. Khan, B.P. Curry and R.J.
    Fortner, Phys. Rev. Let. $\underline{21}$, 1731 (1968)

11. S. Dvoretsky, R. Novick, W.W. Smith and N. Tolk, Phys. Rev.
    Let. $\underline{18}$, 939 (1967)

12. F.J. De Heer, L.W. Muller and R. Geballe, Physica $\underline{31}$ 1745 (1965)

13. F.T. Smith, Phys. Rev. $\underline{179}$, 111 (1969)

14. A. Russek, Phys. Rev. A$\underline{4}$, 1918 (1971)

15. J. Von Neumann and E.P. Wigner, Phys. Z. $\underline{30}$, 467 (1929)

16. See for instance E.E. Nikitin in Chemische Elementarprozesse.
    Springer Verlag $\underline{43}$ (1968)

17. W. Lichten, Phys. Rev. $\underline{131}$, 229 (1963)

18. B. Andresen and S.E. Nielsen, Mol. Phys. $\underline{21}$, 523 (1971)

19. See for instance V. Sidis, invited lectures and Progress
    Reports IX° ICPEAC Seattle (1975)

20. D.R. Bates and R.H.G. Reid, Advances in atomic and Molec.
    Physics $\underline{4}$, 13 (1968)

21. See for instance C.A. Coulson and A. Joseph, Int. J. Quant.
    Chem. $\underline{1}$, 337 (1967)

22. J.C. Houver, J. Fayeton and M. Barat, J.Phys. B $\underline{7}$, 1358

23. R. McCarroll and R.D. Piacentini, J.Phys. B $\underline{3}$, 1336 (1970)

24. D.R. Bates and D.A. Williams, Proc. Phys. Soc. $\underline{83}$, 425 (1964)

25. C. Bottcher, J.Phys.B $\underline{4}$ 1140 (1971), $\underline{6}$, 2368 (1973). J. Pascale
    and J. Vandeplanque,   J.Chem.Phys. $\underline{60}$, 2278 (1973)

26. U. Fano and W. Lichten, Phys. Rev. Let. $\underline{14}$, 627 (1965).
    W. Lichten, Phys. Rev. $\underline{164}$, 131 (1967)

27. M. Barat and W. Lichten, Phys. Rev. $\underline{6}$, 211 (1972)

28. S.S. Gershtein and V.D. Krivchenkov, Sov. Phys. JETP $\underline{13}$, 1044
    (1961)

29. For a review see Ref. 5 and Q.C. Kessel and B. Fastrup, Case
    Studies in Atom. Phys. $\underline{3}$, 137 (1973)

30. N. Stolterfoht, P. Ziem and D. Ridder, J. Phys. B $\underline{7}$, L409 (1974)

31. W.E. Meyerhof, Phys. Rev. Let. $\underline{31}$, 1341 (1973)

32. Yu.N. Demkov, Sov. Phys. JETP $\underline{18}$, 138 (1964)

33. F.W. Saris, W.F. Van der Weg, H. Tawara and R. Lambert, Phys.
    Rev. Let. $\underline{28}$, 717 (1972)

34. For a review : F.W. Saris and F.J. De Heer, in Atomic Physics
    (4) Plenum Press 287 (1974), see also :

35. P.H. Mokler, S. Hagmann, P. Armbruster, G. Kraft, H.J. Stein,
    K. Rashid and B. Fricke, in Atomic Physics (4) Plenum Press
    301 (1974)

36. J.C. Brenot, D. Dhuicq, J.P. Gauyacq, J. Pommier, V. Sidis,
    M. Barat and E. Pollack, Phys. Rev. A$\underline{11}$, 1245 (1975), A$\underline{11}$,
    1933 (1975)

37. J. Fayeton, M. Barat and N. Andersen, Abstract of Papers IX°
    ICPEAC Seattle (1975)

38. D. Coffey, D.C. Lorents and F.T. Smith, Phys. Rev. $\underline{187}$, 201
    (1969). M. Barat, J.C. Brenot, D. Dhuicq, J. Pommier, V. Sidis,

R. Olson, E.J. Shipsey and J.C. Browne, J. Phys. B (in press)

39. V. Sidis and H. Lefebvre-Brion, J. Phys. B 4, 1040 (1971)

40. M. Barat, D. Dhuicq, R. François, R. McCarroll, R.D. Piacentini and A. Salin, J. Phys. B 5, 1343 (1972). V. Sidis, J. Phys. B 6, 1188 (1973)

41. V.V. Afrosimov, Yu. S. Gordeev, M.N. Panov and N.V. Fedorenko, Sov. Phys. Techn. Phys. 9, 1248, 1256, 1265 (1965).

IONIZATION BY NUCLEAR TRANSITIONS

Melvin S. Freedman

Chemistry Division, Argonne National Laboratory

Argonne, Illinois 60439

I. INTRODUCTION

Ionization and excitation in an atom induced by nuclear processes [1,2] within that atom exemplify the so-called single step
and two-step mechanisms. The former is characterized as a sudden
shaking transition of the composite nucleus-atom system, analogous
to shaking in photoemission. In the latter process the charged
particles emitted in a nuclear decay transfer energy by a relatively
slow "direct collision" final state interaction with atomic electrons. These are not truly distinct mechanisms; rather they are
labels for the extremes of the time spans involved in different
types of nuclear decays. In some theories they are even coherent
interactions. For either the shaking or direct collision processes,
we can give a phenomenological description of an ionization event
that will apply to both types.

In any given nuclear transition, alpha, beta or electron capture decay, or internal conversion of the decay of a nuclear excited
state, a definite energy release occurs, and nuclear radiation is
emitted. These decays are each associated with some change in the
charge or charge distribution in the central core of the atom. If
an orbital electron is simultaneously ionized, the same total energy
release is shared by the orbital electron and the nuclear particle
or particles in a statistical way, but the energy division occurs
with the orbital electron most probably taking only a small fraction
of the total as its kinetic energy, usually an amount of the order
of magnitude of its initial shell binding energy. The nuclear part-
icle(s) take most of the energy in the usual energy partition. Thus
the energy distributions of emitted electron and nuclear particle
are continuous spectra, which in principal overlap over the entire

energy range from zero to maximum, but are largely concentrated at
the extremes. Figure 1 shows an idealization of the electron spectra emitted from the K and L shells associated with a nuclear decay
in which a single particle is emitted, e.g., an alpha particle or
the neutrino in K electron capture or an inner shell internal conversion electron. Note that the K electron spectrum extends beyond
the L spectrum, as is expected from its larger binding energy, al-
through its intensity is very much weaker; ejection probability increases rapidly with shell number. Complementary to each is the
mirror image spectrum of the alpha or neutrino or conversion electron. Each of these has its maximum energy displaced below the normal nuclear transition energy by the shell binding energy. Just
below the strong line spectrum of the normal nuclear radiation one
sees the weak lines of the nuclear radiation corresponding to excitation of electrons to unoccupied bound states. The composite spectrum of line and continua satellites is the analogue of shaking processes seen in the photoelectron spectrum.

    Orbital electron emission can occur from any shell. It's al-
ways a weak process in inner shells for any type of nuclear decay;
typically K ejection occurs with a probability per nuclear decay in
the range 10^{-6}-10^{-3} at medium Z values, depending on the transition
energy and type of decay. For outer shells the probability is much
higher, the total emission being about 20-30% for all shells, with a
major contribution from valence shell ionization for all Z values
for beta decay or internal conversion: for alpha decay the probability is nearly unity since it varies as $(\Delta Z)^2$. In contrast, the
total ejection probability is negligibly small for nuclear electron
capture, because in this case the effective change in central atomic
charge is nearly zero.

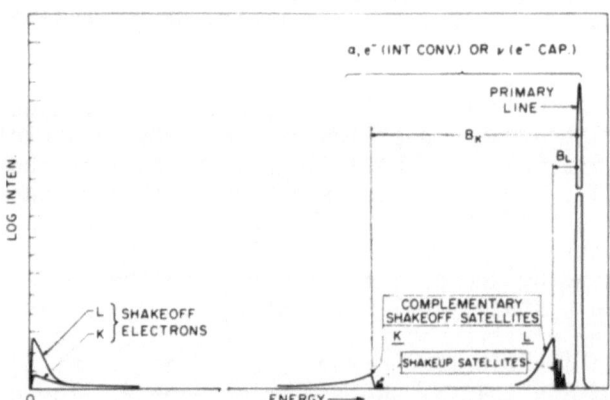

Fig. 1. Schematic energy spectra of electrons emitted from K and
L shells, and complementary "nuclear" particle satellite continua.
Satellite lines corresponding to excitation of K or L electrons
and K and L binding energies B_K and B_L shown.

Most of the experimental studies have been done on inner electron ionization despite its relative rarity. Inner shell electrons usually have energies that can be detected whereas the energies of outer shell electrons will mostly be too low to observe. Similarly, the complementary spectra of the nuclear particle or internal conversion electron can be resolved from the normal transition line only if it is displaced by a sufficient (inner shell) binding energy. The specific shell ionized is characterized by the shell x-ray; this is difficult for outer shells.

II. SHAKING IN NUCLEAR DECAYS

The principal mechanism contributing to ionization in beta and electron capture decay and in internal conversion is the shaking process. In fact, there is no evidence of any direct collision component in any of these decays, at a level much below its expected intensity. The familiar description of any nuclear decay process is that of a bare nuclear event. If the process involves emission or redistribution of electric charge on a fast time scale within the nucleus or the inner core of the atom, then this description of nuclear events in real atoms is incomplete; the only valid description is that of a one-step transformation of the system nucleus + atom as an entity, including both atomic and nuclear variables in initial and final states. The usual picture is that the shaking process is induced by the sudden change in central atomic charge, which leads to a conceptualization which seems rather to imply a sequence of events; first the nuclear event emits a charged particle with a certain initial energy, then an interaction results in the excitation or ionization of atomic electrons, which derive their energy from the emitted charged particle. Most of the errors of interpretation that one finds in the literature arise in this misconception of the process, which is actually a valid description only of the usually much weaker direct collision process.

A better picture of the process follows the recognition that the atomic wave functions of the initial atom are not eigenfunctions of the final state Hamiltonian, since this corresponds to a different effective central atomic charge. The atom must relax into one of the infinite number of available final eigenstates. The initial eigenfunctions cannot persist for a finite time as non-stationary states; the fast charge change does not permit a slow adiabatic adjustment of the wave function. The nuclear and atomic transitions are simultaneous processes in the composite nucleus-atom system. The overall transition rate is given by Fermi's Golden Rule No. 2 for the sudden approximation of time-dependent perturbation theory by

$$\lambda = 2\pi |M|^2 \rho \quad ; \tag{1}$$

ρ is the density-of-final-states factor expressing the conservation of energy and the statistical sharing of momentum among all final state emitted particles, and the matrix element of the composite system can be written

$$|M|^2 = |<\phi_{fin}\psi_{fin}U_{fin}|H'_{NUC}|\phi_{in}\psi_{in}U_{in}>|^2 \; ; \tag{2}$$

ϕ denotes the nuclear wave function, ψ is the atomic wave function, and U is the wave function of any particles incident onto or emitted by the nuclear process. H'_{NUC} is the perturbation term in the nuclear Hamiltonian that generates the overall process. The familiar "bare nucleus" transition matrix element which appears in the normal nuclear transition rate without shaking is:

$$|M_{NUC}|^2 = |<\phi_{fin}U_{fin}|H'_{NUC}|\phi_{in}U_{in}>|^2 \; . \tag{3}$$

If $|M_{NUC}|$ is independent of the energy of the emitted particles, which is the case in allowed beta decay and in K electron capture, then equation (2) can be factored into a product of nuclear and atomic factors

$$|M|^2 = |M_{NUC}|^2 \cdot |M_{AT}|^2 \tag{4}$$

Here the atomic matrix element

$$|M_{AT}| = \int \psi_{fin}(Z') \cdot \psi_{in}(Z)d\tau \tag{5}$$

is the wave function overlap integral of initial and final atomic states. This procedure is a whole-atom-decay description, and is applicable if the charge perturbation is fast enough; this leads to the criterion that the speed of the emerging charged particle should be large compared to the orbital speeds of the bound electrons, so that they cannot rearrange adiabatically as the charge changes. Expressed in terms of energies, this is equivalent to the rule that the transition energy E_0, which comes mainly from the nucleus, must be large compared to the binding energy B_i of the electron under consideration; $E_0 \gg B_i$. The suddenness condition is fulfilled for all measured nuclear decay cases except only in the ejection of K, L or M electrons in alpha decay; in this case the alpha velocity is less than the bound orbital electron velocity, and only the direct collision mechanism operates.

The atomic matrix element M_{AT} has the simple form of an overlap integral, with an atomic operator of unity, as a result of the (infinitely) sudden approximation involved in the derivation. As in the case of photoionization, it represents a monopole electronic transition, for which the selection rule is $\Delta L = \Delta J = 0$. In a higher

approximation when the finite rate of charge shift is taken into
account, the contribution of the monopole shaking transition decreases, and added terms appear representing dipole or higher multipole transition operators; such terms contribute in inner shell
ionization in alpha decay.

The transition rate for orbital electron ionization depends on
the ejected particle energies in the statistical factor ρ, and also
in the overlap integral; this integral decreases very rapidly with
increasing emitted orbital electron energy, about as $(E_e/B_i)^{-7/2}$.
The latter dependence explains the concentration at low energy, near
B_i, for the shakeoff electrons, and correspondingly at high energy
for the complementary satellite spectra of the nuclear particles.

One is usually interested in the electron shaking rate relative
to the rate of normal nuclear decay without shaking, rather than in
the absolute rate. Since the nuclear perturbation operator appears
only in $|M_{NUC}|$ in equation (4), and since this factor appears also
in the normal nuclear decay rate, it cancels in the relative shaking
rate. Thus this normalized rate is approximately independent of the
detailed nature of the nuclear operators causing the decay, their
multipolarity or forbidden character. This does not, of course,
imply that shakeoff probability is the same for all types of nuclear
decay; the nuclear transition characteristics also determine the
effective charge change ΔZ_{eff}, and shaking varies as $(\Delta Z_{eff})^2$.
Moreover the shaking rate still depends on the transition energy and
final energies of the nuclear particles.

II A. Shakeoff in Internal Conversion

Shakeoff in internal conversion is an almost perfect analogue of
shakeoff-multiple-ionization in photoemission; it similarly produces
two inner shell vacancies and either an electronic shakeup excitation or a shakeoff continuous spectrum plus a matching internal-
conversion-electron continuum. But there is a competitive process
in the decay of a nuclear excited state that can also eject two orbital electrons with continuous energy distributions, namely, the
rare second order double conversion process. Thus all measurements
which determine the probability of producing a pair of inner shell
vacancies per normal internal conversion necessarily measure the
summed contributions of the double conversion and shakeoff processes
and one must resort to rather unreliable theory to separate them.
Theory indicates that the shakeoff process usually dominates but by
only a small factor. Such measurements consist of double K x-ray
coincidence or K x-ray hypersatellite experiments.

One can distinguish the separate contributions because they
yield very different shapes for the electron spectra; the shakeoff
and complementary satellite conversion electron continua are

concentrated at the low and high energy extremes of the range,
whereas double conversion gives a continuous two electron spectrum
that is in general broadly distributed between these limits. Figure
2 shows the first observation of a satellite continuous spectrum below the K-conversion line of the 122 keV transition in ^{57}Fe as observed in the Argonne high resolution beta spectrometer. It rides
on the tail of the K line, with a tiny fraction of the line intensity
and its high energy threshold is displaced from the conversion line
centroid by just the calculated L_3 subshell binding energy in an Fe
ion with one K vacancy. It represents the K conversion electron
continuous spectrum associated with the shakeoff of an L electron.
It is a continuous spectrum extending down about 1% in momentum, in
comparison with the narrow K line of .05% width shown fitted to the
upper limit of the continuum. The L shakeoff electron spectrum lies
within a kilovolt of zero; this satellite spectrum is at $\sim$ 114 keV.
The intensity of the double conversion spectrum in this region would
be negligibly small. The 136 keV transition in ^{57}Fe gives an almost
identical result, with exactly the same relative intensity per normal internal conversion. This transition is electric quadrupole
whereas the 122 keV transition is magnetic dipole, demonstrating
that shaking is independent of the intrinsic character of the nuclear transition; in contrast, their internal conversion probabilities
differ by tenfold.

    The spectrum area corresponds to an L shakeoff probability of
0.9% per K conversion. The self-consistent-field (SCF) wave function overlap calculations of Dr. Carlson's Oak Ridge group

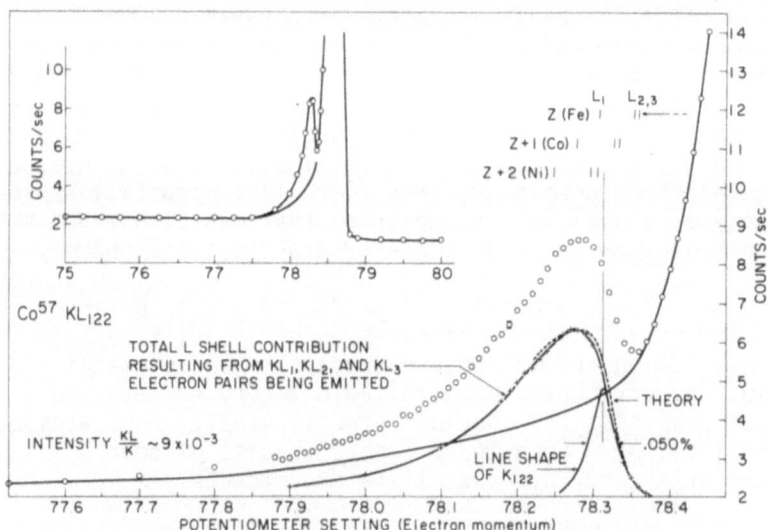

FIG. 2.   Complementary L-shakeoff satellite continuum on tail of
K conversion line of 122 keV transition in ^{57}Fe.

corrected by a Slater screening factor to take account of the
smaller effective change in central charge for K conversion compared
to beta decay, gives 0.95%. Figure 2 shows the excellent agreement
of the continuum shape with the theoretical shape for the satellite.
A few similar cases now establish the phenomenon; in each, the
shakeoff electron is from a shell outside the conversion shell.
Even the satellite due to valence electron shakeoff has been seen.
For shakeoff outside the conversion shell the measured shakeoff
intensity lies within a factor of 2 of the Oak Ridge predictions.

The experimentally much more difficult case of shakeoff in the
same shell as that in which conversion occurs is especially interesting as an indication of the need for taking account of correlations between the two closely associated bound electrons. The very
weak K shakeoff in K conversion was definitely seen for transitions
in ^{57}Fe and ^{137}Ba, but the weak continua could not be as accurately
delineated as for L shakeoff. The assumption of the theoretical
continuum shape gave shakeoff yields per K conversion between 3 and
8 times higher than predictions from the same calculations. A
similar result applies to the L shakeoff in L conversion of the 89
keV transition in ^{109}Ag. Perhaps this larger disparity is indica-
tive of the need to use a many body calculation for such cases,
although accounting for correlations in the similar case of K shakeoff in nuclear K capture actually lowers the predicted intensity.
However, Professor Shimizu's Kyoto group [3] showed that relativistic screening constants applied to the SCF wave function overlap integrals of the Oak Ridge group brought much nearer agreement in
these cases of the same shell shakeoff. The Kyoto groups also
calculated the shakeoff probability in internal conversion using
relativistic hydrogenic wave functions with screening constants obtained from SCF wave functions. For K shakeoff their predictions
are somewhat lower, and hence in poorer agreement with the meager
data, than the earlier SCF calculation; for shakeoff in a higher
shell, the Kyoto predictions are about 1/3 of the SCF predictions,
again in poor agreement with the data. In summary one can only
conclude that the question of the adequacy of a single electron
model is not clear at present; better data is needed, particularly
for shakeoff in the same shell, and by the unambiguous but very
difficult electron-spectroscopic technique.

The final state has two electrons in the continuum so we write
a form differential in each energy, and keep only energy dependent
factors. The relative probability is

$$N(W_S,W_K)dW_SdW_K = |\{F(Z_f,W_S)B(W_S)\}|^2 p_S W_S p_K W_K \delta(A-W_S-W_K)dW_SdW_K. \quad (6)$$

The subscript K denotes the K conversion electron and S, the shakeoff electron; p is momentum and W is energy. The curly bracket is

the atomic overlap integral of the continuum final state Coulomb-field Fermi function $F(W_S)$ with the initial bound state wave function, $B(W_S)$. The density of states factors follow, including the energy-conserving δ function, the constant A being the decay energy minus the sum of both electron binding energies. By integrating the formula over W_K one obtains the shakeoff spectrum and integration over W_S gives the mirror image conversion electron satellite spectrum of Fig. 2. Note that the nuclear matrix element which governs the internal conversion probability has been factored out of equation (6) in writing the shakeoff probability per internal conversion. This has removed the sensitive dependence on multipolarity that internal conversion exhibits. The bound electron wave functions $B(W_S)$ were evaluated with non-relativistic hydrogenic wave functions to produce the shapes shown in Fig. 2; the Kyoto group improved these with relativistic wave functions.

II B.   Shakeoff in Electron Capture

Nuclear capture of an atomic electron is a beta-decay process alternate to positron emission. If the transition energy exceeds the K electron binding energy, K electron capture is most probable. The K electron charge is transferred into the nucleus from near the nuclear surface leaving a K vacancy ion, and a neutrino of unique energy is emitted. The effective change in central atomic charge is much smaller in K capture than the unit charge change in positron decay since no charge is removed from the atom; for the remaining K electron the decrease in nuclear charge is nearly compensated by the removal of the partial screening afforded by the captured K electron; for outer shells this componsation is almost complete since the screening constant of a K electron for outer shells is nearly unity. Thus K shakeoff is about an order of magnitude less probable than for positron decay, being typically 10^{-4}-10^{-5} per decay; L and higher shell shaking is still much weaker. Recall that shakeoff probability varies as the square of the effective change in central charge.

When K shakeoff occurs, the energy spectrum resembles that of the shakeoff component in internal conversion, with its maximum at zero energy and rapidly decreasing in intensity with energy above zero. The complementary satellite continuous spectrum is that of the unobserved neutrino with which the emitted electron shares the available energy. In consideration of the low interaction cross-section of neutrinos, there is no direct collision process; this is the purest example of shaking. The residual atom is doubly ionized in the K shell, so the definitive evidence of the process is the emission of two K vacancy deexcitations in coincidence, either K x-rays or Auger electrons, or the identification of a K x-ray hyper-satellite as the first K vacancy is filled. K shakeoff has been observed in six isotopes, ^{7}Be, ^{37}Ar, ^{55}Fe, ^{71}Ge, ^{131}Cs and ^{165}Er, and L shakeoff in none.

The atomic matrix element overlap integral in the shakeoff rate has two bound K electrons in the initial state, one of which is destroyed, and one free electron in the final state. The principal problem in its evaluation is how to account for the initial state interaction or correlation between the two electrons; all published theories do so, by widely varying means, and since they enjoy comparable agreement (or lack of it) with experimental results, selection among them is as present a matter of theoretical preference. Because of the very small change in effective central charge in K capture, the initial and final wave functions of the remaining cortege change only very slightly. This means that the overlap integral between initial bound state and final continuum state wave functions is between nearly orthogonal functions, and thus it is very sensitive to their small details. It seems certain that omitting the correlation interaction would yield a grossly different result.

Three different approaches are currently competitive. All are two-initial-electron models; the remaining electrons are regarded as static during the transition. All theories antisymmetrize the initial wave function for exchange of the two K electrons and all include a density-of-states factor. The calculation of Primakoff and Porter used a variational product of two hydrogenic 1s wave functions, with extra factors adjusted to account for the effect of their mutual Coulomb interaction on their spatial correlation and for their mutual screening of the nuclear charge. The constants were evaluated by matching the Hylleraas wave function for helium. They calculated the overlap with a free electron Coulomb wave function as a function of its energy.

The Kyoto group also calculated an overlap integral, but of a relativistic hydrogenic single bound electron with a continuum electron. They accounted for the initial interelectron interaction in terms of screening constants σ for initial and final wave functions that determine the effective nuclear charge $Z_{eff} = Z-\sigma$ that each electron sees. They calculate σ from relativistic self consistent field wave functions, and take account of the change of screening in the final K-vacancy state.

Intemann uses a very different idea. His initial wave function is without interelectron interaction; instead the K-K interaction is put into the Hamiltonian as a perturbation along with the beta decay operator, and thus the wave function can be evaluated with any desired accuracy by perturbation methods. The initial basis set without interaction is obtained as solutions of a "semi-relativistic" Hamiltonian. The result is obtained without having to calculated screening constants or an effective nuclear charge. For

this reason it is probably the preferred approach. Intemann's
method also includes significant corrections for dipole contributions to the dominant monopole ejection; these arise from the relativistic mixing of orbital angular momentum.

    In summary of the theoretical results, the effect of making the
calculation relativistic is to reduce the shakeoff probability by
30-50%, and to change the shape of the shakeoff electron spectrum
by a modest amount compared to the non-relativistic calculation.
Figure 3 shows the only truly valid measurement of the K electron
spectrum in K capture, done on ^{55}Fe by the Kyoto group [4] in coincidence with two K x-rays. The solid curve is their prediction of
the shape, which is close to Intemann's calculation; the dashed
curve is that of the non-relativistic theory of Primakoff and
Porter. The fit to the relativistic form seems better, and constitutes the only clear basis for choice among the theories. However,
the agreement with experimental shakeoff probabilities is better
for the non-relativistic predictions in three low atomic number isotopes, while it is better with both relativistic predictions in two
higher Z cases, as one might expect. The Kyoto group also predict
a large contribution of K shakeup to unoccupied bound states, about
equal to their shakeoff values; the ensemble of experimental values
is indecisive evidence pro or con. Overall the average agreement

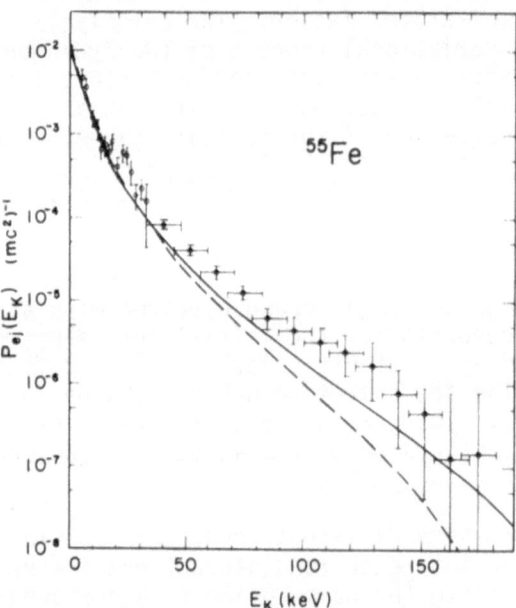

FIG. 3. K shakeoff energy spectrum in K capture in ^{55}Fe. Solid
curve; relativistic theory of Kyoto group. Dashed curve; non-
relativistic theory of Primakoff and Porter.

is poor, but this is at least in part owing to large variance among the few measurements in these difficult experiments.

II C. Shakeoff in Beta Decay

Here the situation is less ambiguous than for electron capture; only the theory of Law and Campbell for K shakeoff in beta transitions with allowed shapes appears to be very well supported by the ensemble of recent experiments of better quality. These are measurements of shakeoff probability done with high resolution x-ray detectors and measurements of the composite energy spectra of shakeoff-electron plus-nuclear-beta done by magnetic spectrometry in coincidence with K x-rays. All of the dozen or so published calculations of K electron emission in beta decay have used the sudden (shaking) approximation, but the complication of a three particle final state, two free electrons and an antineutrino, has been handled by way of various approximations that are often inadequate. Law and Campbell's treatment is the culmination of several attempts of increasing sophistication spanning 30 years; it is the first to use a fully antisymmetrised, relativistic form incorporating nuclear and atomic variables. Its approximations involve: a) the use of a one electron atom model, so the initial state is described with a Dirac wave function without screening and of course without correlations; and b) apart from the antisymmetrization of the two electron wave function, the final state interactions of the two free electrons with each other (or with other electrons) are not yet formally included, though their interaction with the residual nucleus is accounted for.

Law and Campbell use the formalism of second quantization to express the beta decay transition for an initial atom with nuclear charge Z and one K electron to a final atom with charge Z ± 1, plus two free electrons and a neutrino. The perturbation operator in the transition matrix element which induces the transition is the V-A beta decay Hamiltonian H_β. In a standard notation for Fermi's golden rule the K shakeoff rate in $\beta^\pm$ decay is written:

$$\lambda_\pm = 2\pi |<f|H_\beta|i>|^2 \, \delta \, (E_i - E_f) \; ; \tag{7}$$

where the wave functions f and i include nuclear and atomic variables, the delta function conserves energy and the density of states is implicitly included. This means that all three emitted particles including the neutrino share the energy statistically. If ρ is the momentum of an electron and q is that of the neutrino, the differential shakeoff rate dependence on p, which is the electron momentum spectrum is

$$\lambda_\pm \, (p)dp = Ap^2 dp \int q^2 dq \, M \; ; \tag{8}$$

this resembles a standard allowed beta spectrum. Constant A includes the β decay coupling constants and 2π's. The second quantized form of the overall transition matrix element M brings in the
third momentum, s, of the ejected electron, and also the essential
difference between the $β^+$ and $β^-$ cases. In the $β^+$ decay the two
electrons in the free state can be distinguished by their charges;
but not in $β^-$ decay. The necessary antisymmetrization of the final
state wave function with respect to the two electrons in the $β^-$ case
results in a complete formal symmetry of the final expression for
the electron momentum spectrum; this symmetry arises from the coherent addition of the amplitudes of the wave functions of the free
electrons. The double differential electron momentum spectrum shows
this symmetry in all functions of s and p:

$$\lambda_{\pm}(p,s)dpds = Ap^2s^2(E_0-E_p-E_s)^2 \, dsdp\{|\langle\psi_s|\psi_K\rangle|^2F(Z',E_p) +$$

$$|\langle\psi_p|\psi_K\rangle|^2F(Z',E_s) - |\langle\psi_s|\psi_K\rangle\langle\psi_K|\psi_p\rangle| \sqrt{F(Z',E_s)F(Z',E_p)}\} \qquad (9)$$

E denotes energy and the factor $(E_0-E_s-E_p)$ is the neutrino energy;
F is the Fermi function of beta decay theory that accounts for the
interaction of each electron with the nuclear Coulomb field; the
angular brackets are the wave function overlap integrals of the s
and p free electrons with the initial bound K electron. Inside the
curly brackets are three terms, two squared terms and a negative
interference term. Only the first term appears in the $β^+$ shakeoff
calculation; this leads to a reduction of the predicted shakeoff
probability by $\sim 1/2$ compared to that in $β^-$ decay at the same energy
and atomic number. By integrating over either electron momentum the
spectrum of the other electron is obtained, but the momentum dependence of the overlap integrals results in different spectrum shapes
for the three terms. The two leading terms, which give equal total
contributions to the intensity, are relatively emphasized at high
and low energy regions of the spectrum, and in analogy to the idealized shakeoff spectrum (Fig. 1) these are called the nuclear and
orbital components, respectively. The interference term is of equal
intensity in low energy decays with a low ratio of transition energy
to K binding energy, but it becomes relatively small for high decay-
energy cases.

The inner orbitals of heavy atoms and high energies involved in
beta decay both require the use of relativistic wave functions in
the overlap integrals; for simplicity, Law and Campbell used hydrogenic type. The calculated shakeoff rate is doubled to account for
two initial K electrons. The calculation was extended to include
shakeup excitation to unoccupied bound states, which adds significantly to the total shaking, particularly in cases with low ratios
of transition energy to K binding energy. From the nuclear standpoint the theory is appropriate only for allowed beta transitions
or for those forbidden transitions with allowed shapes.

Experimental comparisons to the theory are most reliable with 33 recent improved measurements of the K shaking probability P_K, covering 15 β^- decaying isotopes from Z = 20-83, and ranging in P_K values from 2×10^{-6} to 2×10^{-3}; the measurements were made by a variety of independent techniques. Only two cases of highly forbidden-shape transitions were in slight disagreement with the predictions; the averaged agreement for all isotopes lies within a few percent with Law-Campbell theory, but tenfold or more worse with any other calculation.

Accurate measurement of spectrum shapes again turns out to be a more sensitive test of the theory. The Argonne group has measured five representative cases covering a wide range in transition energy and in isotopes with both allowed and forbidden shapes for the singles spectrum. The composite nuclear beta- K electron spectrum focused in a magnetic spectrometer was registered in coincidence with K x-rays observed in a GeLi detector. In Fig. 4 is the momentum

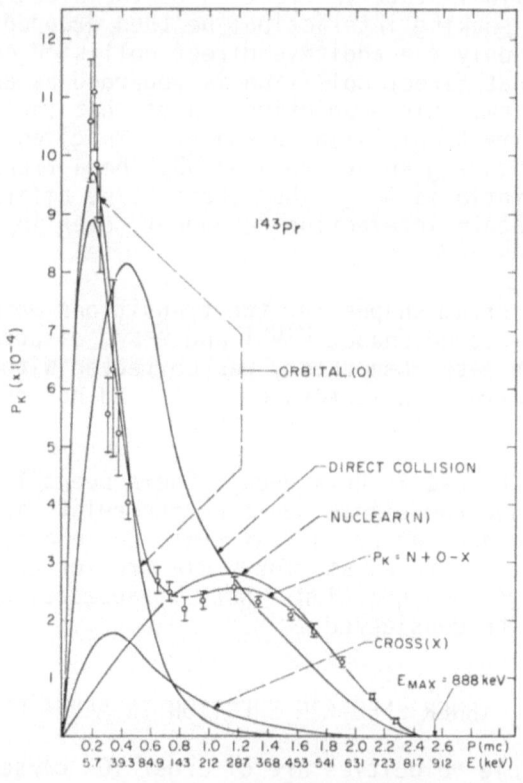

FIG. 4. Composite β^- + K-orbital shakeoff momentum spectrum in β^- decay of ^{143}Pr. "N + O - X" curve is theory of Law and Campbell, summing nuclear (N), orbital (O) and interference (X) terms. Direct collision shape of Feinberg theory shown.

spectrum of ^{143}Pr, a 930 keV allowed shape β^- transition. The separate nuclear, orbital and interference terms of Law-Campbell theory are shown, and their sum clearly fits the observed spectrum very well over the full energy range from 6 keV up; for comparison, the K binding energy is 44 keV. Similar excellent fits are seen in the 227 keV decay of ^{147}Pm and even in ^{151}Sm, which has a transition energy of 77 keV, only 1.5 times the K binding energy of 49 keV. In this case the low energy shakeoff spectrum which has a maximum energy of only 28 keV serves as a severe test of the applicability of the sudden approximation treatment. The good fit of the shape to the Law-Campbell theory and the exact agreement of the K emission probability (obtained from the spectrum area) at 2.3×10^{-6} per decay attests the validity of the sudden approximation in this limit.

The agreement of the shakeoff value also sets the lowest limits on the contribution of the direct collision mechanism, to $\sim 1/5$ of that predicted by the Feinberg theory. Feinberg analyzed direct collision as a final state interaction of the electrons which is coherent with the shaking interaction; he then dropped the interference term, retaining only the additive direct collision component. It is in this sense that direct collision is regarded as an independent additive mechanism. His prediction states that the relative intensity ratio of direct collision to shakeoff is given approximately by the ratio of K binding energy to available beta transition energy. For ^{151}Sm this ratio is 38%. Thus there is no affirmative evidence for this final state interaction component, even in this most favorable case.

Shakeoff spectrum shapes for two transitions with large deviations from the allowed shape, ^{210}Bi and ^{89}Sr, disagreed with L-C theory, but much less when theory was corrected with the measured singles shape correction factor, per L-C's ad-hoc prescription. This is the principal deficiency of the theory.

To sum up the case in beta decay, there seems little indication of need to improve the simple one-electron relativistic wave functions or to take account of initial state correlations in the K shell. Not enough data exists for L electron or outer shell shakeoff on which to judge this question. Nor, it appears, need final state energy exchange be considered.

III. INNER ELECTRON EJECTION IN ALPHA DECAY

Alpha particle velocities are of order 10^9 cm/sec, which is below mean orbital electron velocities in K, L or M shells in heavy elements. Thus the α passage is essentially adiabatic so electron shakeoff due to the relatively slow reduction of the central field should contribute little to the ejection probability. The electron

emission arises instead from the direct collision process. In the
first stage the α is ejected by the nucleus through the potential
barrier past the outer classical turning radius of about 4 Fermis
with a definite total energy. This is transformed to over 99% of the
final α kinetic energy inside the mean radius of the K orbit. The
second stage energy exchange with the inner electron mainly occurs in
the vicinity of the orbit, since the Coulomb interaction potential
between α and electron is maximum as the α passes the shell.

All but one recent calculation use time dependent perturbation
theory in a higher order approximation than the sudden approximation
to treat this problem. Thus, in the original theory of Migdal, the
probability for ejection of an electron from the n shell of the
ground state configuration g, to a final state, j, is given by

$$P_n(j) = \frac{i}{h} \int_0^\infty e^{\frac{-i(E_j - E_g)t}{h}} \; |<\psi_j|H'(t)|\psi_g>|^2 dt \tag{10}$$

Here E is energy, and $H'(t)$ is the time-varying Coulomb perturbation
operator given by:

$$H'(t) = -2e^2/r_\alpha = -2e^2/v_\alpha t \tag{11}$$

where r_α and v_α are the alpha radial position and velocity. In a
later version by Levinger, this interaction is greatly diminished by
an opposing contribution from the potential of the slowly recoiling
nucleus with velocity v_r. The integral in equation (10) was developed in a series resembling a multipolar expansion. The monopole term
corresponding to shaking was shown to be negligible; the dipole term
was evaluated with hydrogenic wave functions to given an ejection
probability for the n shell

$$P_n = \frac{(v_r - 2v_\alpha/Z)^2}{Z^2} \; C_n \tag{12}$$

where the C_n are squares of hydrogenic dipole matrix elements for the
n shell. In Migdal's version the recoil velocity term is omitted.
Various refinements such as relativistic and nuclear screening corrections and excitations to unoccupied bound levels have been added.

The only isotope with a simple ground-to-ground state α decay
scheme that is feasible for measurement of the weak ejection probabilities is [210]Po, on which essentially all experiments have been
made. The probability is obtained from the intensity of the shell
x-rays, corrected for fluorescence yield, measured either in singles
or in coincidence with alphas or with ejected electrons. For the
inner shells the probabilities are given in Table 1, averaged over
nine K shell, six L shell and three M shell measurements. One sees

TABLE 1.   EMISSION PROBABILITIES PER ALPHA DECAY IN ^{210}Po

	Exp't	Hansen	Migdal	Levinger	Rubinson
K($\times 10^{-6}$)	1.88 ± 0.15	2.0	2.5	0.10	0.14
L($\times 10^{-4}$)	7.9 ± 0.4	5.9	1.13	0.6	0.028
M($\times 10^{-2}$)	2.3 ± 0.3	1.90	0.16		0.0056

from the comparison to the Migdal predictions that the agreement with the K shell values is fair, but for the L shell theory is at least a factor of seven low, and for the M shell, a factor of twenty; even worse for the later improved versions of the theory.

A recent calculation by Hansen [5], has achieved fairly good agreement with the data in all three inner shells. This theory is an adaptation of the binary encounter approximation used in the evaluation of electron ejection in ionic bombardment; ionization occurs when the coulomb interaction transfers at least the electron binding energy from the incident ion to a single electron. The cross section for this transfer is integrated over the infinite range of impact parameters in the ion beam. For α decay only the outgoing wave is considered, and only zero impact parameter. The energy-transfer cross section is evaluated with an exact two-body treatment of a classical α trajectory and a particular electron treated quantum mechanically; the cross section is a function of the α and averaged electron velocities. The electron velocities are calculated using screened relativistic hydrogenic wave functions, and the α acceleration is taken into account. (However, there appears to be an inconsistency in the relativistic velocity transformation, which seriously affects the results, particularly for the K shell.)

Again the shapes of the spectra of α and electron turn out to be more severe tests of the theory than the integrated probabilities. The α complementary satellite spectrum shape associated with L electron ejection was measured [6] with high resolution in the Argonne magnetic α spectrometer, in coincidence with K or L x-rays. Figure 5 shows these satellites corresponding to K and L electron ejection, displaced below the α line by the K and L binding energies. Near zero energy is the electron spectrum which was measured in the Argonne magnetic electron spectrometer in coincidence with L x-rays; the mirror image of this spectrum, broadened with the line shape of the alpha spectrometer, is the dashed curve (b), which fits the L shell α complementary spectrum quite well, as expected. But great disparities with the predictions of any theory appear in the K and L electron spectra extended to 100 keV. The binary encounter theory gives no match whatever to the measured shapes, despite the fairly good agreement with the ejection probabilities, and Migdal's predicted shape for the K spectrum also fails below 30 keV.

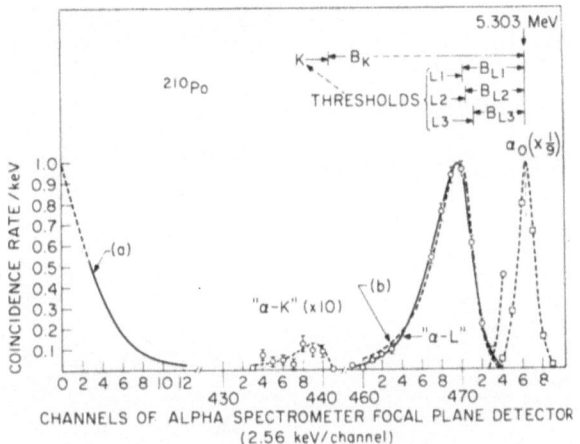

FIG. 5. α complementary satellite continua associated with ejection of K and L electrons in ^{210}Po. L electron spectrum (a) shown at left, and fit of its complement, convoluted with α_0 normal line, to the "α-L" continuum is shown dashed (b). B_K and B_L binding energies at top.

An unanticipated result emerged that is a further challenge to ejection theory; namely, it appears that in ∿ 2/3 of the interactions of the emerging α with an L electron, (but not with a K electron), the electron is captured into a bound state of the α particle, which must then emerge as a singly charged He$^+$ ion. It would seem that an attempt to account for this result must consider the interaction of the developing level structures of the α particle and the residual lead nucleus, as the α moves through the shells, and must also reproduce the measured energy spectra.

1. Melvin S. Freedman, Ann. Rev. Nuc. Science, 24, 209-47 (1974). See for all pre-1974 references.

2. R. Walen and C. Briançon, in Atomic Inner Shell Processes, ed. B. Crasemann, Academic Press, Inc., Vol. 1, p. 233.

3. Takeshi Mukoyama and Sakae Shimizu, Phys. Rev. C, 11, 1353-63 (1975).

4. Tetsuo Kitahara and Sakae Shimizu, Phys. Rev. C, 11, 920-6 (1975).

5. J. S. Hansen, Phys. Rev. A, 9, 40-43 (1974).

6. H. J. Fischbeck and M. S. Freedman, Phys. Rev. Lett. 34, 173-6 (1975).

THEORY OF ATOMIC DECAY FOLLOWING INNER-SHELL IONIZATION

T. Åberg [*]

Institut du Radium and Université Pierre et Marie Curie

11, rue Pierre et Marie Curie, 75231 Paris Cedex 05, France

1. INTRODUCTION

We assume that the target consists of free non-interacting atoms. Inner-shell ionization by photons and charged particles leads to the emission of x-rays and electrons [1]. This emission seems to be characteristic of the atoms in the sense that the energies of the individual lines are generally found to be independent of the excitation mode although the intensity distribution may be very different. Notable exceptions are the Auger electrons which are emitted with Doppler-shifted energies in heavy-ion collisions and molecular x-rays which are characteristic of the colliding complexes rather than of the target atoms or the projectiles [2].

The appearance of a characteristic emission spectrum is consistent with the assumption that the decay process of a particular inner-shell hole state is independent of the excitation process and possible preceeding emission events. However, this assumption may not always be valid. For example, there may even be observable changes of the shape and energy of an Auger line close to an inner-shell photoionization threshold due to the interaction between the slow photoelectron and the Auger electron [3]. In this case the Auger effect should be treated as a resonance in the double photo-ionization cross section [4] rather than as an independent decay mode.

In the usual theory of radiative and radiationless inner-shell transitions the hole states are treated as stationary states of the Hamiltonian of the ionized atom.

The emission is taken into account by assuming an initially prepared discrete state which decays without " memory " of the excitation process. One or several photons or electrons or even both particles are emitted simultaneously [1].

We shall only discuss the non-relativistic theory but include the spin-orbit interaction in the Hamiltonian. Whether this theory is adequate or not depends on the inner-shell property to be studied. For example, the non-relativistic theory seems to be adequate for calculations of Auger rates of atoms with Z below 20 [5]. On the other hand, an analysis of the correlation energy corrections of the inner-shell binding energies requires a proper treatment of the relativistic energy contributions even in very light atoms [6]. Note also that in heavy atoms the relativistic motion of the innermost electrons affects the screening of the nucleus [7] which may especially influence the probabilities of the radiationless transitions of the " non-relativistic " outer-shell electrons.

By the assumptions, mentioned above, the theory of the atomic decay following inner-shell ionization is to a large extent reduced to the ordinary theory of atomic spectra which is described in many text books [8]. In the following we shall try to examine some of these assumptions and to establish the most important steps which lead to the conventional non-relativistic one-electron frozen-core theory of inner-shell transition probabilities.

2. DESCRIPTION OF VARIOUS DECAY MECHANISMS

The inner-shell hole states have very short lifetimes corresponding typically to linewidths of the order of 1 eV. In general, the widths of the radiationless transitions dominate with the exception of the K widths of very heavy atoms which are mostly due to radiative transitions [9]. The inner-shell binding energies range from a few eV to hundreds of keV. These properties demonstrate that the inner-shell hole states are highly excited autoionizing states of the ionized atom. This is illustrated by the energy diagram of Ne in Fig. 1 which is sketched by following Parratt's suggestions [10].

Several decay modes are indicated in Fig. 1 such as single and double photon emission, single Auger electron transitions, and radiative Auger emission. Double Auger transitions are not shown. Absent are also the Coster-Kronig and Super Coster-Kronig transitions which do not occur in the Ne atom. The former transition refers to a decay process of the type $X_i \rightarrow X_j Y$ where X_i and X_j pertain to holes in the same shell but different subshells and the latter to a process of the type $X_i \rightarrow X_j X_k$ where

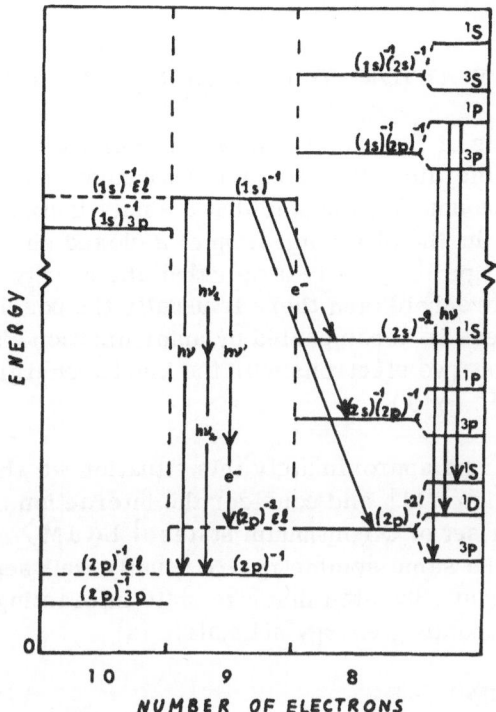

Fig. 1 Schematic part of the energy diagram of Ne ionized in the 1s, 2s and 2p subshells. Only a few energy levels (solid lines) with associated continua (dashed lines) are shown. The energy levels are not in scale . The zero level corresponds to the ground-state energy of the Ne atom. The thresholds of the continua are indicated by the configuration of the ion and the ejected electron. Hence the number of electrons refers to the system under consideration. Examples of single x-ray photon (hν), single Auger electron (e$^-$), radiative Auger (hν', e$^-$), and two x-ray photon (hν_1, hν_2) transitions are shown.

all the holes are in the same shell. Note that the inner-shell transitions are usually identified by specifying the initial and final holes with respect to the ground state configuration of the neutral atom. In addition to that the multiplets are identified by the LS values of the initial and final terms or the transitions by the J values of the initial and final levels. Depending on the strength of the spin-orbit interaction a subshell is described by ℓ or $j = \ell \pm 1/2$ (j = 1/2, ℓ = 0). In the latter case only J is a good quantum number. In most cases it is possible to associate an inner-shell transition with a specific initial and final state configuration.

3. CONCEPT OF AUTOIONIZING INNER-SHELL HOLE STATES

A proper analysis of the radiationless decay of the inner-shell hole states requires that the joint discrete and continuous features of these states are taken into account. We assume that a singly ionized hole state $|\gamma JM\rangle$ is created in the photoionization of a closed shell atom. If the single ionization energy $I^+ (\gamma J)$ is larger than the energy $I^{++} (\alpha J')$ of a double ionization threshold then there is usually the possibility that the photoionization process is accompanied by autoionization. This results in the ejection of at least two electrons with the kinetic energies $I^+ (\gamma J) - I^{++} (\alpha J)$ and $\omega - I^+ (\gamma J)$.

In order to describe approximately this situation we shall use Fano's theory of autoionization [11] and consider the interaction between a discrete state $|\gamma JM\rangle$ and a set of Q continuum states $|E\alpha JM\rangle$. Since the inner-shell hole states of the same symmetry are usually well separated, it is sufficient to consider an isolated discrete state interacting with the Q continua. The corresponding energy submatrix is

$$\langle \gamma JM | H-E | \gamma JM \rangle = E_\gamma$$
$$\langle \alpha E'JM | H-E | \gamma JM \rangle = V_\alpha (E',E) \qquad\qquad \alpha = 1 \ldots Q \quad (1)$$
$$\langle \alpha E'JM | H-E | \beta E''JM \rangle = \delta_{\alpha\beta} \delta (E''-E') (E'-E),$$

where the matrix elements are independent of M. Note also that it is assumed that the submatrix of the continuum wave functions is diagonal. The interaction between the discrete state and the continua is described by the coefficients a_μ of the eigenfunctions

$$|E\mu JM\rangle = a_\mu(E) | \gamma JM \rangle + \sum_{\alpha=1}^{Q} b_\mu^\alpha(E,E') |\alpha E'JM\rangle dE' \quad (2)$$

which are normalized per unit energy range and orthogonal. The coefficients a_μ fulfil the relation [11]

$$|A(E)|^2 = \sum_{\mu=1}^{Q} |a_\mu(E)|^2 = \frac{\frac{1}{2\pi}\Gamma(E)}{(E-E_r)^2 + \frac{\Gamma(E)^2}{4}} , \quad (3)$$

where the resonance energy E_r is given by

$$E_r = E_\gamma + P \int \frac{dE' \frac{1}{2\pi} \Gamma(E')}{E_\gamma - E'} \qquad (4)$$

and the resonance width $\Gamma(E)$ by

$$\Gamma(E) = 2\pi \sum_{\alpha=1}^{Q} \left| V_\alpha(E,E) \right|^2. \qquad (5)$$

In Eq. (4) P means the " principal part ".

The final state wave functions $|E \mu \varepsilon J'M'\rangle$ correspond to $J'=1$, $M'=0$ in the photoionization process. They are given by linear combinations of antisymmetrized products of $|E \mu JM\rangle$ and one-electron continuum wave functions $|\varepsilon \ell jm\rangle$. In each wave function the sum of the products $|\alpha E'JM\rangle|\varepsilon \ell jm\rangle$ describes a doubly ionized core and two ejected electrons. Since the photon-electron interaction operator is a one-electron operator the corresponding part of the cross section will vanish if the initial and final state wave functions are described by the one-electron frozen-core approximation. This suggests that, in general, the part $\sigma^+(\omega)$ which describes transitions to the discrete states $|\gamma JM\rangle$ and the continuum states $|\varepsilon \ell jm\rangle$ dominates. It follows by neglecting the interference term also that

$$\sigma_t(\omega, E) \cong |A(E)|^2 \sigma^+(\omega), \qquad (6)$$

where $|A(E)|^2$ is given by Eq. (3). According to Eq. (6) $|A(E)|^2 dE$ gives the relative number of atoms which emit electrons into any of the energy ranges $(E_\alpha, E_\alpha + dE_\alpha)$, where $E_\alpha = E - E_0 - I^{++}(\alpha J)$. The corresponding photoelectrons are emitted with the energies $\omega - E + E_0$. Here E_0 is the ground state energy. According to Eq. (6) it is also possible to interpret the resonant behaviour of the double photoionization cross section $\sigma_t(\omega, E)$ as a radiationless decay process which occurs via $|\gamma JM\rangle$.

The probability distribution $|A(E)|^2$ is approximately Lorentzian if $\Gamma(E)$ is weakly dependent on E in the neighbourhood of the resonance energy $E_r \cong E_\gamma = I^+(\gamma J) + E_0$. This approximation as well as the omission of the integral in Eq. (4) should be good for inner-shell hole states. However, in a very accurate analysis of transition energies, the energy corresponding to the integral may be significant.

In order to describe the radiative decay as well, the photon field and the operator representing the interaction between the photons and the electrons should in principle be included in the Hamiltonian of Eqs. (1) [12]. The basic set would then consist of products of electron and photon wave functions. With this model it should be possible to demonstrate the circumstances which permit the usual addition of the radiative and radiationless decay probabilities.

A heuristic argument of the additivity is based on the theory of the radiative decay by Weisskopf and Wigner [13]. In their theory it is assumed that the probability amplitude of the initial state decreases according to the exponential law exp (- 1/2 Pt). It is then shown by first-order time dependent perturbation theory that the relative number of atoms per unit energy range, which decay by emission of photons of energy E is given by Eq. (3). In this case E_r is the energy difference between the initial and final state and $\Gamma = \Gamma (E)$ is related to the decay probability P by

$$\Gamma = \hbar P .$$

Consequently the energy distribution (3) is Lorentzian with a full width at half maximum (FWHM) which is Γ. Since this is approximately true also for the radiationless decay, the convolution of the two probability distribution functions is also a Lorentzian with a FWHM which is $\Gamma_{nr} + \Gamma_r$ i.e. the sum of the radiationless (non-radiative) and radiative widths. Similar arguments lead to the result that the width of an x-ray or Auger (Coster-Kronig) line is the sum of the total decay widths of the initial and final states.

We have assumed for simplicity that the autoionization is a consequence of photoionization of inner-shell electrons of closed shell atoms. The results can easily be generalized to excitation processes which are described by one-electron operators and to initial open-shell configurations.

4. PROBABILITY OF RADIATIONLESS AND RADIATIVE TRANSITIONS

In a radiationless transition the initial and final state configurations differ by two sets of one-electron quantum numbers (nℓjm) because there is at least one outgoing electron in the final state. If it is assumed that $\langle \alpha EJM | \gamma JM \rangle = 0$, then the orthogonality of the one-electron orbitals yields

$$P = 2\pi \sum_{\alpha = 1}^{Q} \left| \langle \alpha EJM | \sum_{i > j} \frac{1}{r_{ij}} | \gamma JM \rangle \right|^2 \qquad (7)$$

in atomic units. We shall neglect the spin-orbit splitting in the following and consider a transition between an initial-state configuration term $(n_1\ell_1)^{-1}\ {}^2L_1\ (L_1 = \ell_1)$ and a final-state configuration term $(n_2\ell_2)^{-1}$ $(n_3\ell_3)^{-1}\ {}^{1,3}L\varepsilon\ell\ {}^2L_1$. Note that the parity is conserved i.e. $\Pi_i = \Pi_f$ or $(-1)^{\ell_1+\ell_2+\ell_3+\ell} = 1$. Usually only a few $\alpha = \ell$ values are possible. By using the properties of the 3j-symbols we obtain

$$P = 2\pi\frac{(2L+1)\ (2S+1)}{2\ (2\ell_1+1)}\sum_\ell\ \left|\left\langle n_1\ell_1\ \varepsilon\ \ell\ \text{LSJM}\left|\frac{1}{r_{12}}\right|n_2\ell_2 n_3\ell_3\text{LSJM}\right\rangle\right|^2 \quad (8)$$

in the one-electron frozen-core approximation. In Eq. (8) the initial and final state wave functions are the antisymmetrized two-particle LS-coupling wave functions belonging to the quantum numbers LSJM. The component transition rate (8) gives the probability that an initial $(n_1\ell_1)^{-1}$ hole state will decay into any of the $(n_2\ell_2)^{-1}$ $(n_3\ell_3)^{-1}$ ${}^{1,3}L$ hole states. The rate can be expressed as a function of the direct and exchange radial matrix elements R^k $(n_1\ell_1\varepsilon\ell,\ n_a\ell_a n_b\ell_b)$ (a = 2, b = 3 and a = 3, b = 2) with the aid of the 3j - and 6j - symbols [14]. The summation of Eq. (8) over the possible quantum numbers LS yields the $(n_1\ell_1)^{-1}\longrightarrow (n_2\ell_2)^{-1}$ $(n_3\ell_3)^{-1}$ group transition rate which is independent of the coupling scheme.

The one-electron frozen-core LS coupling model is usually improved by the inclusion of spin-orbit interaction in the final state by the intermediate coupling scheme. Some mixing between double-hole configurations is also introduced. If the initial hole and the ejected electron are characterized by their j values the coupling scheme is called mixed. In this case

$$P = 2\pi\frac{(2J+1)}{(2j_1+1)}\sum_j\ \left|\sum_{\beta LS}C(\beta LS,\alpha J)\left\langle n_1\ell_1 j_1\varepsilon\ \ell\ jJM\left|\frac{1}{r_{12}}\right|\beta LSJM\right\rangle\right|^2 \quad (9)$$

where $|\beta\text{LSJM}\rangle$ is an antisymmetrized two-particle LS-coupling wave function which represents the LS term of one of the double-hole configurations $\bar{\beta}$ to be mixed. The coefficients C (βLS, αJ) form a unitary M-independent matrix which diagonalizes a Hamiltonian submatrix for each J. Since the Hamiltonian includes the spin-orbit interaction operator H', each submatrix contains usually both off-diagonal spin-orbit interaction matrix elements $\langle\bar{\beta}\,\text{LSJM}\,|\,H'\,|\,\bar{\beta}\,L'S'JM\rangle$ and off-diagonal Coulomb interaction matrix elements $\langle\,\beta\text{LSJM}\,|\,\sum_{i>j}r_{ij}^{-1}\,|\bar{\beta}'\,\text{LSJM}\rangle$. The mixed matrix elements in Eq. (9) can be evaluate by using the known unitary transformation which connects the two-particle jj- and LS- coupling wave functions which belong to same configuration and to the same quantum numbers JM [8].

Examples of works which apply the analysis, outlined above, are given

by McGuire [14]. The calculation of the Ne KLL rates by Bhalla [15]
and the analysis of the $L_{2,3} M_{4,5} M_{4,5}$ spectrum of free Zn atoms by
Aksela et al. [16] are also good examples. Recently Krause has compared
his measurements of the Ar KLL spectrum [17] with the calculations of
Chen and Crasemann [5]. The theory, based on Eq. (9), seems to predict
relative Auger line intensities quite well. The calculations are based on
Hartree-Slater type one-electron wave functions and only a few configurations are mixed. The Auger electron energies usually include the relaxation [1] contribution. This makes the theory somewhat inconsistent. Note
also that the continuum wave functions which are used do not prediagonalize
the corresponding Hamiltonian submatrix as implied by Eq. (1).

The theory, described above, only applies strictly to the decay of
inner-shell hole states of closed shell atoms. McGuire [14, 18] has recently
developed the formalism for the calculation of radiationless transition rates
in cases where there is an arbitrary number of open shells in the initial
and final configurations. He considers both the LS- and mixed coupling
schemes in the one-electron frozen-core approximation. His results are
important for the proper analysis of fluorescence yields of multiply ionized
atoms which are produced in ion-atom collisions [19] and for the analysis
of the radiationless part of the cascades which follow deep inner-shell ionization [20].

Since the radiationless transitions are determined by a two-electron
operator the one-electron frozen-core approximation does not account for
the double Auger effect. Similarly, since the photon-electron interaction
operator is a one-electron operator this approximation does not account
for the radiative Auger effect. For a review of these second order effects
which involve the simultaneous emission of two electrons and of an electron
and a photon, see Ref. 1. A complete theory of these and the ordinary transitions would require an extension of the theory of autoionization. The
photon-electron and electron-electron interaction should be treated on equal
basis as was suggested in Sec. 3.

The radiative probability concept in the non-relativistic single-electron
theory is based on the first order perturbation theory result [8]

$$P_{ab} \, d\Omega = \frac{\alpha^3 \omega}{2\pi} \left| \langle b | \sum_{j=1}^{N} e^{-2\pi i y_j / \lambda} \frac{\partial}{\partial z_j} | a \rangle \right|^2 d\Omega \qquad (10)$$

which gives the probability (in atomic units) that a photon with polarization
z is emitted in the solid angle $d\Omega$ in the direction of y by an atom which
undergoes a transition from state $a = | \gamma JM \rangle$ to state $b = | \gamma'J'M' \rangle$.

If it is assumed that there is no special alignment of the emitting atoms in space, the probability that a photon is emitted in any direction with an arbitrary polarization direction is given by

$$P(\gamma J, \gamma'J') = \frac{4\alpha^3\omega^3}{3} \frac{1}{(2J+1)} \sum_{qMM'} |\langle \gamma'J'M'| \sum_{j=1}^{N} T_{qj} |\gamma JM\rangle|^2 \quad (11)$$

in the dipole approximation. Here

$$T_{oj} = z_j \quad \text{and} \quad T_{\pm 1j} = \frac{x_j \pm iy_j}{\sqrt{2}}, \quad (12)$$

The selection rules are $J-J' = \pm 1, 0$, $J+J' \geq 1$, $M'-M = q$ ($q = +1, 0, -1$), and $\pi_b = -\pi_a$. Equation (11) represents the transition rate from the level γJ to the level $\gamma'J'$. The rate $P(LS, L'S)$ of transitions between two terms LS and L'S is obtained by summing $(2J+1) P(\gamma J, \gamma'J')$ over J and J' and by dividing by $(2L+1)(2S+1)$. This rate reduces to

$$P(LS, L'S) = \frac{4\alpha^3\omega^3}{3} G D_1 (n\ell, n'\ell')^2 \quad (13)$$

in the one-electron frozen-core approximation. The geometrical factor G can be evaluated for arbitrary initial and final configurations and expressed in terms of 3j-, 6j-, and fractional parentage coefficients [8]. In the radial dipole matrix element $D_1 (n\ell, n'\ell')$ $n\ell$ denotes the initial and $n'\ell'$ the final hole. For a closed-shell atom $G = \ell_>/(2\ell+1)$, where $\ell_>$ is the larger of ℓ and ℓ'. The calculations of the intensity of x-ray satellites and of the fluorescence yield of multiply ionized atoms are partly based on Eq. (13) [19,21].

The magnitude of the radial matrix elements $R^k (n_1\ell_1 \varepsilon\ell, n_a\ell_a n_b\ell_b)$ and $D_1 (n\ell, n'\ell')$ depends on the initial and final state one-electron radial wave functions. Usually these functions are calculated by a self-consistent-field (SCF) method which is based on a spherical average potential. Several schemes are available such as the restricted Hartree-Fock, the X_α, and the transition operator method [22]. However, relatively little attention seems to have been paid to the question of the sensitiveness of the radial matrix elements to the radial wave functions [23]. If they are obtained by separate SCF calculations of the initial and final state (the ΔE_{SCF} method) then the decay rate becomes also a function of monopole matrix elements [24]. The influence of the monopole interaction on the radiative emission rates has been examined [1] but very little is known about their influence on the radiationless decay rates.

5. INTENSITY OF X-RAY, AUGER, AND COSTER-KRONIG LINES

The separation of the excitation and decay probabilities makes it possible to express the integrated intensity in terms of the initial state population and the transition rate. Suppose that a collimated monoenergetic beam of particles which has the flux density I_0 hits a thin target. If the cross section of the excitation to a particular final state $|\gamma JM\rangle$ is $\sigma(\gamma JM)$, then $N'(\gamma JM) = \sigma(\gamma JM) I_0$ is the number of states which is produced in unit time per target atom. Under steady state conditions, i.e. in dynamical equilibrium, the population which gives the relative number of γJM states present is

$$N(\gamma JM) = \frac{N'(\gamma JM)}{P_t(\gamma JM)}, \qquad (14)$$

where $P_t(\gamma JM)$ is the total probability of decay from the state γJM to all possible final states. Note that $\sigma(\gamma JM)$ may also contain contributions from excitations to states higher in energy than γJM via the decay processes.

Suppose that $P(\gamma JM, \gamma'J'M')$ is the probability of the radiative or radiationless transition $\gamma JM \rightarrow \gamma'J'M'$. The number of particles, I, emitted per unit time and target atom in a transition between the corresponding levels γJ and $\gamma'J'$ is

$$I = \sum_{MM'} N(\gamma JM) P(\gamma JM, \gamma'J'M'). \qquad (15)$$

Usually it is assumed that $N(\gamma JM)$ is independent of M (natural excitation). In that case Eq. (15) reduces to

$$I = N(\gamma J) P(\gamma J, \gamma'J'), \qquad (16)$$

where $N(\gamma J) = (2J+1) N(\gamma JM)$ and $P(\gamma J, \gamma'J')$ is the probability of the transitions between the levels γJ and $\gamma'J'$. In the case of the radiationless transitions $P(\gamma J, \gamma'J')$ is usually given by Eq. (9) and in the case of radiative transitions by Eq. (11). Note the different meaning of J in Eq. (9).

If it is further assumed that $N(\gamma JM)$ is level independent and that the term splitting is small the number of transitions between the terms LS and L'S' is given by

$$I = N(LS) P(LS, L'S'), \qquad (17)$$

where $N(LS) = (2L+1)(2S+1)N(\gamma JM)$ and where $P(LS, L'S')$ is given in the one-electron frozen-core approximation respectively, by Eq. (8) or (13). Note that $J=j_1$ should be used in connection with Eqs. (9) and (16) and $L=\ell_1$, $S=1/2$ with Eqs. (8) and (17). It may even be reasonable to assume that $N(\gamma JM)$ is term independent in which case it is meaningful to consider the number of particles which is emitted in transitions between the states of two configurations. The multiplet intensity ratios are determined solely by the strengths $(2L+1)(2S+1)P(LS, L'S')$ for a given initial and final configuration. For comparison of intensities of multiplets which belong to different pairs of configurations the ratio of the populations (14) must also be known. The analysis of the Ne K [21] and the Zr L [25] spectrum is based on the concept of term independent populations.

It is not true in general that $N(\gamma JM)$ is independent of M. Since $P_t(\gamma JM)$ is dominated in most cases by the radiationless transition rate it can be considered independent of M but not of the initial term or of the initial configuration in particular [14]. However, the cross section $\sigma(\gamma JM)$ depends on M even in the case of photon impact because the ionized target atoms align themselves, on the average, in the direction of the photon beam. The condition of the alignment is $\ell \geq 1$ for the initial hole $n\ell$ [26]. The alignment effect has observable consequences such as the polarization of x-rays and the unisotropic angular distribution of Auger or Coster-Kronig electrons [27]. The cross section depends on M also for the impact by charged particles [28]. However, the effect has been found to be rather small, which suggests that Eq. (16) is a reasonable approximation. Whether the approximation of term independent populations is useful or not requires further examination.

6. SUMMARY

We examine briefly the background of the one-electron theory of the atomic decay following inner-shell ionization. Attention is paid to assumptions which permit the factorization of the autoionization cross section into an inner-shell ionization cross section and a decay probability. The concept of an autoionizing inner-shell hole state, based on the separation of the excitation and decay processes, is introduced. In the frozen-core approximation the radiationless and radiative decay probabilities are expressed as a function of a few radial matrix elements. The modifications which are introduced by the configuration and monopole interactions are mentioned. The intensities of the radiative and radiationless transitions are related to the the population of initial states and to the transition rate. The non-statistical nature of the initial state population is pointed out.

ACKNOWLEDGEMENTS

It is a great pleasure to thank Dr. O. Goscinski, Dr. G. Howat and Professor W. Mehlhorn for helpful and stimulating discussions on the theory of autoionization.

REFERENCES

*Permanent address : Laboratory of Physics, Helsinki University of Technology, 02150 Espoo 15, Finland.

1. For a review, see Atomic Inner-Shell Processes, Vols. I and II, eds. B. Crasemann (Academic Press, New York, 1975).
2. See e.g. J.D. Garcia, R.J. Fortner, and T.M. Kavanagh, Rev. Mod. Phys. 45, 111 (1973).
3. F.H. Read, in the Proc. of the IX International Conference on the Physics of Electronic and Atomic Collisions, Seattle, 1975, to be published.
4. R.V. Vendrinskii and V.V. Kolesnikov, Bull. Acad. Sci. USSR, Phys. Ser. 31, 904 (1968).
5. M.H. Chen and B. Crasemann, Phys. Rev. A 8, 7 (1973).
6. T. Åberg, Phys. Rev. 162, 5 (1967).
7. Y.K. Kim and J.P. Desclaux, J. Phys. B 8, 1177 (1975).
8. As listed e.g. by J.H. Scofield in Ref. 1.
9. O. Keski-Rahkonen and M.O. Krause, Atomic Data and Nuclear Data Tables 14, 139 (1974).
10. L.G. Parratt, Rev. Mod. Phys. 31, 616 (1959).
11. U. Fano, Phys. Rev. 124, 1866 (1961) ; see also F.H. Mies, Phys. Rev. 175, 164 (1968).
12. A. Nitzan, Molec. Phys. 27, 65 (1974).
13. V. Weisskopf and E. Wigner, Z. Phys. 63, 54 (1930).
14. See e.g. E.J. McGuire in Ref. 1.
15. C.P. Bhalla, Phys. Lett. 44A, 103 (1973).
16. S. Aksela, J. Väyrynen, and H. Aksela, Phys. Rev. Lett. 33, 999 (1974).
17. M.O. Krause, Phys. Rev. Lett. 34, 633 (1975).
18. E.J. McGuire, Phys. Rev. A 12, 330 (1975).
19. M.H. Chen, B. Crasemann and D.L. Matthews, Phys. Rev. Lett. 34, 1309 (1975) ; C.P. Bhalla, Phys. Rev. A 12, 122 (1975) ; M.H. Chen and B. Crasemann, Phys. Rev. A 12, 959 (1975).
20. E.J. McGuire, Phys. Rev. A 11, 1889 (1975).
21. T. Åberg, in Proc. Intern. Symp. X-ray Spectra and Electronic Structure of Matter, Vol. I, eds. A. Faessler and G. Wiech, München, 1 (1973).

22. See e.g. F.P. Larkins in Ref. 1 and O. Goscinski, G. Howat and
    T. Åberg, J. Phys. B 8, 11 (1975).
23. H.P. Kelly in Ref. 1.
24. See e.g. T. Åberg, lecture on " Shake theory of multiple photoexcita-
    tion ", this issue.
25. M.O. Krause, F. Wuilleumier, and C.W. Nestor Jr, Phys. Rev. A 6,
    871 (1972).
26. S. Flügge, W. Mehlhorn, and V. Schmidt, Phys. Rev. Lett. 29, 7
    (1972).
27. S.C. McFarlane, J. Phys. B 8, 895 (1975).
28. B. Cleff and W. Mehlhorn, J. Phys. B 7, 593 (1974).

CORRELATION EFFECTS IN X RAY EMISSION SPECTROSCOPY

Jean Pierre Briand

Université Pierre et Marie Curie and Institut du Radium

11, rue Pierre et Marie Curie, 75231 Paris Cedex 05, France

INTRODUCTION

X ray emission spectroscopy was, in the past, the unique method of study of atomic inner shells. In the last ten years, many new methods have been developed to study as well atomic inner shell properties as atomic collision mechanisms : x ray photoelectron spectroscopy (XPS), precision Auger electron spectroscopy (AES), energy loss spectrometry, heavy ions bombardment, charge spectroscopy ... The development of these new techniques, namely the photoelectron spectroscopy, led in 1965 to the direct observation, by Carlson and Krause (1) (2), of the multiexcitation and multiionization processes in photon absorbtion. At the same time the multiionization theory (shake theory) was developed by Carlson et al. (3) and by Åberg (4) (5) and there was then a renewal of interest for inner shell atomic properties. The x ray spectrometry received then a new impulse ; the old data about x ray satellites were reviewed in the scope of the shake theory and numerous new weak lines corresponding to new processes were discovered : Radiative Auger satellites (RAE) (6), hypersatellites (7), Radiative Electron Rearrangment satellites (RER) (8), heavy ions satellites (9), multiplet splitting " satellites " ... Most of these new results allowed a better understanding of multielectron interactions in atoms.

Because we are mostly interested in multiionization processes in atoms and by their manifestation in x ray spectrometry we shall first define what we mean by an x ray satellite. Numerous weak lines are generally observed in any x ray spectra, mostly on the high energy side of the diagram lines,

they were all called   satellite lines. In the past, they were soon interpreted as due to multiionized atoms but now we have to define them more accurately. These multiionized states can be attained either after an Auger or
a Coster-Kronig transition or directly in a collision process. We shall be
interested in these lectures only by these direct multiionization processes.

The first problem in this field is to have an accurate identification of
these satellite lines and to recognize what processes are responsible for
each of them. Such identification can be achieved either by doing coincidence experiments between x ray satellites and any direct proof of the multiionization process and by doing HF calculations. Very few direct identifica-
tions have been done nowadays. Three different types of experiments can
be considered (i) coincidence between an Auger or a Coster-Kronig electron
and the x ray satellites (can only be used for L or M ... shells), (ii) coincidence between a K hypersatellite and a K satellite, (iii) coincidence
between a photoelectron (or any charged ejected particle) and any
x ray satellite.

In fact most of the satellites are identified using HF calculations. Such
calculations fell in the past to well identify the x ray satellites but it is now
possible to have accurate results using HF or DF calculations like the
Desclaux's ones (J.P. Desclaux, this volume).

We shall define as a satellite the x ray emitted by an atom having an
initial and/or a final state composed by an inner hole (the x ray hole) plus
one (or more) other holes corresponding to an additional ionization or excitation. A few lines that we shall briefly discuss, are of different origins
and cannot be strictly   called   x ray satellites. Most of the observed satellite lines appear on the high energy side of their corresponding diagram
line but we can only observe those which have an energy shift that is higher
than the natural width of the parent line. It has been pointed out, either by
using HF calculations (10) or direct coincidence experiments (11) that most
of the observed satellites come from atoms doubly ionized in the same shell
(hypersatellites) or in two adjacent shells (KL satellite for instance).
All other satellites coming from doubly ionized atoms in two not adjacent
shells are hidden by diagram lines. In some cases of excitation, e.g. internal conversion, the diagram line is so contaminated by its satellite compa-
nions that the monoionized state line is quite negligible (11). During a K
ionization most of the shake processes arise in the outermost shells and
the corresponding satellite lines can never be observed in x ray spectra.

The x ray satellites can also be perturbed by the
molecular or solid state bonding. In Fig. 1 we present, for instance, the

splitting of the K_α (K → L) line of molecular nitrogen. The final hole of the transition is in the valence shell, and the molecular orbitals (and also some vibrational states)are then observed in this case. Such a splitting due to the environment of the atom can also be observed in K_β lines (1s → 3p) of the elements of the third row of Mendeleiff Table (In fact, this is mainly due to solid state effects in transition metals and the K_β line is generally only broadened. Another interesting case of chemical effect on x ray spectra is the so-called " cross transitions ". Let us consider a molecule composed by two different atoms. If an x ray hole is created in the K shell of one of them, it is possible that,in some cases, this x ray hole is filled by an electron coming from an outer shell of the other atom. Such a " cross transition " has a different energy from that of the diagram line and appears as an extra x ray " satellite ". Numerous cross transitions have been observed in the last years and accurately identified (13).

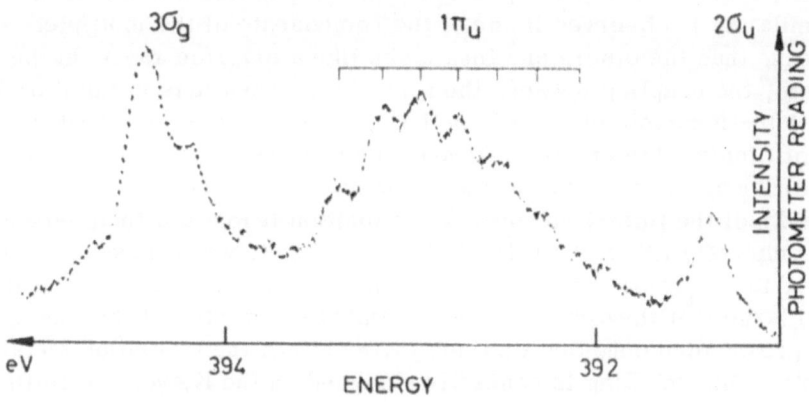

Fig. 1. Splitting of the K_α line of molecular nitrogen from Ref. (12).

If we exclude these previous examples the effect of the chemical bonding on the energy of the x ray satellite is generally negligible. That means, in fact, that the chemical shift is the same for both initial and final states. However, we should notice that the effect of chemical bonding is very important for the relative intensity of the x ray satellite. An example of the importance of this bonding will be given at the end of the second part of these lectures (Fig. 8).

We shall now study three examples showing how important correlation effects are in x ray spectra.

1. Multiplet splitting of x rays emitted by atoms having incompletely filled shells (In this case, the observed satellites do not fulfill the previous definition).

2. The shake theory and the Kα satellites.

3. The shake theory and the Kα hypersatellites.

1. MULTIPLET SPLITTING OF X RAYS EMITTED BY ATOMS HAVING INCOMPLETELY FILLED SHELLS

We shall study, in this section, the x rays emitted by atoms which have shells incompletely filled as the transition metals or rare earths, in which single ionization occurs. There is, in principle, a coupling between the x ray hole and the incompletely filled shell which can split the diagram ray in various components. There is then no diagram ray in this case but a more or less well resolved complex spectrum which looks like a satellite spectrum (This similarity is observed if one of the components of the multiplet is more intense than the others and then looks like a diagram ray). In the initial state, the coupling between the x ray hole, when it is in the K or L shell, with the incompletely filled shell (M or N) is generally weak (In the case of titanium the energy difference between the main components is of the order of magnitude of a few tenths of eV, i.e. smaller than the natural width of the lines). If the x ray transition leads to a final vacancy in the incompletely filled shell itself, the coupling between these two holes is expected to be strong, and the various final states to have energy differences so large that they can be higher than natural widths of the lines. If K x rays are studied only the Kβ lines (K $\rightarrow$ M) of the transition metals can be split (this splitting is generally designed as the K$\beta'\beta_{1,3}$ satellite spectrum). If L x rays are studied, most of the 2p $\rightarrow$ 3d transitions can be split (Lα_1, Lβ_1).

In most cases this multiplet splitting is expected to look as low energy satellite spectrum. This result can be explained in a simple way. Let us consider, for instance, the K$\beta'\beta_{1,3}$ group ; the initial ionization state is a 1s hole, the final state a 3p hole which is coupled to the 3d subshell . In such a transition the multiplet strengths are proportional to the multiplicity of the final state. It is a general rule that the terms of the highest multiplicity lies at the lowest energies. Therefore the K$\beta_{1,3}$ line has a characteristic low energy structure. This general consideration was the basis of the the explanation of the observed β' low energy satellite spectra, given in 1927 by Coster and Druyvesteyn (14).

Many calculations have been done of this multiplet structure which conclude to a possible observation of such satellite spectra. We present in

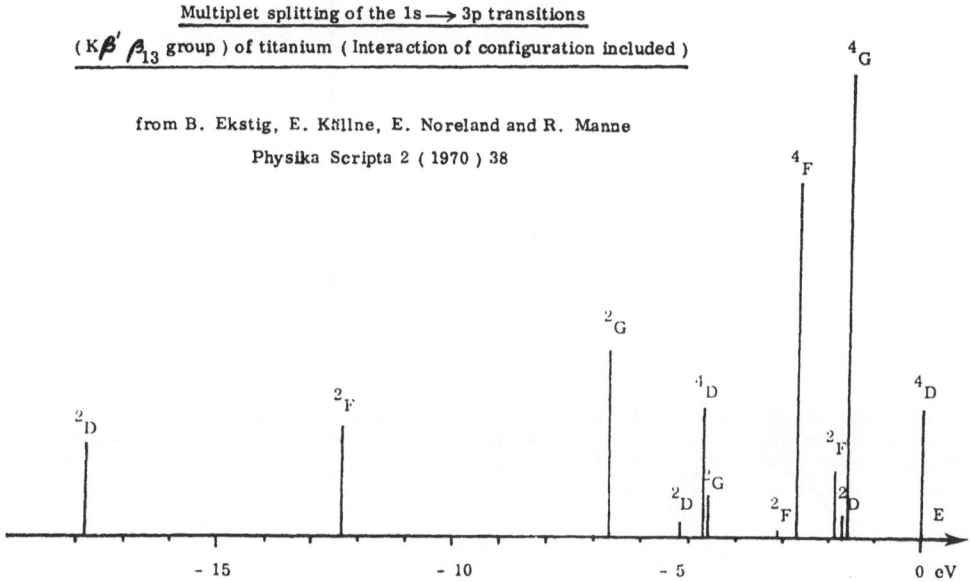

Fig. 2. Calculated multiplet splitting of Ti.

Fig. 2 the calculated splitting of the K$\beta'\beta_{1,3}$ group of titanium (B. Ekstig). It can be seen that most of the lines are concentrated in the same energy region (few eV) constituting the most intense structure of the spectrum which looks like a " diagram ray ", when few other ones have an important energy shift and look like a continuous spectrum. In general, the agreement between calculated structures and the observed spectra is qualitative . In any case, the β' structure of the K$\beta_{1,3}$ line is a continuum on the low energy side of the main line. Recently, a new K$\beta'\beta_{1,3}$ spectrum of titanium has been observed by Pessa et al.,   with a high resolution double flat crystal spectrometer. It is presented in Fig. 3 and compared to their own calculations. The observed fine structure seems then in a rather good accordance with c and d calculations (configuration interaction included).

In the case of the L spectra, such low energy structures have been observed and have been qualitatively explained as multiplet splitting resulting from hole interactions in final state.

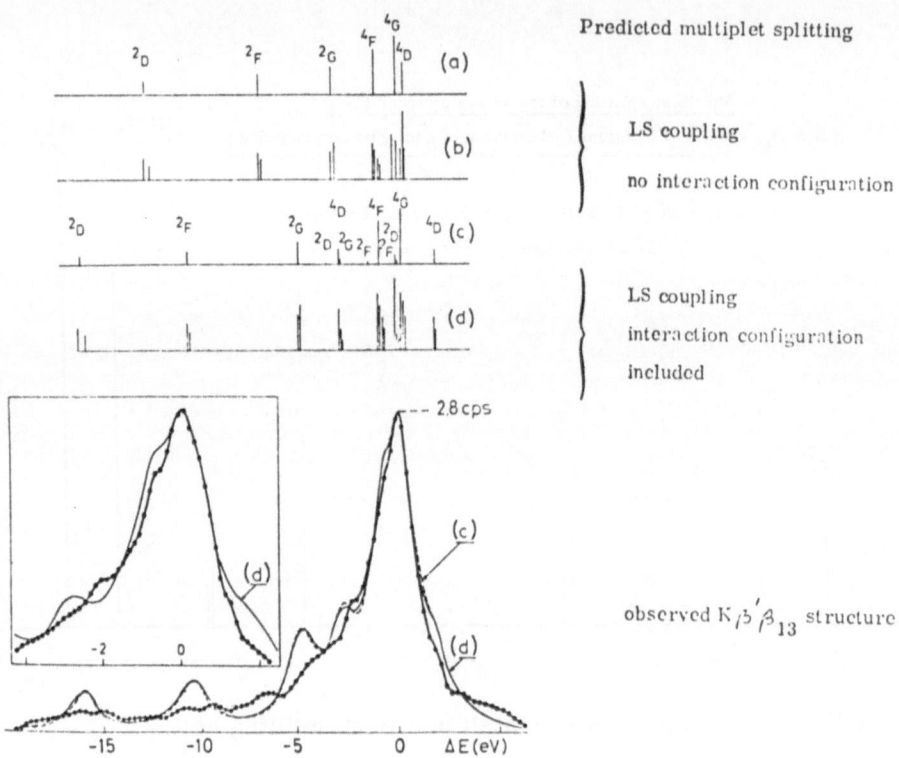

Fig.3. Comparison between experiment and predicted multiplet splitting of Ti. (from M. Pessa, E.K. Kortela, A. Suikkanen and E. Suoninen, Phys. Rev. A8 (1973) 48.

These " satellites " should then, in principle, give information about configuration interaction but as previously described the agreement with calculations is still qualitative and such results are today somewhat inconclusive.

2. THE SHAKE THEORY AND THE K$_\alpha$ SATELLITES

The first attempt of a theory of multiionization was given by nuclear physicists. In 1941 Migdal (15) and Feinberg (16) first introduced the sudden approximation to explain the atomic autoionization following β decay. Migdal demonstrated that the additional ionization in inner shells of the residual atom

which appears, with a small probability, during β decay, is not due to collision induced by the outgoing β electron in the atomic cloud, but to the
" sudden change " of the nuclear charge. The only requirement for such a
theory is that the β electron fastly leaves the atom, i.e. its velocity will
be much higher than the velocity of the considered electron ; this condition
is generally fulfilled in nuclear β decay. The atomic electrons feel a
sudden change of the nuclear charge and have to rearrange to be in a stationary state of the new atom. The " autoionization " in an (n, ℓ) shell is
then caused by an imperfect rearrangement of this (n, ℓ) electronic state.
In the language of nuclear physicists, atomic rearrangement means that
the atomic orbitals of a Z atom becomes that of a (Z+1) atom. The autoionization probability can be calculated in a simple way (which is not the
original Migdal's approach) : the probability that a (n, ℓ) electron
of the initial atom remains in the same shell of the final atom is equal to
the square of the integral overlap of the initial and final individual wave-
functions of the (n, ℓ) electron in the Z and Z+1 atom : $|\langle \psi_f^{n,\ell}(z+1)| \psi_i^{n,\ell}(z)\rangle|^2$
The probability of autoionization i.e. the probability that the electron does
not remain in the same shell as in the initial atom is then simply equal to
$1 - |\langle \psi_f^{n,\ell}(z+1)| \psi_i^{n,\ell}(z)\rangle|^2$; $|\langle \psi_f / \psi_i\rangle|$ is then the probability that the initial wavefunction will be an eigenfucntion of the Hamiltonian of the final atom. Such
a calculation takes directly into account as well shake-off process (ejection
of the " shaked " electron in the continuum) as shake-up process (promotion of the shaked electron to an unoccupied discrete state of the final atom).
The previous formula has then to be corrected when outer states are occupied. The transition which occurs, when a (n, ℓ) electron of the initial
atom becomes the same (n, ℓ) electron of the final atom (rearrangement),
an electron of the continuum (shake-off process) or a (n', ℓ') electron
of the final atom (shake-up process) is a monopolar transition and follows
the $l_i = l_f$ and $m_i = m_f$ selection rules. The wavefunction of a single electron
state in an inner shell of an atom Z, being very close to that of the neighbouring Z+1 atom, the shake-off probability is very small (10^{-3} to 10^{-4})
for medium and heavy mass atoms. If outermost shells are considered the
wavefunctions for Z and Z+1 atoms are very different ; the shake process
has then a very high probability (typically 10%). The use of this simple
model leads in general to a good accord for β decay, between experimental
results and theoretical predictions (see for instance, the Freedman's lecture). When the outgoing particle has a low velocity compared to that of
the atomic electrons, the "correct" rearrangement becomes more probable
(adiabatic transition) and the autoionization probability falls rapidly to
zero (In the case of α decay, where the α particle goes out very
slowly, the " autoionization " probability is two orders of magnitude less
than in the case of β decay). This " shake model " in nuclear decay is
the basis of the present theory of atomic multiionization processes.

In 1966, Sachenko and Demekhin (17) considered that the fast' removal
of a K electron is, in fact, similar, for the outer electrons, to a sudden change
of the effective (screened) nuclear charge. They used this " autoionization "
theory to explain, satisfactorily as we shall see later, the satellite appearance during K ionization. In 1968, Carlson et al (3a) using HF wavefunctions calculated systematically such shake-off + shake-up probabilities for every atom in the case of β decay. As a first approximation
the removal of an inner electron is then considered as equivalent to a change
of the nuclear charge by one unity. This " $\Delta Z=1$ " approximation seems
then to be valid essentially when the " shaked " electron is in the outest
shells and the " initial " hole in the innest shell. In the case where both electrons are in the same shell, it was proposed that a $(\Delta Z)^2$ correction
should be used taking into account the relative ΔZ screening of the two
electrons. In 1973, these authors did more refined calculations taking properly into account the origin of sudden change of the central charge (3b). A
comparison of the total shake probability for K and L shell in both cases :
$\Delta Z=1$ sudden change of nuclear charge and K initial ionization (in this
case, due to the screening of the K electron, the sudden change of charge is
less than the unity) is presented in table 1 (similar calculations were done
for few elements by Åberg in 1967 (4) and 1969 (5).

Removal of a K electron				'Sudden $\Delta Z=1$ change of nuclear charge'			
Carlson (1973)				Carlson (1968)			
	K	L_I	L_{II}	L_{III}	K	L_I	L_{II}	L_{III}
Ne	0.03	1.71	5.44	10.63	Ne 1.03	4.77	5.55	11.10
Ar	0.001	0.3	0.54	1.07	Ar 0.264	0.914	0.703	1.41
Kr	0.002	0.06	0.089	0.18	Kr 0.07	0.185	0.124	0.230
Xe	0.001	0.023	0.032	0.061	Xe 0.034	0.0791	0.0494	0.0819

Table 1. K and L shake-off + shake-up probability (in%) from ref. (3a, b).

The comparison between both calculations well illustrate the model which
is used : when a K electron is removed, the change of effective central
charge is less than in the case of $\Delta Z=1$ change of charge in nuclear decay,
because the screening of both K electrons, and all shake processes have
a smaller probability : the final wavefunction is, in this case, closer to that
of the initial one. An extreme illustration of this model is the ionization in
electron capture decay. In this case the change of central charge is near
zero because the K electron is not ejected but goes inside the nucleus and
the probability of a shake process becomes then negligible for outermost
shells (18).

To conclude this brief description of the theory, we must notice that this shake model implies (i) a single particle model (correlation effects not included) (ii) that the probability of a shake process does not depend on the nature of the initial ionization (particle bombardment or photoionization ...). We shall now discuss the origin of the K_α satellite lines according to this shake theory.

Most of the K_α satellites appear on the high energy side of the diagram lines. They are well known only in the case of light or medium mass atoms ($Z \leq 42$). Figure 4 shows, for instance, the K_α satellite spectrum of neon, observed by Keski-Rahkonen (19). All the K_α satellite spectra observed have a similar structure. The satellite lines are generally designated by by K_α'', K_α', $K_{\alpha 3}$, $K_{\alpha'3,4}$, $K_{\alpha 4}$, $K_{\alpha 5}$, $K_{\alpha 6}$ following an approximate increasing energy law. K_α is generally a small bump on the $K_{\alpha 1}$ diagram line tail ; the K_α'' and $K_{\alpha'3,4}$ have only been observed in some cases and their exact positions are uncertain. The $K_{\alpha 3}$-$K_{\alpha 4}$ and $K_{\alpha 5}$-$K_{\alpha 6}$ satellites

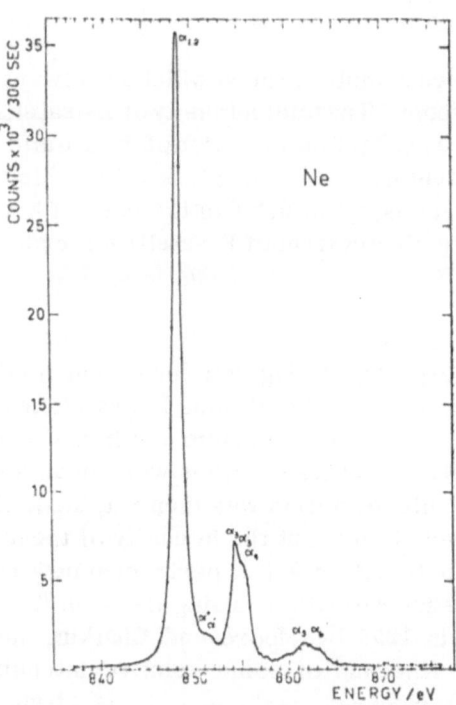

Fig.4. The K neon spectrum (19).

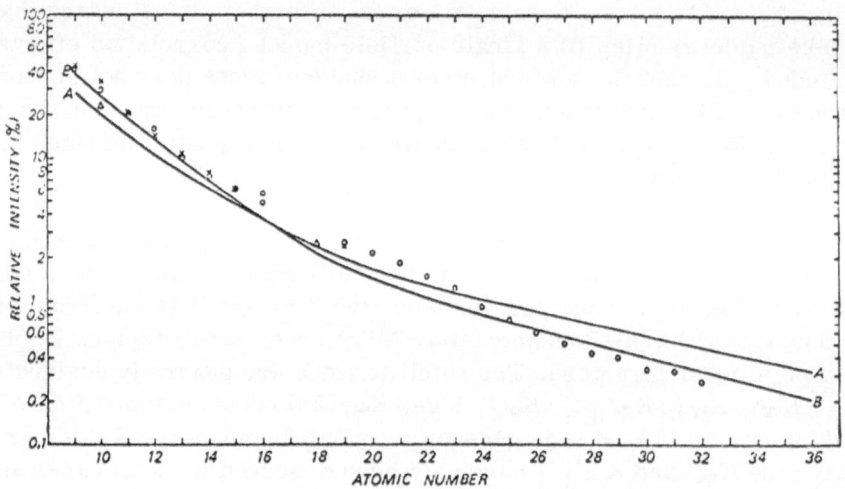

Fig. 5. The $K^{-1}L^{-1}$ ($\alpha'\ \alpha_3\ \alpha_4$) satellite group intensity relative to $K_{\alpha 1}$, according to shake-off theory, (Curve A : shake probabilities, Curve B : corresponding satellite intensity. T. Åberg, invited talk at Atlanta Conf.).

form two well defined complex groups which are always well separated in high resolution spectroscopy. The total intensity of K_α satellites, relative to the diagram line, is plotted in Fig. 5 as a function of the atomic number Z. A rapid decrease of this intensity is evident ; this value, which is of the order of 26% for neon, becomes equal to 0.3% in the case of krypton (Z=36). In fact, there is no accurate observation of K satellite for atoms having a Z value higher than 42.

The first correct identification of the K satellites was given in 1934 by Kennard and Ramberg (20). Using self consistent field methods they calcu-lated exactly the energies of the various K_α satellites which should appear in $K^{-1}L^{-1} \longrightarrow L^{-2}$ transition in sodium and in approximate way for various other atoms. The K_α satellite energies were in agreement with their calcu-lated values, their interpretation was then retained. However, these authors did not give any prediction about the intensity of these lines with respect to the diagram one. In 1960, Horak (21) performed new refined calculations leading to the same identification ; using the neon K_α complete satellite spectrum obtained in 1955 by Moore and Chalkin, he gave an explanation of every observed line and of some relative intensities. The first group of satellites : α', α_3, α_4, α'', $\alpha'_{3,4}$ is attributed to $K^{-1}L^{-1} \longrightarrow$ L^{-2} transitions, the second one α_5, α_6 being due to $K^{-1}L^{-2} \longrightarrow L^{-3}$ radiative decay. Some other groups (α_7, α_8...) have been mainly observed in heavy ion collisions and have been attributed to transitions arising from

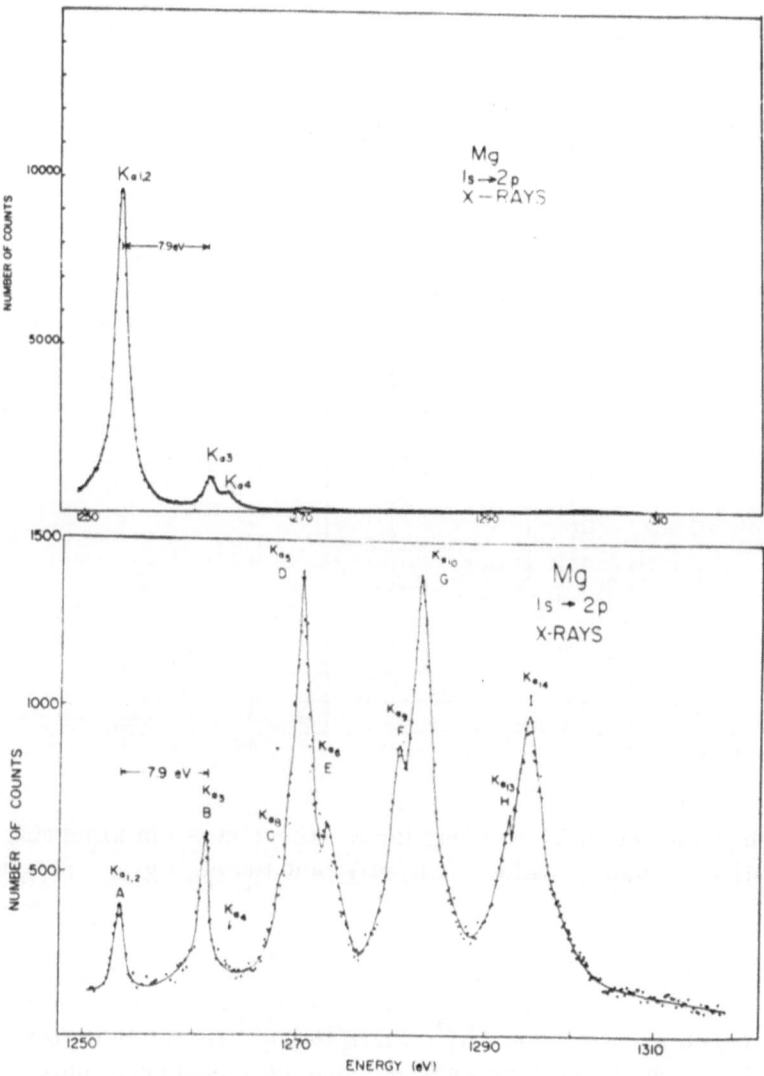

Fig. 6. MgKα spectrum produced by 2 MeV protons (upper half of Figure) and by 21 MeV oxygen (lower half of figure) bombardment. (B : initial state $K^{-1}L^{-2}$; C.D.E. : $K^{-1}L^{-3}$; F.G. : $K^{-1}L^{-4}$; H.I. : $K^{-1}L^{-5}$). (from P. Richard, Proceedings of Atlanta Conf. p.1641 (1972)).

initial states having more than 2 L holes (in heavy ion bombardment the probability of having numerous K and L ionization is very high and these lines are easily observed as shown in Fig. 6.

The first group of satellites (originating from $K^{-1}L^{-1}$ states) may

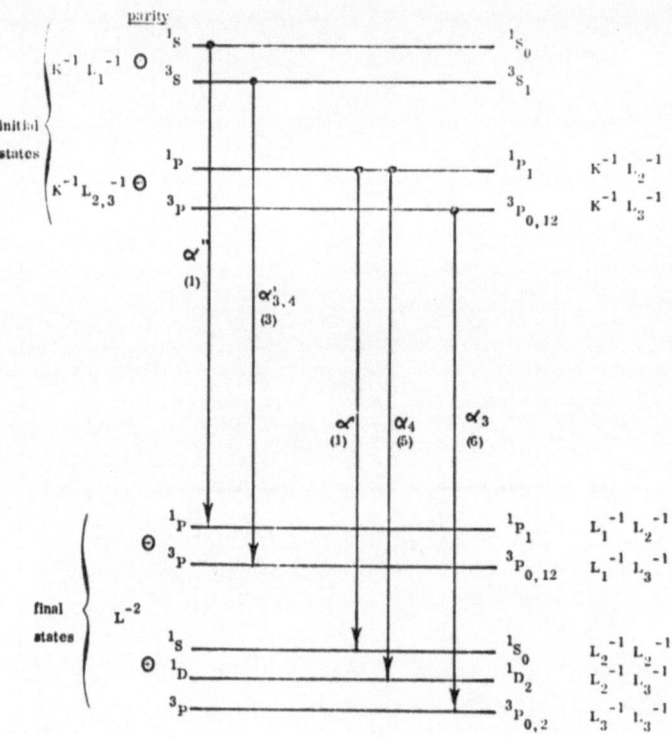

Fig. 7. Decay scheme in LS coupling for $K^{-1}L^{-1}$ states (in brackets, predicted relative intensities within the $K^{-1}L_{\bar{1}}^{1}$ and the $K^{-1}L_{2,3}^{-1}$ groups).

be divided in two parts : the $K^{-1}L_{\bar{1}}^{-1} \longrightarrow L_{\bar{1}}^{1}L_{2,3}^{-1}$ transitions , the α'', α'_3, α'_4 lines, which are very weak and have been observed only in few cases, and the $K^{-1}L_{2,3}^{-1} \longrightarrow L_{2,3}^{-2}$ transitions which correspond to the α', α_3 and α_4 lines ; these lines have been observed for all atoms below $Z = 42$ and are generally well separated from the diagram line. They correspond to all lines predicted in LS coupling decay scheme, as shown in Fig. 7. The relative strengths of α' and α_4 lines which originate from the same ^1P initial level , can be predicted by the general rules of multiplet strength : $\alpha_4/\alpha' = 5$. Such intensity ratio is observed for light atoms.

    It is now possible to compare the predictions of the shake theory and the relative intensity of $K\alpha$ satellite. Either relative intensity of various lines or relative intensity of total satellite spectrum with respect to diagram one versus Z are in good agreement with the predictions of shake theory as shown in Fig. 5 and tables 2 and 3. One of the main features of the shake theory : the independence of the mode of the "primary" process of ionization has been also experimentally confirmed in some experiments of Graeffe (22) who observed the same relative intensity of K_α satellite as well in photoionization as in electron bombardment. However, some other experiments (23) seem to lead to the opposite conclusion.

In conclusion we can say that, when a double ionization occurs in two different shells (the K and L shells in our example), the conventional shake theory i.e., shake theory using pure single particle model, seems to work quite well.

However, some notable disagreement seems to appear in two cases, between the K_α satellite spectra observed and the predictions of the shake theory. First, the relative intensity of $K^{-1}L^{-2}$ satellite group observed (see Table 4) is generally higher than predicted. This effect is probably due to the neglect of correlation between the two L electrons. This effect increases the ionization probability, as we shall discuss it later. Unfortunately, there is no complete calculation today about the probability of this ionization state. Second, in Auger electron spectrometry some satellites appear corresponding to a $^3P/^1P$ ratio ($KL_{2,3}$) inversed with respect to that observed in K_α x ray satellite spectra. In fact, the relative intensity ratio of K x ray satellite for light elements is well known to be very sensitive to chemical bonding. Such chemical effect is well illustrated in Fig. 8 where $K_{\alpha 3}/K_{\alpha 4}$ intensity ratio appeared to be very different in pure Al and Al_2O_3 targets.

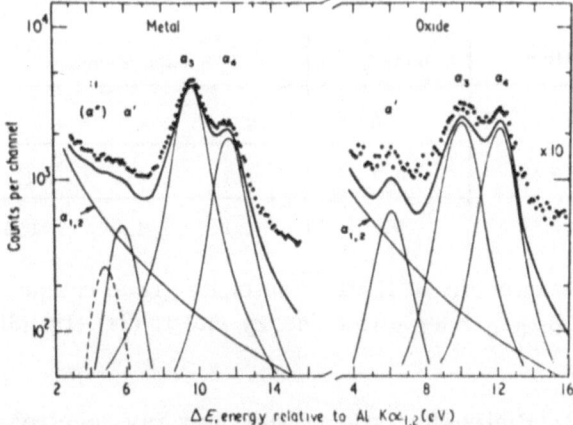

Fig. 8. $K_{\alpha 3}/K_{\alpha 4}$ x ray satellite intensity ratio emitted by metallic alluminium target or oxided aluminium target (24).

	α''	$\alpha'_{3,4}$	α'	α_3	α_4	
Theory	1	3	1	9	5	
10 Ne	1	3.1	1	8.9	4.9	(19)
12 Mg	0.3	-	1	9.1	5.1	(24)
13 Al	0.3	-	0.7	7.3	3.1	(24)

Table 2. Relative intensity of (α' α_3 α_4 α'' $\alpha'_{3,4}$) satellite group with respect to $K\alpha_{1,2}$ diagram line. Comparison for some light elements between theory (13) and recent experiments. (19) (24)

	Theory	Various experiments		
10 Ne	26.7	30	33	34
12 Mg	13.6	15.5	13.6	16.1
13 Al	10.5	11.7	10.7	10.1

Table 3. Relative intensities of $K\alpha$ satellites. Some comparisons between theory (13) and recent experiments for light elements. (24) (19) (30)

Atom	Theory	Experiment		
10 Ne	3.5	9.2	8.3	3.9
12 Mg	1.1	1.7	1.8	1.1
13 Al	0.7	1.1	1.0	0.9

Table 4. Relative intensity of $K^{-1}L^{-2}$ satellite group compared to the shake-off predictions ($K\alpha_{5,6}/K\alpha_{1,2}$). (Theory (13)). (24) (19) (30)

However, the $^3P/^1P$ intensity ratio as seen in x ray spectroscopy, despite these chemical effects, seems to be equal to the statistical ratio for every light elements and is then in contradiction with that observed in Auger spectroscopy. This important discrepancy which is not explained today is discussed in the Mehlhorn's lecture.

3. THE SHAKE THEORY AND THE K$_\alpha$HYPERSATELLITE SPECTRA

In the previous section we have seen that conventional shake-off theory i.e. independent particle model, led to a rather good agreement with experimental results. The gross features of observed spectra as well as the order of magnitude of the multiionization probabilities by photon can be explained by such a simple theory. In fact, some discrepancies between conventional shake theory and experimental results still remain, namely to explain the double ionization probability in the same shell (1). This last effect is observable in x ray spectrometry studying hypersatellites under photon excitation, for instance. The shake theory cannot predict what happens near threshold but such experiments need continuously variable excitation light and are then relevant to the future experiments around the electron collision rings. We shall then be only concerned in this section by double ionization processes in the same shell, where correlation effects are of prime importance. This importance of correlation effects was first demonstrated by Carlson (1) when he discovered the very high probability value of double ionization of neon in the 2p shell. This value was much higher than it was predicted by the conventional shake-off theory which seems to apply only when the double ionization occurs in two different shells. Such similar results have been also obtained more recently measuring double K ionization probabilities (19) during internal conversion, by x ray spectroscopy. In 1967, Åberg (5) proposed a general theory of multiionization taking correlation effects into account. In this theory the general correct formalism for double photoionization is shown to give, as a limiting case, the conventional shake theory under few assumptions. This theory is described in the Åberg's lectures but we can briefly summarize the most characteristic predictions of this theory. Using the most general wavefunction for multielectron systems (i.e. without any separation of variables), he wrote the time dependent Schrödinger equation for a system composed by an atom ionized in an (n, ℓ) shell and an outgoing electron as :

$$i\hbar\frac{\partial\Psi}{\partial t} = \hat{H}(N-1)\Psi + \hat{K}(1)\Psi \tag{1}$$

The general solution of this time dependent Schrödinger equation is a linear combination of products of eigenfunctions of $\hat{H}$ and $\hat{K}$ operators with energy dependent coefficients of which squared value is the probability of finding the system in a particular state. If it is assumed (sudden approximation) that the Hamiltonian of the initial atom (neutral atom) $\hat{H}$ (N) changes

suddenly to the Hamiltonian $\widehat{H}$ (N-1) + $\widehat{K}$ (1) of the final system,
the ground state wavefunction has to be equal to the linear combina-
tion of eigenfunctions, general solution of equation (1), (that is the
continuity condition). Then, it is possible using this equality,to find
the general expression of the probability of a given transition which
is equal to the square of the energy dependent coefficient of the expansion
of the general solution of equation (1). This result leads to a formula :
the generalized shake-off + shake-up probability in which the most complete wavefunctions including correlations can be used. If the used
wavefunctions are linear combinations of Slater determinants, the general
theory leads to the previous conventional shake-off theory which then gives
an understanding of the used approximation.

We shall now discuss our present experimental knowledge about double
ionization in a same shell as seen in x ray emission spectroscopy. The
first observation of x ray hypersatellites(x ray transition emitted by an
atom doubly ionized in the same shell) was presented for L shell (25) in
1970. This observation was obtained in a coincidence experiment between
a KLL Auger line and the L x ray spectrum. The K x ray hypersatellite
was observed for the first time (26) in 1971 for gallium ; the origin of the
double K hole was in this case a direct " K shake process " during a K
electron capture   nuclear decay.   The intensity of hypersatellites is
generally very weak,with respect to diagram lines,when conventional excitation is used (10^{-4} or 10^{-5}) and about two orders of magnitude less than
the conventional satellites we previously discussed. Their energy shift with
respect to the corresponding diagram line is so important (about 10 times
higher than those of conventional satellites) that they can be easily separated from other lines. Numerous x ray hypersatellites are now well known
for various atoms. We present in Fig. 9 the decay scheme of K^{-2} states
in intermediate coupling. For light elements,the singlet to triplet
($^1S_0 \longrightarrow {}^3P_1$) transition is forbidden in the LS coupling scheme and the
K^{-2} state mainly decays via a $K\alpha_2^h$ hypersatellite transition. For heavier
elements the $K\alpha_1^h$ becomes allowed in intermediate coupling and its intensity relatively to $K\alpha_2^h$ grows to be approximately that of the diagram lines for
the heaviest elements (i.e. near twice more intense than the $K\alpha_2^h$ line).
This decay scheme in intermediate coupling seems to be now well known
as a function of Z as well theoretically (T. Åberg to be published) as
experimentally in a large range (Z=28 to Z=82) (J.P. Briand et al. (7)
to be published). In Fig. 10 we present for instance the $K\alpha_1^h$ line of Pb
(Z=82) observed in a coincidence experiment between K L satellite and
K hypersatellites. These hypersatellites are now well identified as well
in coincidence experiments as   in   Desclaux's D.F. calculations, which
fit very well with experimental results.

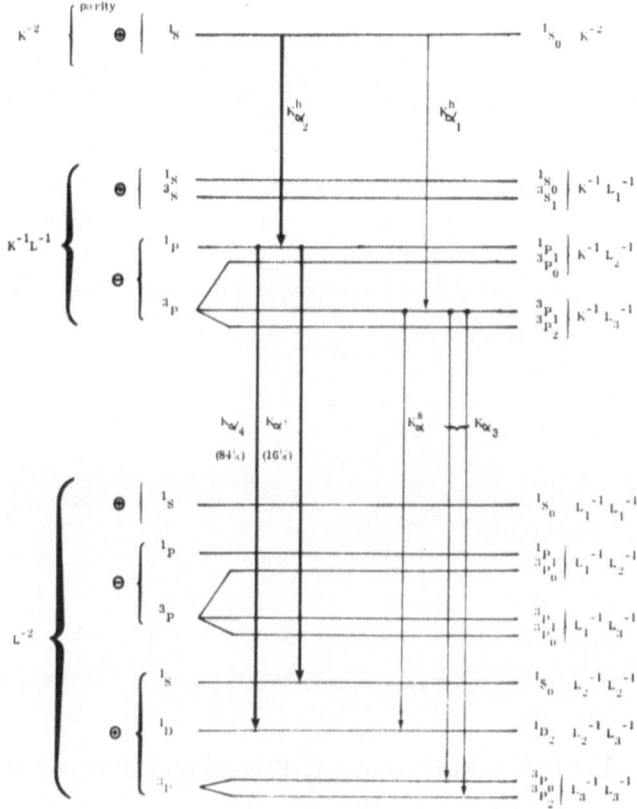

Fig. 9. K^{-2} radiative decay scheme in intermediate coupling (strong lines : allowed lines in pure LS coupling).

We shall now talk about the probability of double K ionization comparing theoretical predictions, including correlation, to experimental data obtained by measuring hypersatellite line intensities. Double K ionization can occur in various processes : photoionization, internal conversion, electron capture, electron bombardment. The generalized shake model may apply to predict probability values, of these double ionization processes except in the case of electron impact.

It is expected that the double K ionization probability in electron capture is different from that in internal conversion or photoionization. In order to compare these probability values in various cases, we shall define a reference level for the probability of double K ionization which will be

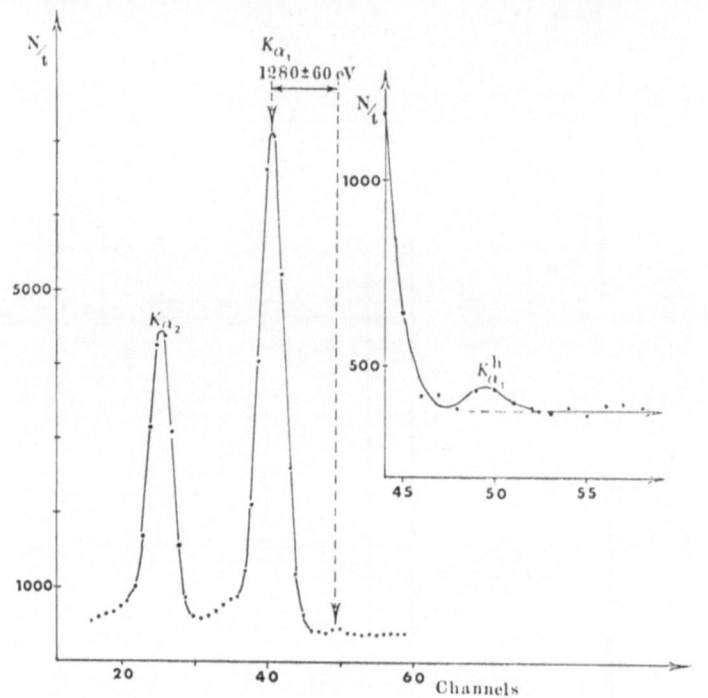

Fig. 10. $K_{\alpha 1}^h$ line of Pb observed in a K-K coïncidence experiment (27).

the conventional shake probability of single K ionization following an allowed β decay (the "Carlson level"). Let us first consider a photoionization process (or the internal conversion process involving virtual photon). The shake process in the K shell (in the independent particle model) has in this case a probability value which is less than the "Carlson level" discussed previously in section 2, because the change of central charge is less than unity (in this case) (Fig. 11). If we now consider the effect of correlation (Coulomb correlation effects), the probability of this effect is increased because both K electrons which have an opposite spin cannot be at the same place in real space as it is allowed in single particle model. The two K electrons are effectively more separated when correlation effects are taken into account. The overlap of the initial and final ground state wavefunctions is then less than in the conventional shake probability and therefore the generalized shake probability is higher. In the case of electron capture for instance, the change of central charge has an opposite sign, but an higher absolute value, nearer $\Delta Z = 1$ than in the previous case, and the decrease of the shake probability is less than in the case of internal conversion.

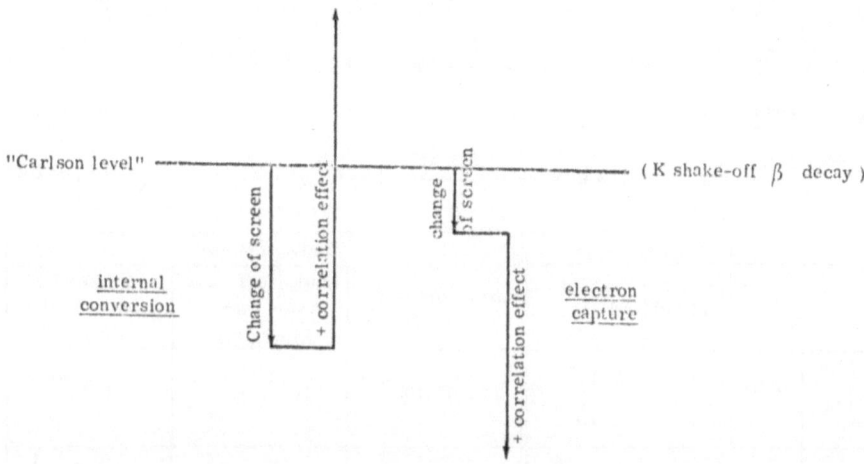

Fig. 11. Comparison of double K ionization probabilities in various processes and under various assumptions.

conversion level. However, the correlation effects decrease this probability of double K ionization in this case, (which is opposite to the previous
case) and then decrease further the " Carlson $\Delta Z=1$ " probability value.

Except in the case of helium which is discussed in the Schmidt's seminar,
very few results are presently known for double K ionization probability in photoionization or internal conversion processes. We present in the first column
of table 5 all the presently known values of P_{KK} probability in the case of
internal conversion, and in the second column, the conventional shake value
probability for double K ionization in photoionization or internal conversion
processes. These values are interpolated taking into account both Carlson's
paper (3a) and (3b) about β decay shake probability and K ionization
shake probability (shake-up + shake-off probability). In the third column
we present the shake-off + shake-up probability for double K ionization
which was obtained by Åberg (4) with correlated wavefunctions in the
generalized shake theory. This probability value is not corrected for relativistic effects.

The probability value corrected for shake-up processes, which are forbidden for real atoms of high Z because all electronic states are occupied, is
presented in the fourth column. As it can be seen from this table the correlation effects increase strongly the probability values and the experimental

Nuclear transitions used

Z = 49 ^{113}In (1/2 -) 394 keV $\xrightarrow{\text{M4}}$ (9/2 +) G.S. B_K = 27,94 keV

Z = 49 ^{114}In (5 +) 191 keV $\xrightarrow{\text{E4}}$ (1 +) G.S.   "   "

Z = 56 ^{137}Ba (11/2 -) 661 keV $\xrightarrow{\text{M4}}$ (3/2 +) G.S. B_K = 37,44 keV

Z = 81 ^{203}Tl (3/2 +) 279 keV $\xrightarrow{\text{E2 + M1}}$ (1/2 +) G.S. B_K = 85,531 keV

P_{KK}/P_K		experimental	calculations		
		Briand and al. (28)	corr. not incl. Carlson (3a, b)	shake-off + up corr. incl. Åberg (5)	shake-off corr. incl. Åberg *
In ^{113} In ^{114}	Z=49	i $\pm 0.3.10^{-4}$ 1.1 $\pm 0.2.10^{-4}$	$0.15.10^{-4}$	10^{-4}	$0.4.10^{-4}$
Ba	Z=56	$0.7 \pm 0.3.10^{-4}$	$0.12.10^{-4}$	$0.79.10^{-4}$	$0.31.10^{-4}$
Tl	Z=81	$0.4 \pm 0.1.10^{-4}$	$0.07.10^{-4}$	$0.37.10^{-4}$	$0.15.10^{-4}$

Table 4. Double K ionization probability. Comparison between experimental results and various theoretical predictions. (* Åberg private communication). The Tl experimental value has also been measured by Desclaux et al (29). We only present accurate data obtained in hypersatellite measurements.

values are in better agreement with theoretical values of Åberg than with those obtained in conventional shake theory. The experimental values are still twice higher than the theoretical one. When atoms of high Z value are considered, we should have to take into account magnetic effects between electrons. It was pointed out by Desclaux that the Breit interaction tends to be important for elements higher than Z=30. This Breit interaction and the spin-spin interaction particularly has in fact the same effect as the one of coulombic correlations : repulsive effect at intermediate distances. This " Breit correlation effect " could then increase the probability value.

In conclusion we can say that these correlation effects in atoms, which are of prime importance when a double excitation occurs in a same shell, can be studied in x ray spectrometry when innermost shells are considered and that they are very sensitive to the probability values. In the case of outermost shells, such effects are much more easily studied in electron spectroscopy.

ACKNOWLEDGEMENTS :

I want to thank Prof. T. Åberg for helpful discussions about the shake model and for his very kind and efficient contribution during writing of this manuscript.

REFERENCES

(1) T.A. Carlson, Phys. Rev. 156, 142 (1967).

(2) M.O. Krause, T.A. Carlson and R.D. Dismukes, Phys. Rev. 170, 37 (1968).

(3a) T.A. Carlson, C.W. Nestor and J.C. Tucker, Phys. Rev. 169, 168 (1968).

(3b) T.A. Carlson and C.W. Nestor Jr., Phys. Rev. A8, 2887 (1973).

(4) T.Åberg, Phys. Rev. 156, 35 (1967).

(5) T. Aberg, Ann. Acad. Scient. Fenn. 308, 1 (1969).

(6) T. Aberg, Atomic Innershell Processes, B. Crasemann ed. vol. 1 Academic Press, New York (1975).

(7) J.P. Briand, A. Touati, M. Frilley, P. Chevallier, A. Johnson, J.P. Rozet, M. Tavernier, S. Shafroth and M.O. Krause, J. Phys. B, (1976) to be published.

(8) K.A. Jamison, J.M. Hall and P. Richard, J. Phys. B8 (1975) to be published.

(9) P. Richard, Atomic Innershell Processes, B. Crasemann ed. vol. 1 Academic Press, New York (1975).

(10) M.O. Krause, F. Wuilleumier and C.W. Nestor, Phys. Rev. A6, 871 (1972).

(11) J.P. Briand et al., to be published (described in oral talk).

(12) L.O. Werme, B. Grennberg, J. Nordgren, C. Nordling and K. Siegbahn, Phys. Rev. Lett. 30, 523 (1973).

(13) T. Åberg, International Symposium : X ray spectra and electronic structure in matter, ed. Faessler and Wiech, München 1972, p.1.

(14) D. Coster and M.J. Druyvesteyn, Z. Phys. 40, 765 (1927).

(15) A. Migdal, J. Phys. USSR 4, 449 (1941).

(16) E. Feinberg, J. Phys. USSR 4, 423 (1941).

(17) V.P. Sachenko and V.F. Demekhin, J.E.T.P. 22, 532 (1966).

(18) H.W. Schnopper, Phys. Rev. 154, 118 (1967).

(19) O. Keski-Rahkonen, Physica Scripta 7, 173 (1973).

(20) H.E. Kennard and E. Ramberg, Phys. Rev. 46, 1040 (1934).

(21) Z. Horak, Proc. Phys. Soc. A77, 980 (1961).

(22) G. Graeffe, J. Sivola, J. Utriainen, M. Linkoaho and T. Åberg,

Phys. Lett. 29A, 464 (1969).

(23) N. Cue and W. Scholz, Phys. Rev. Lett. 32 , 1397 (1974).

(24) M.O. Krause and J.G. Ferreira, J. Phys. B8, 2007 (1975).

(25) J.P. Briand, P. Chevallier, M. Tavernier, J. de Phys. Paris 32
    C4, 165 (1971).

(26) J.P. Briand, P. Chevallier, M. Tavernier, J.P. Rozet, Phys. Rev.
    Lett. 27, 777 (1971).

(27) J.P. Briand, P. Chevallier, A. Johnson, J.P. Rozet, M. Tavernier
    and A. Touati, Phys. Lett. 49A , 51 (1974).

(28) J.P. Briand, P. Chevallier, A. Johnson, J.P. Rozet, M. Tavernier
    and A. Touati, Phys. Fenn. 9 S1, 409 (1974).

(29) J.P. Desclaux, C. Briançon, J.P. Thibaud and R.J. Wallen, Phys.
    Rev. Lett. 32, 447 (1974).

(30) J. Utriainen, M. Linkoaho, E. Rantavuori, T. Åberg and G. Graeffe,
    Z. Naturf. 23a, 1178 (1968).

CORRELATION EFFECTS IN AUGER ELECTRON SPECTROSCOPY

W. Mehlhorn

Fakultät für Physik, Universität Freiburg

78 Freiburg, Germany

1. INTRODUCTION

Electron correlation effects manifest themselves in two ways in Auger spectroscopy.
a) Correlation effects occuring during the Auger transition and in the final state. These can be studied by way of the intensities of diagram Auger transitions or the occurence of many particle Auger transitions (double Auger transitions, radiative Auger transitions). For example, deviations between experimental and theoretical values of diagram Auger line intensities, where the latter have been calculated with HF independent particle wave functions, are due to correlation effects. Many particle transitions, where more than 2 electrons participate in a transition, occur either through electron core rearrangement or through electron correlation. In an independent particle frozen-core approximation both are strictly forbidden.
b) Correlation effects in the primary excitation process by photons or particles. This leads to single particle excitation or ionization cross sections in disagreement with independent particle model values and to multiple ionization and/or excitations. These processes can be studied by the satellite Auger transitions.

2. CORRELATION EFFECTS IN DIAGRAM AUGER SPECTRA [1]

Diagram (or normal) Auger lines result from transitions in atoms having a single initial vacancy and two final vacancies in the ground state configuration.

2.1. Influence of Solid or Chemical State

In comparing experimental quantities (energies, intensities and widths of lines) with theoretical values we must remember that most of the experiments on diagram spectra have been made on atoms in solids, except for the noble gases and some metal vapors (Na, Mg, Ca, Zn, Cd, Hg). On the other hand, theoretical line intensities have been calculated only for free atoms assuming closed shells. The influence of the chemical and solid state on intensities and line widths is negligible as long as Auger transitions involving only inner shells are investigated. The only effect is that the lines are shifted by a constant amount of energy relative to those of the free atom. Multiplet splitting caused by an incomplete outermost shell is in general small for inner shell transitions and manifests itself through line broadening. In this case the transition probability is unaffected by the passive open shells [2]. The situation is very different if electrons from the outermost shell or the next inner shell play an active role in the transition. Figure 1 shows the strong shift of KLL lines of Mg in going from Mg metal [3] to MgO [3] and Mg vapor [4]. Within 0.1 eV each line is shifted by the same amount:

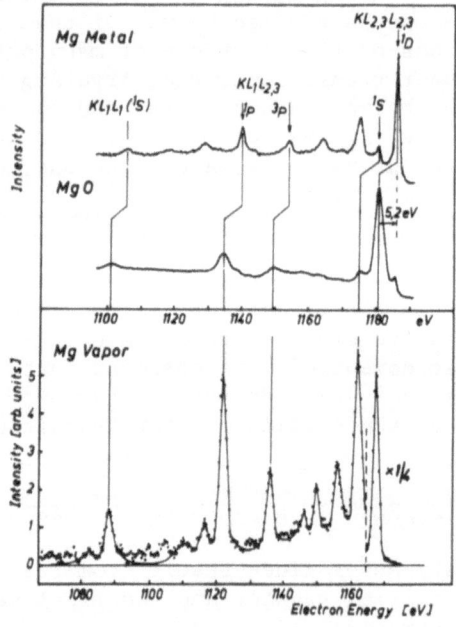

Fig. 1. KLL Auger spectra of Mg metal [3], MgO [3] and Mg vapor [4] showing the large energy shift of Auger lines. The lines of MgO are considerably broader than those of Mg metal or Mg vapor.

$$E(KLL, metal) = E(KLL, atom) + 18.1 \text{ eV}, \tag{1}$$
$$E(KLL, MgO) = E(KLL, atom) + 13.1 \text{ eV}. \tag{2}$$

Ley et al. [5] calculated the shift in Eq. (1) to be 18.9 eV. The change of line intensities between MgO and Mg vapor can be as much as 50 % (e.g. $KL_{2,3}L_{2,2}(^1D_2)$) [4,6]. The same has been found for the KLL lines of atomic Na [7] and Na_2O [6]. No intensities have been reported for solid Na or Mg.

2.2. Nonisotropic Angular Distribution of Auger Electrons

Almost all experiments with solid state targets used the radioactive decay (γ-conversion, electron capture) as excitation source, under these conditions Auger emission is isotropic. For the excitation of gaseous targets particle or photon beams are used and the problem of nonisotropic emission of Auger electrons relative to the primary beam direction arises. Let us consider the process

$$A + e^- \rightarrow A^+(SLJ) + 2e^- \tag{3}$$
$$\hookrightarrow A^{++}(S'L'J') + e^-_{Auger}(\ell_i j_i).$$

For electron ionization the angular distribution of Auger electrons was shown to be [8a]

$$I(\theta,E_o) = I_o(E_o) \{1 + \sum_{n=1}^{L} A_{2n}(E_o)P_{2n}(\cos\theta)\} \,, \tag{4}$$

where the $P_{2n}(\cos\theta)$ are Legendre polynomials. The anisotropy coefficients A_{2n} in Eq. (4) are functions not only of the cross sections $Q(SLJ|M|)$ but also of the Auger amplitudes $<Auger>_i$ and the relative phases δ_{ℓ_i} of Auger partial waves ℓ_i.

The coefficients A_{2n} vanish if $Q(SLJ|M|)$ is independent of $|M|$ and an isotropic distribution results. This is always the case for $J = \frac{1}{2}$. For the following we assume closed shell atoms A. Then ionization in an inner shell $n\ell j$ with $j \geq 3/2$ leads generally to a nonisotropic distribution of Auger electrons. In the special case of $J' = 0$ in Eq. (3) only one Auger partial wave $\ell = L$ is emitted and the A_{2n} are only functions of $Q(SLJ|M|)$. For inner shell ionization in a $p_{3/2}$ shell, the expected distribution of Auger electrons is then given by [8a]

$$I(\theta_o,E_o) = I_o(E_o) \{1 + A_2(E_o)P_2(\cos\theta)\} \,, \tag{5a}$$

$$\text{with} \quad A_2(E_o) = \frac{Q(^2P_{3/2}|\frac{1}{2}|,E_o) - Q(^2P_{3/2}|\frac{3}{2}|,E_o)}{Q(^2P_{3/2}|\frac{1}{2}|,E_o) + Q(^2P_{3/2}|\frac{3}{2}|,E_o)} \,. \tag{5b}$$

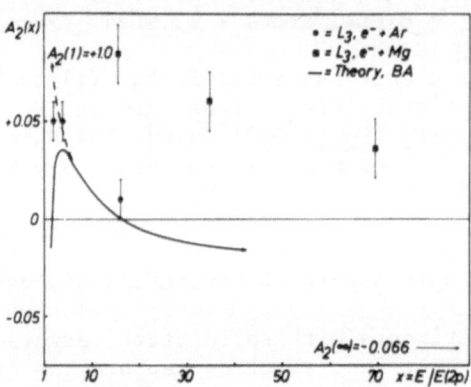

Fig. 2. Experimental anisotropy coefficients $A_2(x)$ for electron impact ionization in the L_3 shell of Ar [8b] and Mg [9] as function of reduced electron energy in units of the binding energy $E(2p)$ and comparison with theory [10].

From the anisotropy of $I(\theta, E_0)$ the alignment of inner shell ionized atoms can be determined directly via Eq. (5b).

The coefficients $A_2(E_0)$ have been measured for electron impact ionization in the $2p_{3/2}$ of Ar [8b] and Mg [9] by way of the angular distribution of the $L_3M_{2,3}M_{2,3}(^1S_0)$ electrons of Ar and the $L_3M_1M_1$ electrons of Mg. In Fig. 2 the experimental results are compared with theoretical values calculated by McFarlane [10] who used the Born approximation and screened hydrogenic wave functions. There is good agreement for Ar but not for Mg, most probably due to the approximate wave functions used. The results of Fig. 2 and additional experiments on the $3d_{3/2}$- and $3d_{5/2}$-shell ionization in Kr [11] indicate that, at least for electron impact, the intensity distribution of Auger electrons deviates only little (< 15 %) from isotropy.

2.3. Correlation Effects in the KLL Auger Diagram Intensities

Correlation effects in diagram intensities are expected to be large when the outermost shell or next inner shell is involved in the Auger transition. If only deep inner shells are involved, then correlation effects are small. This has been shown for the $M_{4,5}NN$ Auger spectrum of Xenon [12], where the theoretical values calculated with independent HFS wave functions and corrected for intermediate coupling agree well with experimental intensities.

A comparison of experimental intensities with theoretical

HFS-values for bound and continuum states as function of atomic number Z would clearly demonstrate the occurance and the importance of correlation. The KLL spectrum is best suited for such a comparison: Here calculations with HFS-functions have been done for $Z \leq 54$ (McGuire [13], Walters and Bhalla [14], Chen and Crasemann [15]) and many body perturbation theory was introduced by Kelly for neon [16]. Earlier calculations [17a] have not been included because the wave functions used were not sufficiently accurate (e.g. analytical functions for the continuum states). Experimental intensities have been obtained from Z = 10 to 94; and in the region of small Z, where correlation effects are expected to be large, free Ne [17b,18], Na [6], Mg [3] and Ar atoms [19] have been investigated.

In Fig. 3 the experimental relative group intensities $I(KL_1L_1)/(I(KLL)$, $I(KL_1L_{2,3})/I(KLL)$ and $I(KL_{2,3}L_{2,3})/(KLL)$ are compared with theoretical values calculated without configuration interaction (CI) by Walters and Bhalla [14] (dash lines). Introduction of CI between final states $2s^02p^6\ ^1S$ and $2s^22p^4\ ^1S$ [20] improves the agreement considerably (solid lines) and is almost perfect for argon [19]. The HFS-values of Chen and Crasemann including CI are given by the dotted lines. McGuire's values are not shown in Fig. 3 because of their unrealistic oscillations at small Z. For atomic numbers as small as 10 there still remains a discrepancy, here only the results of Kelly's many body calculation [16] are in perfect agreement with the intensities of neon. There is also very good agreement for the individual line intensities (see Table 1). The total neon KLL transition probability of $8.055 \cdot 10^{-3}$ atomic units ($\hat{=}$ 0.219 eV) agrees nicely with the K level width $\Gamma(K)$ found by photoelectron spectroscopy [21] to be (0.23 $\pm$ 0.02) eV. Systematic deviations of the group intensities at higher Z are clearly due to relativistic

Table 1. Comparison of experimental relative KLL intensities of neon with theoretical values

Transition	Experiment		Theory		
	[17b]	[18]	HF [16]	HF + MBPT [16]	HFS + CI [14]
$KL_1L_1(^1S)$	1.0	1.0	1.0	1.0	1.0
$KL_1L_{2,3}(^1P)$	2.73(4)	2.87(5)	2.14	2.79	2.98
(^3P)	1.00(2)	1.06(5)	0.83	1.00	0.99
$KL_{2,3}L_{2,3}(^1S)$	1.55(3)	1.5(1)	0.48	1.57	0.99
(^1D)	9.31(15)	10.0(2)	5.98	10.07	8.30

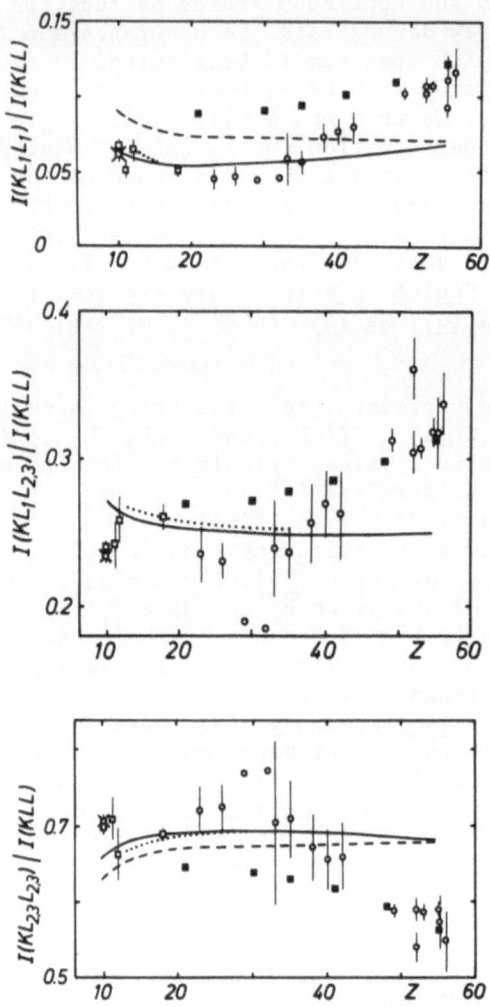

Fig. 3. Comparison of experimental intensities $I(KL_1L_1)/I(KLL)$, $I(KL_1L_{2,3})/I(KLL)$ and $I(KL_{2,3}L_{2,3})/I(KLL)$ with theory. Experiment: ⌀ = gaseous atomic target, ⌀ = solid state target. Nonrelativistic theory: --- = Walters and Bhalla [14] without CI, —— = Walters and Bhalla [14] with CI, ··· = Chen and Crasemann [15] with CI, X = Kelly [16]. Relativistic theory: ■ = Bhalla and Ramsdale [22a] without CI.

effects; this can be seen from the results of relativistic calculations [22a] (squares in Fig. 3). Inclusion of CI in the latter

calculations would certainly improve the agreement with experiment also for medium Z [22b].

An important result emerged from the calculations by Kelly [16]. He considered also the CI between the final total states of Auger transitions, which are continuum states, and showed that for neon the mixing of continuum states, e.g. $1s^2 2p^6 \varepsilon s$ ^2S with $1s^2 2s 2p^5 \varepsilon' p$ ^2S, is of equal importance as the mixing of final Ne^{++} atomic states $1s^2 2p^6$ ^1S and $1s^2 2s^2 2p^4$ ^1S. The mixing of final total states is probably less important if the two final atomic state vacancies are not in the outermost shell, as is the case for the KLL transitions in Ar. This then would explain the rather perfect agreement of KLL intensities of argon [19] with HFS-values of Chen and Crasemann [15], where only CI of final Ne^{++} ^1S atomic states have been included.

2.4. Correlation Effects in L- and M-Shell Diagram Auger Spectra

We expect the largest correlation effects, when the outermost shell is involved in the transition, e.g. in the L-MM spectrum

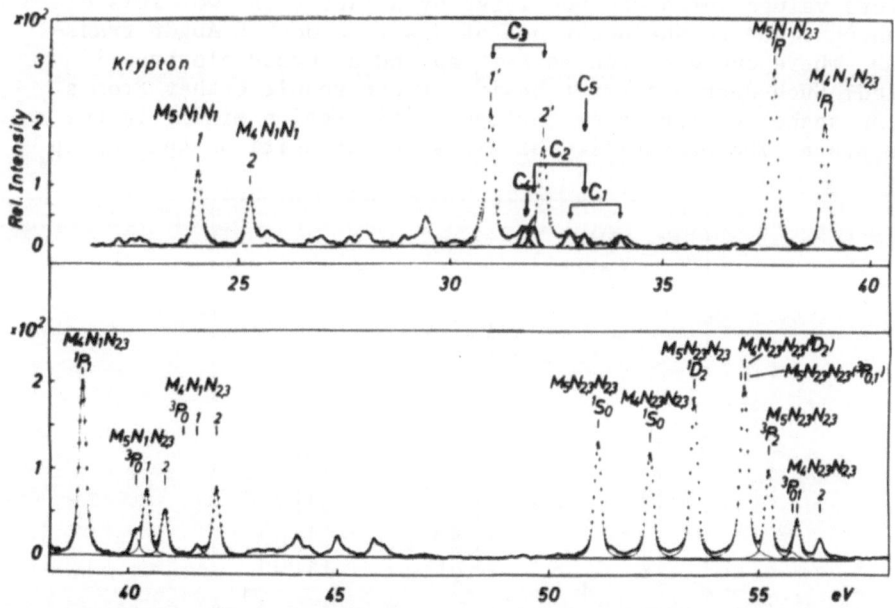

Fig. 4. $M_{4,5}$NN Auger spectrum of Kr [22c]. The discrete double Auger transitions identified and associated with the $M_{4,5}N_1N_{2,3}(^1P_1)$ parent doublet are designated by the doublets C_1, C_2 and C_3; those associated with the $M_5N_1N_{2,3}(^3P_1)$ parent transition are marked by C_4 and C_5.

of argon and the M–NN spectrum of krypton. As an example, the $M_{4,5}NN$ spectrum of krypton is discussed. The experimental spectrum [22c] is shown in Fig. 4 (see also [23]). All $M_{4,5}NN$ diagram lines, except the $M_{4,5}N_1N_1$ lines, could be identified by the known final state energies from optical data [24]. The identification of the $M_{4,5}N_1N_1$ transitions was uncertain for a long time [22c,23,25,26, 27], they were either ascribed to the doublet 1'/2' [22c,23] or to the doublet 1/2 [27,25]. Recently the latter identification has been confirmed indirectly by Hertz [28], who showed by photoemission spectroscopy that the corresponding low energy doublet in the $N_{4,5}OO$ spectrum of Xenon is undoubtedly due to the $N_{4,5}O_1O_1$ transitions. A direct many body calculation of the $4s^04p^6$ ^1S final state energy of krypton is most desirable, the adiabatic HF-value is 72.99 eV [26] compared to the experimental value of 69.77 eV [22c 23].

In Fig. 5 the experimental $M_{4,5}NN$ intensities of Kr are compared with theoretical values, which have been calculated with HFS-functions assuming jj-coupling for the initial M_4 or M_5 vacancy and intermediate coupling with configuration interaction for the two final vacancies [27]. The overall agreement is fairly good except for the $M_{4,5}N_1N_{2,3}(^1P_1)$ transitions, for which theory predicts values which are too large by a factor of two. This discrepancy is due to the occurence of discrete double Auger transitions, where one electron is emitted and a second electron is excited. Such double Auger transitions can result either from a $4p \rightarrow np$ shake up process or through configuration mixing in the final state. The probability of all shake transitions $4p \rightarrow np,\varepsilon p$

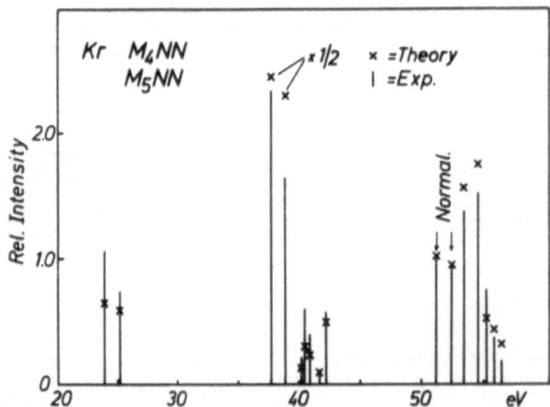

Fig. 5. Comparison of experimental line intensities [22c] of $M_{4,5}NN$ transitions of Kr with theory [27]. The theoretical values have been normalized to the experimental intensities of the $M_{4,5}N_{2,3}N_{2,3}(^1S_0)$ transitions. The theoretical values of the $M_{4,5}N_1N_{2,3}(^1P_1)$ transitions exceed the corresponding experimental intensities by more than a factor of 2.

accompanying the $M_{4,5}N_1N_{2,3}(^1P_1)$ transition is only 0.02 of the diagram intensity [29] and cannot account for the intensity loss of approximately 0.5. On the other hand, it has been shown [22c,23, 27] that several doublets in the experimental spectrum of Fig. 4 can be ascribed to discrete double Auger transitions caused by CI between the final state $4s4p^5$ ^1P of the diagram transition and the final states $4s^24p^3\binom{ns}{nd}$ ^1P of double Auger transitions. In Table 2 all discrete double Auger transitions identified so far in the noble gas spectra are listed.

In comparing the experimental diagram intensities with the theoretical HFS values, the intensities of associated discrete double Auger transitions have to be added. Let us consider for example the above mentioned transitions with the finale states

$$P \equiv 4d^{10}4s4p^5\ ^1P \quad \text{and} \quad S_k \equiv 4d^{10}4s^24p^3\binom{ns}{nd}\ ^1P. \tag{6}$$

In the CI model the parent and satellite final atomic state wave functions are given by

$$\psi_P = C_o\psi_P^o + \sum_k C_k\psi_{S_k}^o$$

and $$\tag{7}$$

$$\psi_{S_j} = C_j\psi_P^o + \sum_k C_{jk}\psi_{S_k}^o\ ,$$

where ψ_P^o and $\psi_{S_k}^o$ are the HFS model wave functions. The initial state wave function Φ_i^o and the total final wave functions Φ_f, including those of the ejected Auger electron $\varepsilon\ell$, are

$$\Phi_i^o = |4d^94s^24p^6\ ^2D >,$$

$$\Phi_{f,P} = C_o\psi_P^o\varepsilon\ell(^2D) + \sum_k C_k\psi_{S_k}^o\varepsilon\ell(^2D), \tag{8}$$

$$\Phi_{f,S_k} = C_j\psi_P^o\varepsilon\ell(^2D) + \sum_k C_{jk}\psi_{S_k}^o\varepsilon\ell(^2D).$$

Then the parent and satellite intensities are given by

$$I_P = K|C_o < \Phi_i^o|V|\psi_P^o\varepsilon\ell > + \sum_k C_k < \Phi_i^o|V|\psi_{S_k}\varepsilon\ell >|^2$$

$$I_{S_j} = K|C_j < \Phi_i^o|V|\psi_P^o\varepsilon\ell > + \sum_k C_{jk} < \Phi_i^o|V|\psi_{S_k}\varepsilon\ell >|^2 \tag{9}$$

with $V = \sum \frac{e^2}{r_{ij}}$. The matrix elements $<\Phi_i^o|V|\psi_{S_k}\varepsilon\ell>$ are identically 0, because Φ_i^o and $\psi_{S_k}\varepsilon\ell$ differ by more than two spin-orbitals. With

Table 2. Identification of discrete double Auger transitions in
noble gases caused by CI of final states. Energies and
intensities are relative to the diagram parent line (P).

Element	Transition	Energy		Intensity
		Moore [24]	Exp.	Exp.
Ne	P: $1s^{-1}-2s2p^5$ $^1P_1$			
	$1s^{-1}-2s^22p^3(^2P)3d$ $^1P_1$	-11.20	-11.2 [18,30]	0.03
	P: $1s^{-1}-2s2p^5$ ^3P			
	$1s^{-1}-2s^22p^3(^2P)3d$ ^3P	-29.10	-29.2 [18,30]	0.015
Ar	P: $2p^{-1}-3s3p^5$ ^3P			
	$2p^{-1}-3s^23p^3(^2P)4s$ ^3P	-11.5	-11.5 [30,31]	0.10
	$2p^{-1}-3s^23p^3(^2P)3d$ ^3P	-12.3	-12.3 [30,31]	0.14
	$2p^{-1}-3s^23p^3(^2D)4d$ ^3P	-19.4	-19.4 [30,31]	0.06
	P: $2p^{-1}-3s3p^5$ ^1P			
	$2p^{-1}-3s^23p^3(^2P)4s$ ^1P	$-$	-12.0 [31,32]	0.3
	$2p^{-1}-3s^23p^3(^2P)3d$ ^1P	$-$	-12.4 [31,32]	~ 0.1
	$2p^{-1}-3s^23p^3(^2D)3d$ ^1P	$-$	-13.1 [31,32]	0.6
Kr[+]	P: $3d^{-1}-4s4p^5$ ^1P			
	C_1: $3d^{-1}-4s^24p^3(^2P)5s$ ^1P	-4.88	-4.89 [22c,23]	0.09
	C_2: $3d^{-1}-4s^24p^3(^2P)4d$ ^1P	-5.75	-5.75 [22c,23]	0.09
	C_3: $3d^{-1}-4s^24p^3(^2D)4d$ ^1P	$-$	-6.51 [27]	0.78
	P: $3d^{-1}-4s4p^5$ $^3P_1$			
	C_4: $3d_{5/2}-4s4p^3(^2P)5s$ $^3P_1$	-7.30	-7.32 [22c,23]	0.26
	C_5: $3d_{5/2}-4s4p^3(^2P)4d$ $^3P_1$	-8.78	-8.79 [22c,23]	0.35
Xe	P: $N_{4,5}-5s5p^5$ $^1P_1$			
	$N_{4,5}-5s^25p^3(^2P)6s$ $^1P_1$	-4.38	-4.40 [23,30]	0.66/0.89

[+] The double Auger transitions C_1 through C_5 are marked also in
Fig. 4

the relation $|c_0|^2 + \sum_j |c_j|^2 = 1$ and Eq.(9) it follows then that

$$\sum_j I_{S_j} + I_P = K|<\Phi_i^{\,o}|V| \psi_P^o \varepsilon \ell |^{\,2} \tag{10}$$

$$= I_{th} \text{ (HFS model)}.$$

In principle one should also include in (6) the continuum states, that is $4d^{10}4s^24p^3(\varepsilon s_{\varepsilon d})^1P$, which are the corresponding final states of continuum double Auger transitions (see section 3). They are not at all very small but have been omitted here for clarity of presentation.

The mixing coefficients can be determined from the intensity ratio of satellite and parent lines by using Eqs. (9):

$$|c_j|^2/|c_0|^2 = I_{S_j}/I_P . \tag{11}$$

In the case of Kr $M_{4,5}N_1N_{2,3}$ ^1P transitions the total intensity according to Eq. (10) and Table 2 is by a factor 2 larger than the diagram intensity I_P and now in good agreement with the HFS value.

Also in the case of the $L_{2,3}MM$ spectrum of argon McGuire [32] has shown, that the existing discrepancies between experimental intensities of the $L_{2,3}M_1M_1$ $^1S_0$ and $L_{2,3}M_1M_{2,3}$ $^1P_1$ transitions and theoretical values can be removed through the interpretation of several hitherto unidentified doublets as double Auger transitions.

3. DOUBLE AUGER AND RADIATIVE AUGER TRANSITIONS

Double Auger transitions have been reviewed by Åberg [33], therefore only the main and some recent results will be presented. In contrast to the discrete double Auger transitions, which were discussed already in section 2.4, the continuum double Auger transitions with two electrons to be emitted into the continuum lead to a continuous energy distribution of Auger electrons and escape detection in the electron spectrum under usual conditions (but see Carlson and Krause [34]). The first evidence [35] of and the following investigations on the double Auger transitions were made using ion charge spectrometry. Table 3 lists experimental double Auger rates found so far. In the independent particle and frozen-core approximation double Auger transitions have zero probability. If core relaxation is allowed Carlson and Krause calculated with HF wave functions a shakeoff probability of 0.5 % for Ne [35] and 0.6 % for Ar [34] which is by an order of magnitude smaller than the experimental values. This discrepancy shows that electron correlation plays a dominant role also in double Auger transitions where two electrons are emitted. The same result was found in the

Table 3. Intensities of double Auger transitions relative to
all radiationless transitions[+]

Element	Transition	Rel. Intensity	Shakeoff
Ne	K-LLL	8(1)	0.5 [35]
Ar	$L_{2,3}$-MMM	10(2)	0.6 [34]
Kr	$M_{4,5}$-NNN	31(2)	
Xe	$N_{4,5}$-OOO	27	
Al	$L_{2,3}$-MMM	6	

[+] For references of experimental values see [33].

case of discrete double Auger transitions, where the three electrons
involved are from the same shell (section 2.4).

This is quite different for discrete double Auger transitions
in which the excited electron stems from a different shell (usually a less tightly bound shell). Then shakeup during core rearrangement accounts for the larger part of the measured intensity. In
Table 4 two examples for such transitions are listed. The theoretical shakeup values have been calculated within the conventional theory using HF wave functions [29]. In the conventional shake
theory the initial and final states are described by Slater determinants with single particle wave functions calculated for the
initial and final configurations.

In radiative Auger transitions an X-ray quantum is emitted
simultaneously with the ejection or excitation of an electron. In
the first case the final atomic state is a two vacancy state
(final state of normal Auger transition). Radiative Auger transitions are observed through the low energy X-ray satellite structures (either continuous or discrete) accompanying normal X-ray

Table 4. Intensities of discrete double Auger transitions (S)
relative to the parent transition (P) in per cent

Element	Transition	Rel. Intensity	Shakeoff
Na	P: $1s^{-1}$-$2s^2 2p^4$(^{1}D)$3s$		
	S: $1s^{-1}$-$2s^2 2p^4$(^{1}D)$4s$	7.8(8) [7]	4.7 [29]
Mg	P: $1s^{-1}$-$2s^2 2p^4$(^{1}D)$3s^2$		
	S: $1s^{-1}$-$2s^2 2p^4$(^{1}D)$3s4s$	7.1(8) [4]	4.3 [29]

transitions. The field of radiative Auger transitions has been reviewed by Åberg [33]. According to Åberg, two mechanisms contribute
in lowest order to the radiative Auger probability: Shakeoff (up)
through rearrangement and CI in the final state; the transition
amplitudes may either add or subtract. If CI can be neglected, then
the total K-LL and K-MM radiative Auger rates have been calculated
with the conventional shakeoff theory [36]. These rates are compared
with experimental results in Fig. 6. It can be seen that the experimental values of K-LL radiative Auger transitions are generally
much smaller than the shakeoff rates indicating the importance of
CI in the final state. For the K-MM radiative Auger rates the experimental results agree much better with the shakeoff rates (at
least for $Z \gtrsim 17$).

It is worthwhile to point out the intimate connection between
the discrete radiative Auger transitions and the shakeup satellites
in photoelectron spectroscopy (PES). Cooper and La Villa [37]
and also Werme et al. [38] found strong discrete radiative Auger
transitions in argon accompanying the $L_{2,3}$-M_1 X-rays due to CI between the final X-ray state $3s^{-1}\,^2S$ and the final states $3p^{-2}(^1D)nd\,^2S$
(n = 3,4) of discrete radiative Auger transitions (Fig. 7a). On
the other hand Spears et al. [39] found in the PES of Ar 3s-shell
intense shakeup lines (Fig. 7b), which they considered to be due
to CI between the final configuration $3s^{-1}\,^2S$ and $3p^{-2}nd\,^2S$.
The low energy structures found in both cases have the same energy
separation from the parent line, which agrees well with optical
data (see Table 5). As CI satellites their intensities relative to
the parent transitions should be independent of ΔZ, the change in

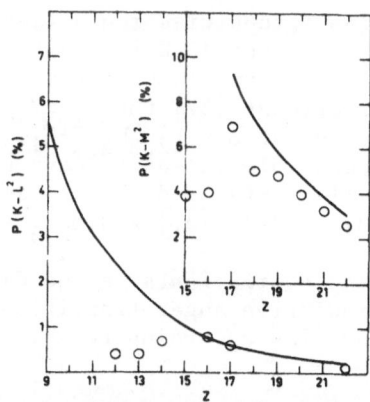

Fig. 6. Relative K-LL and K-MM radiative Auger rates as a function
of atomic number Z [33]. The rates correspond to the total integrated K-XX (X = L or M) intensities divided by the K-X parent-line
intensity. ⎯⎯ = shakeoff calculations [36]. For references of experimental values (o) see [33].

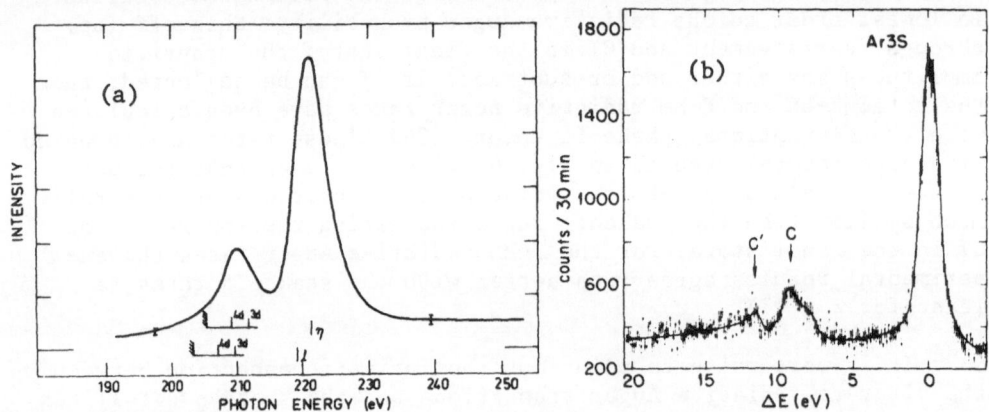

Fig. 7. Comparison of discrete radiative Auger transitions with shakeup transitions in the PES of argon.
a) Discrete radiative Auger transitions associated with the $L_{2,3}-M_1$ parent X-ray transition of Ar [37]. Beneath the spectrum are plotted the positions of the transitions to the final states $3p^{-2}(^1D)nd\ ^2S$ (n = 3 and 4).
b) Photoelectron spectrum of argon ionized in the valence shell by Mg K X-rays [39]. C and C' indicate positions of shakeup lines caused by CI between the final configurations $3s^{-1}\ ^2S$ and $3p^{-2}(^1D)nd\ ^2S$ (n = 3 and 4).

effective Z. The relative intensities found (see Table 5) are not inconsistent with this since the 10 % intensity of radiative Auger transitions is only a lower bound (due to the low resolution in the Cooper – LaVilla experiment [37] the $L_{2,3}M_1$ parent lines have not been resolved from the intense $L_{2,3}M_{2,3}$-$M_1M_{2,3}$ satellites). Analogous structures of comparable intensity have been found also in the PES of the K 3s-shell in KCl and the Kr 4s-shell [39] and in the X-ray spectra $L_{2,3}M_1$ of KCl [37] and $M_{2,3}N_1$ of Kr [40].

Also here the mixing coefficients in the CI final state wave functions of discrete radiative Auger transitions can be determined from the relative intensities according to Eq. (11).

4. MULTIPLE IONIZATION AUGER SATELLITE TRANSITIONS

Multiple ionization in the primary excitation process can be studied by X-ray [41] or Auger satellites. Here only the results of ionization by photons, electrons or protons in an inner shell

Table 5. Relative energies and intensities of CI shakeup lines in
         photo electron spectroscopy (PES) and discrete radiative
         Auger transitions (RAT) in Ar.

Final state	Energy			Intensity	
	PES [39]	RAT [37,38]	Optical [24]	PES [39]	RAT [37,38]
$3s3p^6$ ^2S	0	0	0	100	100
$3s^23p^4(^1D)3d$ ^2S	9.4	9.4	9.33	15	} > 10
4d ^2S	11.9	12.0	11.95	7	

with simultaneous outer shell ionization are discussed. Multiple
ionization in heavy ion-atom collisions is not the subject of this
lecture.

In the case of inner plus outer shell ionization (e.g. KL in Ne,
$L_{2,3}M$ in Ar) the observed double ionization rate for $h\nu$, e^- and
H^+, impact is in reasonable agreement with the conventional shakeoff
calculation [42] , provided the sudden limit is reached in these
experiments. Fig. 8 shows as an example the KL satellite intensity/
total K Auger intensity for excitations by photons, e^- and H^+ as
function of energy. For sufficiently high energies the double ionization rate seems to be independent of the energy and the particle
and agrees reasonably well with the shake theory, although it has
been found [45] that the ratio KL/total K of Ne is smaller by about
20 % for $E(e^-)$ = 40 keV than for $E(e^-)$ = 4 keV. The agreement of
total KL events with the conventional shake theory seems to indicate
that correlation effects are small in simultaneous inner and outer
shell ionization.

This is not the case if one considers the excitation of multiplet states ^3P and ^1P in the 1s2p double ionization of Ne.
Schmidt [46] has shown via the $KL_{2,3}$-$LLL_{2,3}$ Auger satellite intensities of neon [18] that for 4 - 6 keV electron impact the ^3P and ^1P
multiplets of the $KL_{2,3}$ initial configuration have been excited
with the ratio ^3P : ^1P = 1.36 instead with the expected value of 3,
which follows from the conventional shake theory (Fig. 9). Further
evidence for a deviation of the $^3P/^1P$ ratio from the statistical
value was given in experiments by Bhalla et al. [47] but was not
yet found by means of the X-ray satellite intensities [48] . The
reason for this nonstatistical excitation ratio is due to CI of
the continuum states $1s2s^22p^5(^3P)\epsilon p$ ^2S and $1s2s2p^5(^1P)\epsilon'p$ ^2S. This
is a natural extension of the CI between e.g. the discrete shakeup
states $1s2s^22p^5(^1P)np(^2S)$ and $1s2s^22p^5(^3P)np(^2S)$ of neon where the
excited electron is now in a continuum state ϵp. Also the shakeup
excitations corresponding to $1s2s^22p^5(^3P)np$ ^2S and $(^1P)np$ ^2S states
of neon were found in PES [49,50,51] (see also [52]) to deviate

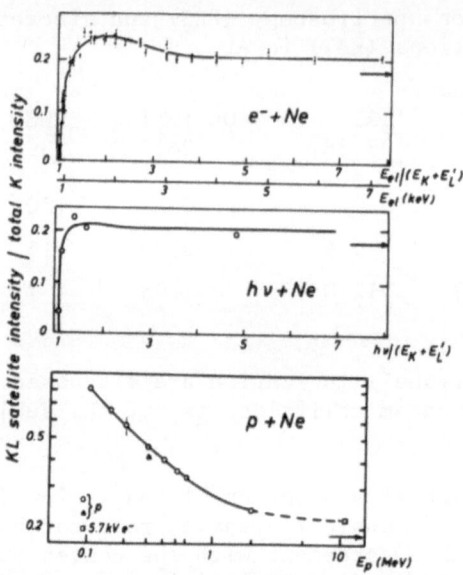

Fig. 8. Ratio of KL double ionization to all K ionizations in neon
as a function of energy of inciedent electron, photon or proton.
As KL/K ratio the ratio of KL Auger satellites to all K Auger
transitions is taken. For high incident energies of electrons and
photons the ratio KL/K is constant, and almost independent of the
ionization mode, the theoretical shakeoff value (indicated by the
arrow) accounts for about 85 % of the observed ratio [43]. For
protons with velocities comparable to those of electrons with
$E \gtrsim 4$ keV the KL/K ratio agrees in the asymptotic limit with the
ratio found for electrons [44]. E_K and E_L^i are the binding energies
of a K shell electron and the shakeoff electron.

strongly from the expected value of 3 and have been calculated
successfully with the shake theory including CI in the final bound
state [49,53].

    For the CI calculation of the above mentioned continuum states
the theory developed by Fano and Prats [54] and applied later to
photoionization by Starace [55] and Lin [56] has been used [57].
In the calculation [57] the mixing of continuum states with discrete states have been neglected and the V-matrix elements instead
of the K-matrix elements have been used. Within these approximations
and using the shake theory the value of $^3P/^1P$ excitation ratio was
found to be 2.28 and thus confirms the importance of final state
interaction also between continuum states. This was already shown
by Kelly [16] in his many body calculation of KLL transition rates

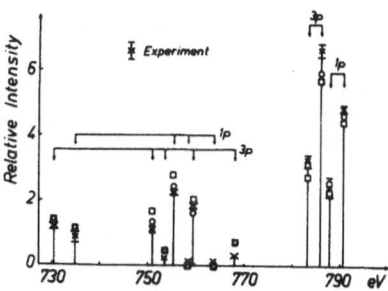

Fig. 9. Relative intensities of $KL_{2,3}$-$LLL_{2,3}$ satellite transitions
of neon excited by 5 keV electrons [18] and comparison with theory
[46]. The initial state of the satellite transitions (either ^3P or
^1P) is indicated. The theoretical values have been obtained by using
either the transition amplitudes calculated by McGuire [13] ($= \square$)
or those extracted from the KLL intensities [46] ($= O$). The excitation ratio of initial multiplet states $^3P/^1P$ must be assumed to be
1.36 in order to achieve good agreement with the experimental intensities.

of Ne (see chapter 2.3). In Fig. 10 the calculated ratio $^3P/^1P$ is
given as a function of energy ε of the shakeoff electron.

Further evidence for nonstatistical population of multiplet
states is given by the 2s-ionization in Na by electron impact:

$$Na(1s^2 2s^2 2p^6 3s) + e^- \rightarrow Na^+(1s^2 2s2p^6 3s\ {}^1S, {}^3S) + 2e^-$$
$$\hookrightarrow Na^{++}(1s^2 2s^2 2p^5\ {}^2P) + e_{CK}$$

From the intensity distribution of Coster-Kronig electrons the intensity ratio $^3S/^1S$ was determined [58] to be 2.5 in contrast to
the expected value of 3.

5. HYPERSATELLITE TRANSITIONS

The term hypersatellite transitions is used here for transitions of the kind KK-LLL or LL-MMM, that is, two electrons make a
transition simultaneously from e.g. the L- to the K-shell and a
third electron is ejected with the combined energy. Following some
controversy whether LL-MMM transitions have been found earlier in
argon [59] there is now firm evidence for $L_{2,3}L_{2,3}$-MMM transitions
to occur in argon [60]. Afrosimov [60] et al. produced the double
vacancy in the $L_{2,3}$-shell of argon in Ar^+-Ar, Cl^+-Ar or N^+-Ar collisions and found an experimental probability of $(4 ... 9) \cdot 10^{-4}$
for the decay via $L_{2,3}L_{2,3}$-MMM transitions. No theoretical calcula-

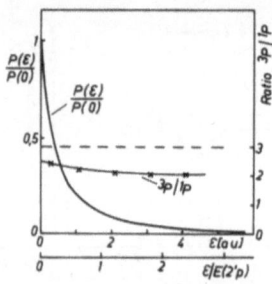

Fig. 10. Shakeoff probability $P(1s2s^2 2p^6\ ^2S \to 1s2s^2 2p^5(^{1,3}P)\varepsilon p\ ^2S)$ and ratio of $^3P/^1P$ excitation in $1s2s^2 2p^5$ ionization as function of the energy ε of the shakeoff electron. The shakeoff probabilities for the excitation of the ^3P- and ^1P initial states have been calculated assuming CI between the final continuum states $1s2s^2 2p^5(^3P)\varepsilon p(^2S)$ and $1s2s^2 2p^5(^1P)\varepsilon'p(^2S)$ [57]. The shakeoff probability has been normalized to 1.0 at $\varepsilon = 0$. $E(2'p)$ is the binding energy of the shakeoff electron.

tions of the probability of such processes have been done to date.

6. AUTOIONIZING TRANSITIONS

Correlation effects are strongest in the outermost shell. Ample evidence was given for this fact in this lecture. Thus the energies, excitation and decay probabilities of autoionizing states are all subject to strong correlation effects. Autoionizing states of the noble gases have been studied extensively by photoabsorption but also via electron ejection. Especially the experimental study of autoionizing states of He has prompted a large number of theoretical investigations of doubly excited states in He and other noble gases. More recently, also autoionization spectra of metal vapors have been investigated and as an example the autoionizing spectrum of the Li-atom, the next simplest case after He, will be discussed [61]. Li-atoms have been excited by electrons [61], H$^+$ and He$^+$ [62], in Li$^+$-He collisions [63] and in beam-foil experiments [64]. The photoabsorption spectrum of Li vapor has been measured earlier by Ederer et al. [65]. The autoionization spectrum excited by 2 keV electrons [61] is shown in Fig. 11. At that energy only optically allowed excitations of ^2P autoionizing states were found. The energies of the ejected electrons are in excellent agreement with those found by photoabsorption [65] and those calculated by Weiss [66], multi-configuration interaction included. Other important quantities are the excitation cross sections of $1s2snp\ ^2P$ states. The absolute and relative cross sections leading

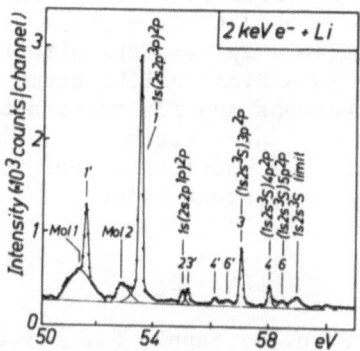

Fig. 11. Autoionization spectrum of Li excited by 2 keV e⁻ and observed at θ = 150° relative to the incident beam. Only ^{2}P states have been excited. The energy loss lines due to the (2s → 2p) inelastic scattering of autoionization electrons with Li atoms are indicated by numbers with primes.

to these states have been determined for 2 keV electron impact [61]. Assuming the Bethe approximation to be valid at this impact energy, then the optical oscillator strengths can be derived from the cross sections and compared with those found in photoabsorption and by theory. In Table 6 the relative excitation cross sections of 1s(2s2p ^{3}P)^{2}P, 1s(2s2p ^{1}P)^{2}P and (1s2s ^{3}S)np ^{2}P, n = 3,4,5, for various excitation modes are listed together with theoretical optical oscillator strengths calculated with multi CI [68]. It should be pointed out that the ratio of collisional cross sections

Table 6. Experimental relative cross sections of ^{2}P autoionizing states of Li for various excitation modes and comparison to theoretical values of optical oscillator strength (OS) and collision cross sections (CCS).

Autoionizing state	Photons [67]	e⁻ [61] 2 keV	H⁺ [69] 100 keV	400 keV	OS [68] with CI	CCS [70] without CI
1s(2s2p ^{3}P)^{2}P	100	100	100	100	100	100
1s(2s2p ^{1}P)^{2}P	5-10	4.0(4)	6.5	4.0	2.2	33
(1s2s ^{3}S)3p ^{2}P		17(1)				
4p ^{2}P		6.2(5)				
5p ^{2}P		1.8(2)				

$1s(2s2p\ ^3P)^2P/1s(2s2p\ ^1P)^2P$ without CI is given $\boxed{70}$ by
$\{3\ Q^d(1s \to 2p) - 2\ Q^e(1s \to 2p)\}/Q^d(1s \to 2p)$,
where $Q^d(1s \to 2p)$ and $Q^e(1s \to 2p)$ are the direct and exchange cross
sections of an electron in a hydrogenlike atom. This ratio reduces
to 3 for high energy electrons and for protons because of
$Q^e(1s \to 2p) = 0$. The experimental ratios in Table 6 clearly show
that the independent particle model that does not include CI
completely fails to predict correct values.

REFERENCES

1. W. Mehlhorn, Phys. Fenn. 9, Suppl. S1, 223 (1974).
   For a broader introduction see e.g. G. Hermann, Auger Line
   Intensities and Correlation Effects, Report, University of
   Aarhus, 1974.
2. E.J. McGuire, in Atomic Inner-Shell Processes, Vol. I, p. 293,
   (ed. B. Crasemann (Acad. Press, New York, 1975)).
3. K. Siegbahn, J. Electr. Spectr. and Rel. Phenom. 5, 3 (1974).
4. B. Breuckmann and V. Schmidt, Z. Physik 268, 235 (1974).
5. L. Ley, F.R. McFeely, S.P. Kowalczyk, J.G. Jenkin, and D.A.
   Shirley, Phys. Rev. B11, 600 (1975).
6. H. Fahlmann, R. Nordberg, C. Nordling and K. Siegbahn, Z. Physik 192, 476 (1966).
7. H. Hillig, B. Cleff, W. Mehlhorn and W. Schmitz, Z. Physik
   268, 225 (1974).
8a. B. Cleff and W. Mehlhorn, J. Phys. B: Atom. Molec. Phys. 7,
   593 (1974).
8b. B. Cleff and W. Mehlhorn, J. Phys. B: Atom. Molec. Phys. 7,
   605 (1974).
9. B. Breuckmann and V. Schmidt, Verhandl. DPG (VI) 9, 411 (1974).
10. S.C. McFarlane, J. Phys. B: Atom. Molec. Phys. 5, 1906 (1972).
11. E. Döbelin, W. Sandner and W. Mehlhorn, Phys. Letters 49A,
   7 (1974).
12. S. Hagmann, G. Hermann and W. Mehlhorn, Z. Phys. 266, 189 (1974).
13. E.J. McGuire, Phys. Rev. 185, 1 (1969); A2, 273 (1970).
14. D.L. Walters and C.P. Bhalla, Atomic Data 3, 301 (1971).
15. M.H. Chen and B. Crasemann, Phys. Rev. A8, 7 (1973).
16. H.P. Kelly, Phys. Rev. A11, 556 (1975).
17a. For references see W. Mehlhorn, in The Physics of Electronic and
   Atomic Collisions, eds. T.R. Govers and F.J. de Heer (North-
   Holland Publ. Co., Amsterdam, 1972).
17b. W. Mehlhorn, D. Stalherm and H. Verbeek, Z. Naturforsch. A23,
   287 (1968).
18. M.O. Krause, T.A. Carlson and W.E. Moddeman, J. Phys. (Paris)
   32, C4-139 (1971).
19. M.O. Krause, Phys. Rev. Lett. 34, 633 (1975).
20. W.N. Asaad, Nucl. Phys. 66, 494 (1965).
21. U. Gelius, S. Svensson, H. Siegbahn, E. Basilier, A. Faxälv
   and K. Siegbahn, J. Electron. Spectr. and Rel. Phen. 2, 405

(1973).

22a. C.P. Bhalla and J.D. Ramsdale, Z. Physik 239, 95 (1970).

22b. W.N. Asaad, private communication, 1975.

22c. W. Mehlhorn, W. Schmitz and D. Stalherm, Z. Phys. 252, 399 (1972).

23. L.O. Werme, T. Bergmark and K. Siegbahn, Physica Scripta 6, 141 (1972).

24. Ch.E. Moore, Atomic energy levels, vol. I-III (NBS, Washington).

25. K. Siegbahn et al., ESCA Atomic, Molecular and Solid State Structure Studied by Means of Electron Spectroscopy (Stockholm: Almquist and Wiksells, 1967).

26. F.P. Larkins, J. Phys. B: Atom. Molec. Phys. 6, 2450 (1973).

27. E.J. McGuire, Phys. Rev. A11, 17 (1975).

28. H. Hertz, Z. Phys. A 274, 289 (1975).

29. B. Breuckmann, private communication.

30. K. Siegbahn et al., ESCA Applied to Free Molecules, (North-Holland Publ. Co., Amsterdam, 1969).

31. L.O. Werme, T. Bergmark and K. Siegbahn, Physica Scripta 8, 149 (1973).

32. E.J. McGuire, Phys. Rev. A11, 1880 (1975).

33. T. Åberg in Atomic Inner-Shell Processes, Vol. I, p. 353, ed. B. Crasemann (Academic Press, New York, 1975).

34. T.A. Carlson and M.O. Krause, Phys. Rev. Lett. 17, 1079 (1966).

35. T.A. Carlson and M.O. Krause, Phys. Rev. Lett. 14, 390 (1965).

36. T. Åberg, Phys. Rev. A4, 1735 (1971).

37. J.W. Cooper and R.E. LaVilla, Phys. Rev. Lett. 25, 1745 (1970).

38. L.O. Werme, B. Greenberg, J. Nordgren, C. Nordling and K. Siegbahn, Phys. Lett. 41A, 113 (1972).

39. D.P. Spears, H.J. Fischbeck and T.A. Carlson, Phys. Rev. A9, 1603 (1974).

40. R.E. LaVilla, Phys. Rev. A8, 1143 (1973).

41. See e.g. J.P. Briand, lecture on "X-ray emission spectroscopy", this issue.

42. T. Åberg, in X-ray Spectra and Electronic Structure of Matter, eds. A. Faessler and G. Wiech (München, 1973), 1.

43. T.A. Carlson, W.E. Moddeman, M.O. Krause, Phys. Rev. A1, 1406 (1970).

44. D. Schneider, D.F. Burch, N. Stolterfoht, Abstracts of Papers, VIII. Conf. on Electr. Atomic Coll., Beograd, 1973, vol. 2, 729.

45. K.-J. Bekk, Diplom Thesis, University of Freiburg, 1974.

46. V. Schmidt in Inner Shell Ionization Phenomena and Future Applications, eds. R.W. Fink, S.T. Manson, J.M. Palms, and P.V. Rao, U.S. Atomic Energy Comm. Rep. Nr. CONF-720404, 548 (1973).

47. C.P. Bhalla, D.L. Matthews, C.F. Moore, Phys. Lett. 46A, 336 (1973).

48. O. Keshki-Rahkonen, Physica Scripta 7, 173 (1973).

49. M.O. Krause, T.A. Carlson, and W.E. Moddeman, Phys. Rev. 170, 37 (1968).

50. K. Siegbahn et al., loc. cit ref. [30].

51. U. Gelius, E. Basilier, S. Svensson, K. Siegbahn, Univ. of

Uppsala, Report UUIP-817 (1973).

52. M.O. Krause, lecture on "Photoelectron Spectrometry: Experiments with Atoms", this issue.

53. R. Martin, private communication, 1975.

54. U. Fano and F. Prats, Proc. Natl. Acad. Sci. (India) A33, 553 (1963).

55. A.F. Starace, Phys. Rev. A2, 118 (1970).

56. C.D. Lin, Phys. Rev. A9, 171 (1974).

57. D. Chattarji, W. Mehlhorn, and V. Schmidt, to be published.

58. E. Breuckmann, Diplom Thesis, University of Freiburg, 1973.

59. G.N. Ogurtsov, I.P. Flaks, and S.V. Avakyan, Zh. Tekh. Fiz. 40, 2124 (1970) [Sov. Phys.-Tech. Phys. 15, 1656 (1971)]. M.E. Rudd, B. Fastrup, P. Dahl and F.D. Schowengerdt, Phys. Rev. A8, 220 (1973).

60. V.V. Afrosimov, Y.S. Gordeev, A.N. Zinoviev, D.H. Rasulov, A.P. Shergin, IX. Int. Conf. Electr. Atom. Coll., Seattle 1975, Book of Abstracts, p. 1068.

61. H. Prömpeler, Diplom Thesis, University of Freiburg, 1975.

62. P. Ziem, R. Bruch and N. Stolterfoht, J. Phys. B, to be published.

63. D.J. Pegg, H.H. Haselton, R.S. Thoe, P.M. Griffin, M.D. Brown and I.A. Sellin, Phys. Rev. A12, 1330 (1975).

64. R. Bruch, G. Paul, J. Andrä, Phys. Rev. A12, 1808 (1975).

65. D.L. Ederer, T. Lucatorto and R.P. Madden, Phys. Rev. Lett. 25, 1537 (1970).

66. A. Weiss in Ref. [65].

67. D.L. Ederer, private communication, 1975.

68. A. Weiss, communicated by Ederer [67].

69. N. Stolterfoht, private communication.

70. A.D. Parks and D.H. Sampson, Ap. J., to be published.

SOME ASPECTS OF INNER SHELL PHENOMENA FOR THE IONS IN A PLASMA [+]

Pierre JAEGLE

E R "Spectroscopie Atomique et Ionique"

Bât.350 - Campus d'Orsay   91405   Orsay (France)

The highly charged ions considered in this paper consist in outershell striped atoms which appear generally in other conditions than innershell ionized atoms produced by electron or photon impact experiments. Outershell striping is the normal situation of an atom immersed in high temperature medium as existing either in matter of astrophysical interest or in plasma for thermonuclear fusion. That is the reason why stellar X-ray measurements aboard satellites as well as plasma diagnostics for fusion urge on a progress of the quantitative knowledge of oscillator strengths, collisional transition probabilities, steady state populations of ionic species or of specified levels of such striped atoms.

From a theoretical point of view, the highly charged ions do not reveal a basic difference with the neutral atoms, but a comparison of theoretical calculations with experiments is generally much more difficult in the case of ions. To a large extent the difficulty arises from the fact that a plasma is not a pure medium but a mixing of ions of different stages, surrounded by a gas of free electrons. On the other hand, as a rule, a high temperature medium produced in laboratory has a short life-time and the specific plasma parameters vary on a short time scale. Thus sophisticated techniques are required for most experiments. A continuous beam of ions with selected electrical charge can also be produced but, in the present stage of art, the ion density in such a beam remains too low for using any usual experimental technique of atomic physics, although some possibilities for electron-ion collision study are shown in Dolder's Lecture[1]. In fact most experiments require a density of ions which occurs only in very dense plasmas, as produced by laser.

Here we do not deal with the collective effects which are specific of the plasma physics but with the processes occuring in plasmas owing to the immersion of atomic systems in a free electron gas.
Generally speaking this leads to emphasize the processes of recombination and its consequences on the population distribution. As a
peculiar process of a great importance in theoretical as well as in
experimental regard, the inverse autoionization, namely the so-called dielectronic recombination[2], will be considered with some details. Furthermore, seeing the role of photoabsorption measurements
in the experimental population studies[3,4], we shall discuss the extension of such measurements to the case of highly charged ions in a
plasma.

I. PLASMA CHARACTERISTICS[5,6,7]

Significant quantities in a plasma are the densities and the
temperatures. It is possible to define several different densities,
generally expressed in cm^{-3}, that characterize the total number of
particles, the free electrons, each ionic species and the populations
of each excited level of the ions. The temperatures, often expressed in electron-volt (1 eV = 11520° K), are defined for the free
electrons and for the ions. Both temperatures are not necessary
equal ; it happens that a temperature can be defined for the electrons but not for the ions.

In laser produced plasmas, the temperature is comprised between
10eV and 1000eV. A few thousand eV is the maximum of the temperature which has been reached up to now in laboratory plasmas. On the
other hand, typical electronic densities are

1 cm^{-3}	in	interstellar matter
10^8-10^9 cm^{-3}	in	solar corona
10^{13}-10^{14} cm^{-3}	in	spark discharges and Tokamak
10^{19}-10^{21} cm^{-3}	in	laser produced plasma

The population density N_i of the i-level of a given species of
ion in a plasma varies against the time in accordance with :

$$\frac{dN_i}{dt} = \sum_{k > i} N_k A_{ki} + N_e \sum_k N_k \langle \sigma_{ki} v \rangle - N_i \sum_{j < i} A_{ij} - N_e N_i \sum_j \langle \sigma_{ij} v \rangle \quad (1)$$

where the terms referring to such processes as photoionization, photorecombination or three body processes have been omitted. In (1),
A_{ki} and $\langle \sigma_{ki} v \rangle$ are respectively the radiative decay probability
and the mean velocity of collisional transition from the k-levels,
A_{ij} and $\langle \sigma_{ij} v \rangle$ are respectively the absorption probability and
the mean velocity of collisional transition from the i-level. A
steady state occurs when, in (1), the right member is nought. Let's

assume that this occurs when two processes are largely predominant, namely a collisional transition for populating and a radiative decay for depopulating the i-level. From (1) we would have :

$$N_i = N_e \ N_k \ \langle \sigma_{ki} \ v \rangle \ /A_{ij} \tag{2}$$

that is N_i is proportional to the electronic density N_e and to the population of the k-level. Calculating a system of coupled equations of the form (1) or using very simplified expressions like (2) represent two extreme cases for population studies in plasmas.

When a thermal equilibrium prevails, the populations verify several familiar relations which are given here with numerical factors for the sake of practical applications. The population ratio of two levels i and j, having the energies E_i and E_j and the statistical weights g_i and g_j, is given by Boltzmann's law :

$$N_i/N_j = g_i/g_j \ \cdot \ \exp \left[-(E_i - E_j)/KT \right] \tag{3}$$

The statistical weight of a level of total angular momentum J is $g = 2J + 1$. For the free electrons surrounding the ions, Bolzmann's law is replaced by Maxwell's distribution. The number dN_e of electrons per cm^3 with an energy lying between E and E + dE, in a electronic gas of density Ne,is given by :

$$dN_e = 0.866 \ N_e \frac{E^{1/2}}{(KT)^{3/2}} \ e^{-\frac{E}{KT}} \ dE \tag{4}$$

(E and KT in eV, N_e in cm^{-3})

An extension of statistical considerations leading to the relations (3) and (4) enables to express the population ratio of two ions, which differ in charge by one unit,by the so-called Saha-Boltzmann equation :

$$\frac{N_{z+1}}{N_z} = 5.898 \ X \ 10^{21} \frac{Q_{z+1}(T)}{Q_z \ (T)} \ \frac{1}{N_e} \ (KT)^{3/2} \ e^{-\frac{i}{KT}} \tag{5}$$

z is the ionization stage, $\dot{I}$, the ionization potential of the ion of charge Z, $Q_z(T)$, the partition function of the same ion. The form of the partition function is

$$Q (T) = \sum_i g_i \ e^{-\frac{E_i}{KT}}$$

namely a sum over all levels of a given ion. In simplest cases, $Q(T)$ is approximated by the statistical weight of the ground state of the ion. (5) is not valid when collisions do not dominate in setting the plasma in a steady state.

From an atomic point of view, the screening of the nuclear charge of a given ion by the other particles is an important feature of
the plasmas. Due to this screening, the potential varies versus the
distance r to the nucleus as :

$$V_r = \frac{Ze}{r} e^{-r/\rho_D} \tag{7}$$

where ρ_D is the Debye radius. In taking into account only the free
electrons, ρ_D can be calculated by

$$\rho_D = 7.45 \ 10^2 \sqrt{\frac{KT_e}{N_e}} \tag{8}$$

The number of particles in the Debye sphere of radius ρ_D is of
importance in evaluating the causes of atomic level perturbation and
line broadening[8] in plasmas. It is given by the relation :

$$n = 1.73 \times 10^9 \sqrt{\frac{(KT)^3}{N_e}}$$

with KT in eV and N_e is cm^{-3}.

II. DIELECTRONIC RECOMBINATIONS AND POPULATION OF AUTOIONIZING LEVELS

Autoionization, like as the Auger effect, is known as a phenomenon
strongly related to electronic correlations in the atomic system.
Formally this is expressed in the structure of the wave function of
the autoionizing state, which is a mixing of discrete and continuous
state wave functions [9]. On the other hand the inverse process of
autoionization is the so-called dielectronic recombination whose the
important role has been recognized for the first time in the solar
corona (2,10) and must be invoked to explain the relative abundances
of highly charged ions of Fe at the temperature of the corona. Generally speaking, one can expect that the presence of an autoionizing
level in the continuum of states of an ion leads to a resonance in
the recombinations with these free electrons that have an energy close to that of the autoionizing level.

Let Ψ be an autoionizing level of an ion of charge Z, E the energy of this level above the ionization limit (fig.1). Two stages are
often considered for calculating the yield of recombination through
the level Ψ .

1) the recombination $X_{z+1} + e^- \longrightarrow X_Z^{\Psi}$ with a cross section σ ,
2) the stabilization $X_z^{\Psi} \longrightarrow X_Z^{x} + h\nu$ with a probability Ar. X_Z^{x} is
any state of the Z-ion below the ionization limit. From ref.(11) the
dielectronic recombination coefficient for electrons of energy between E and E + dE is given by the expression :

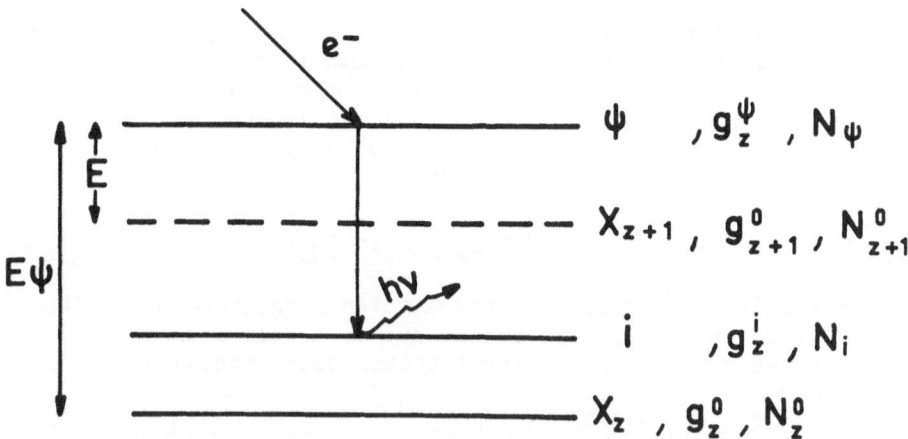

Fig. 1 . Dielectronic recombination from the ion (Z+1) and a free
electron to the level i of the ion Z.

$$\alpha_d = \frac{A_r \sqrt{\frac{4}{2\pi m}} \frac{1}{(KT)^{3/2}} Q\, E\, dE}{A_r + \frac{16\pi m}{h^3} \frac{g^o_{z+1}}{g^\psi_z} Q\, E\, dE} \; e^{-\frac{E}{KT}} \qquad (9)$$

For small A_r, (9) reduces to

$$\alpha_d = A_r \frac{g^\psi_z}{g^o_{z+1}} \frac{h^3}{(2\pi m KT)^{3/2}} \; e^{-\frac{E}{KT}} \qquad (10)$$

where g^o_{z+1} and g^ψ_z are respectively the statistical weights of the
ground state of the ion (Z+1) and of the autoionizing state of the
ion Z. The last expression does not depend on Q but leads to divergent results in summing over all autoionizing and excited final
states.

Recently the calculation of dielectronic recombination has been
treated by a more general method in which the recombination coefficient is expressed as a function of the photorecombination matrix
elements of a generalized S-matrix(12,13). With this treatment it is
no longer necessary to divide the dielectronic recombination in two
processes giving rise to separated probability calculations. Then
dielectronic recombination appears as a resonance in photorecombi-

nation   exactly as autoionization appears as a resonance in photoionization spectrum.

Experimental results involving a negative absorption at a wave-
lenth of 117,4 Å for neon-like ions in aluminium laser-produced plas-
ma(14-20) emphasize a somewhat different dielectronic recombination
mechanism, occuring only in high electronic density medium. It could
be expressed as the result of the two following processes :

1) recombination : $X_{Z+1} + e^- \longrightarrow X_Z^{\Psi}$

2) stabilization : $X_Z^{\Psi} + e^- (\mathcal{E}) \longrightarrow X_Z^{*} + e^- (\mathcal{E}')$

(like previously X_Z^{Ψ} represents the autoionizing state of the ion of
charge Z). The stabilization is now due to a collision instead of a
radiative transition. In joining together both processes we would
have

$$X_{Z+1} + e^- + e^- \longrightarrow X_Z^{x} + e^-$$

which denotes usually a three body recombination. Thus, this new type of dielectronic recombination appears still as a resonance in recombination processes but now in these recombinations that are just
the inverse of ionization by electronic impact.

The yield of this recombination process has not been calculated
up to now. A rough evaluation of its role for populating peculiar
ionic levels could be done with the help of relations like (2), but
slightly modified to account only for dominant collisional transitions , for instance :

$$N_i = N_\Psi \frac{\langle \sigma_{\Psi i} v \rangle}{\sum_j \langle \sigma_{ij} v \rangle} \tag{11}$$

where N_Ψ is the population of the autoionizing level, $\langle \sigma_{\Psi i} v \rangle$
the collisional transition velocity from this level to the i- level,
$\langle \sigma_{ij} v \rangle$ the transition velocity from the i-level to other levels.
Therefore it is relevant to investigate the problem of the population
of an autoionizing level in a plasma(21).

It would be ordinary to consider the population of an autoionizing level as controlled by Boltzmann's law applied to a discrete
state, namely.

$$\frac{N_\Psi}{N_Z^o} = \frac{g_\Psi}{g_Z^o} e^{-\frac{E_\Psi}{KT}} \tag{12}$$

where N_Z^o is the ground state population. As an example, for the
$2s2p^64p$ $^3P_1$ level of the Al^{3+} ions, $E_\Psi = 145$ eV while the ioniza-

tion potential from the $2s^2 2p^6\, {}^1S_0$ ground state is 120 eV ; from
(12), in a plasma at a temperature of 50 eV, the population N_ψ
would be less than 2.10^{-1} times the ground state population.

Instead, autoionizing level is a mixing of discrete and continuous states, leading, when interacting, to states

$$\Psi = a\, \varphi \;+\; \int_E b_E\, \Phi(E)\, dE \tag{13}$$

The statistical weight of the autoionizing level must account
for the continuous part of this function, that is the second term
of the right member of (13). This can be obtained in introducing
there the statistical weight of free electrons. In considering the
integral over the small energy interval ΔE for which the interaction
between discrete and continuous state is strong, one obtains an expression of the statistical weight of the autoionizing level which
enables to replace (12) by :

$$\frac{N_\psi}{N_Z^o} = A^2\, \frac{g_\psi}{g_Z^o}\, e^{-\frac{E_\psi}{KT}} + B^2\, \frac{g_{Z+1}^o}{g_Z^o}\, \frac{8\pi}{N_{Z+1}^o}\, \frac{m^{3/2}(2E_c)^{1/2}}{h^3}\, \Delta E\, e^{-\frac{E_\psi}{KT}} \tag{14}$$

E_c is the difference between E_ψ and the ionization energy of Z-ion
(E_c = 145-120 = 25 eV in the previous example), and A^2 and B^2 are the
probabilities that the system is in a discrete or continuous state
lying in the energy interval ΔE, and :

$$A^2 + B^2 = 1 \tag{15}$$

The meaning of (14) appears clearly in assuming $A^2 \longrightarrow 0$; then :

$$N_\psi \longrightarrow 2\frac{g_{Z+1}^o}{g_Z^o}\, N_e\, f(v)\, 4\pi\, v^2\, \Delta v$$

where v is the electron velocity and f (v) is the Maxwell distribution ; (14) means that population of an autoionizing level consists
in a mixing of bound and free electrons in a proportion which depends on the weights of the discrete and continuous part of the wave
function. For the $2s2p^6\, 4p\, {}^3P_1$ level of Al^{3+} ion, in assuming a density of ground state Al^{4+} ions of 5.10^{18} cm^{-3}, numerical application
of (14) leads to :

$$\frac{N_\psi}{N_Z^o} = 1.7\ 10^{-1} \times A^2 + 1.5 \times 10^2 \times B^2$$

(from experimental data ΔE is taken equal to 0.1 eV).

From this result we see that the population N_ψ of the autoionizing
level will be larger than the ground state population N_Z^o of the Al^{3+}

ion , unless B^2 is very small, which is not expected in a case of large probability of autoionization. Turning back to relation (11) it can now be seen that mechanisms populating via autoionizing levels are able to act very strongly on the population of some levels of ions in dense plasmas. This occurs very likely, for instance, in the case of $2s^22p^54d$ $^3P_1$ level of Al^{3+} ion (18), as well as Si^{4+} ion (22), in laser produced plasma.

III. PHOTOABSORPTION IN POSITIVE IONS

Seeing the large field of hot plasma studies depending on the knowledge of populations, either for the whole of an ionic species or even for a specified excited state of the species, the development of new methods able to provide such experimental informations is now very desirable. For dense complex plasmas, photoabsorption measurements are of special interest because they test either the total population of a given ion, if performed close to an inner shell ionization threshold, or the populations of an excited level, if referred to a discrete transition. With respect to the case of neutrals, multiply charged ions introduce in photoabsorption study two pecularities making the experiments more difficult: i) high densities are obtained only in transient media having a life-time less than 1 μs ;   ii) generally ions are emitting by themselves in the wavelength range of interest for photoabsorption measurements.

The first difficulty is got over[14-19,23,24] in using devices producing simultaneously two plasmas the one as an X-ray source, the other one as an absorber. The synchronization between the lighting of the source and the creation of the ions allows time integrated as well as time resolved measurements.

Interpreting the results of these measurements in taking into account the radiation emitted by the "absorber" (fig.2 and 3) leads to a theoretical study of radiation transfer through the plasma. Basically such a study stands up on an expression of energy conservation written under the differential form :

$$\frac{d(I\,\delta\nu)}{dx} = \left[\frac{A_{21}}{4\pi} N_2 + \frac{I_\nu}{C} B_{12} \left(N_1 - \frac{g_1}{g_2} N_2 \right) \right] h\nu\, P_\nu \delta\nu \qquad (16)$$

$I_\nu \delta\nu$ is expressed in energy per cm^2 and sec ; A_{21} and B_{12} are Einstein's coefficients for spontaneous emission and absorption : P_ν is a profile function depending upon the frequency ν and is assumed to be identical for absorption and spontaneous or stimulated

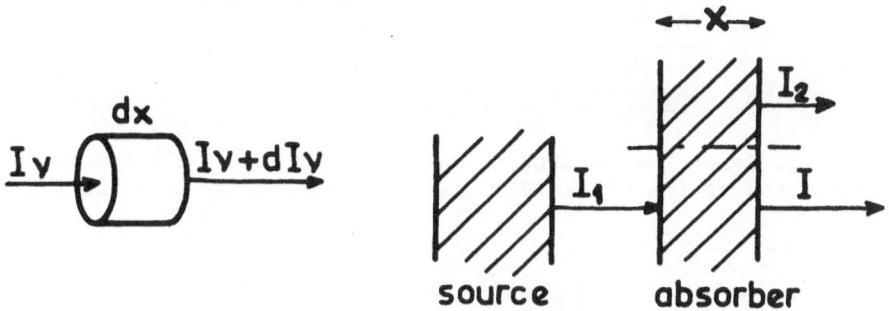

Fig.2 Elementary plasma volume   Fig.3 Sketch of a two plasma
                                          experiment.

emission ; $\delta\omega$ is a small frequency interval.

Two types of calculations can be carried out from expressions
like (16). First, one can study the change in the line profiles due
to the plasma unhomogeneity. This has been extensively treated by
Zwicker in ref.(7) in showing how a strong increasing of continuous
absorption lets move from the profile of fig.4 a to the profile of
fig. 4b. Alternatively, integration of (16) over the line half-width
$\Delta\omega$, then over a finite path x across the plasma, leads to the
expression of mean intensity for a given line

$$I = I_2 + I_1 \; e^{-\frac{h\omega}{c}\frac{12}{\Delta\omega}\left(N_1 - \frac{g_1}{g_2} N_2\right) \; x} \qquad (17)$$

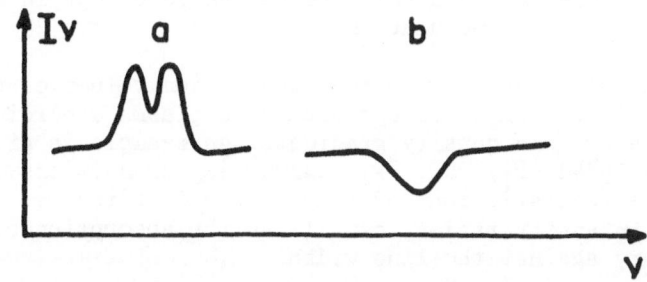

Fig.4 Line profiles due to reabsorption ; profile b occurs only when
continuous absorption is very strong.

with :

$$I_2 = \frac{2h\nu^3}{c^2} \frac{\frac{g_1}{g_2} N_2}{N_1 - \frac{g_1}{g_2} N_2} \left(1 - e^{-\frac{h\nu}{c} \frac{B_{12}}{\Delta\nu}(N_1 - \frac{g_1}{g_2} N_2) x}\right) \qquad (18)$$

(see   fig. 3 for the meaning of I_1 and I_2) . From (17) the mean transmission of the plasma for the line under investigation will be :

$$T = \frac{I - I_2}{I_1} \qquad (19)$$

and the values of T, obtained experimentally in measuring I, I_1 and I_2, are related to the atomic features by the expression

$$T = e^{-\frac{h\nu}{c} \frac{B_{12}}{\Delta\nu}(N_1 - \frac{g_1}{g_2} N_2) x} \qquad (20)$$

This enables to define the absorption coefficient of an homogenous plasma as :

$$\alpha = -\frac{h\nu}{c} \frac{B_{12}}{\Delta\nu}(N_1 - \frac{g_1}{g_2} N_2) \qquad (21)$$

For unhomogenous plasmas, (16) has to be integrated numerically in taking into account the gradients of density N_1 and N_2 which are sometimes related merely each other by a function of the temperature (see (3)). For intensities and absorptions averaged over the line widths, a numerical integration can be performed with the help of (17) and (18) in using a small value of x as an integration step. As an important result of a number of trial calculations, (19) remains valid in very unhomogeneous plasmas in this sense that the T-value is independent of the "probe" intensity I_1 of the external source, for any assumption on the density and temperature gradients of the absorber. Thus (19) can be widely used in experiments, although in many cases (20) and (21) are not valid except in evaluating a mean value of the populations.

Similar calculations can provide expectation values of the intensity ratios in the emission spectrum of a plasma according to various temperature and density gradients. An example is given on fig. 5 for the $2p^5 4d \ ^1P_1$, $^3D_1$, $^3P_1 \longrightarrow 2p^6 \ ^1S_0$ transitions of the Al^{3+} ion. For a realistic ion and temperature distribution, the figure shows how the intensity ratios, as well as the absorption of the lines, should vary against the line width.

Such calculations, as well as photoabsorption experiments like previously mentioned in this section, played an important role in interpreting a strong intensity anomaly at a wavelength of 117.4 Å in an aluminium plasma as due to a departure from the equilibrium

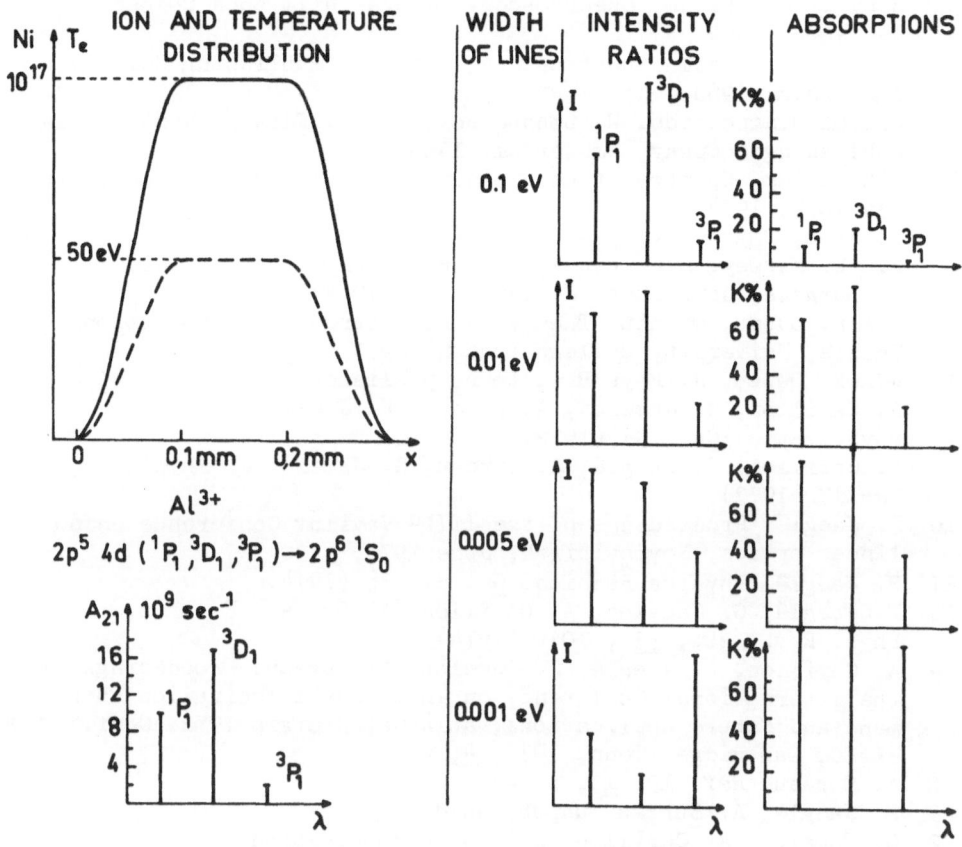

Fig. 5   Influence of the line width on the intensity ratios and line absorption in an unhomogeous plasma.

value of the upper level population for the transition under investigation. Undoubtedly such techniques can be considered as promising for future experiments on highly charged ions in dense plasmas.

REFERENCES

1   K. Dolder , this volume
2   H.S.W. Massey, H.B. Gilbody , Electronic and Ionic Impact Phenomena, Oxford at the Clarendon Press, 1974, Vol. IV, p. 2118
3   A. Carillon , Thèse, 1973 , n° AO 8424, Université Paris VI
4   A. Carillon , P. Jaeglé, P. Dhez , Journal de Physique, 1971, 32, p. C4-48

5   H.R. Griem, Plasma Spectroscopy, McGraw-Hill Book Company,
    New York, 1964
6   G.V. Marr, Plasma Spectroscopy, Elsevier Publishing Company,
    Amsterdam, 1968
7   Plasma Diagnostics, W. Lochte-Holtgreven, Editor, North Holland
    Publishing Company, Amsterdam, 1968
8   H.R. Griem, Spectral Line Broadening by Plasmas, Academic Press,
    New York, 1974
9   U. Fano, Phys. Rev. 124 , 1866 (1961)
10 H.S.W. Massey, D.R. Bates, Rep. Prog. Phys. 9, 6 (1942)
11 A. Burgess, Astrophys. J. 139 , 776 (1964)
12 J.A.R. Dubau, Quantum Theory of Dielectronic Recombination
    Thesis, University College London, 1973
13 J.A.R. Dubau, J. Phys.B. , to be published
14 A. Carillon, G. Jamelot, A. Sureau, P. Jaeglé
    Phys. Lett. 38A , 91 (1972)
15 A. Carillon, P. Jaeglé, A. Sureau, G. Jamelot J. Physique, 34
    C2-117 (1973)
16 P. Jaeglé, Proceedings of the IIIrd Vavilov Conference on non-
    linear optics, Novossibirsk, June 1973, p.30
17 P. Jaeglé, Physica Fennica, 9 , S1, 38 (1974)
18 P. Jaeglé, G. Jamelot, A. Carillon, A. Sureau, P. Dhez,
    Phys. Rev. Lett. 33 , 1070 (1974)
19 A. Carillon, P. Jaeglé, G. Jamelot, A. Sureau, Proceedings of
    the International Conference on Inner Shell Ionization Pheno-
    mena and Future Applications, Atlanta, Georgia 1972, Conf.720404,
    USAEC, Oak Ridge, Tenn, 1973, 4, p. 2350
20 A. Sureau, Ref. 19, 1 , p. 430
21 P. Jaeglé, A. Sureau, unpublished work
22 G. Jamelot, A. Carillon, private communication
23 A. Carillon, P. Jaeglé, P. Dhez, Phys. Lett. 25, 140 (1970)
24 J.M. Esteva, These 1974, n° A09976, Université Paris-Sud

[+]Work partly supported by the D.G.R.S.T. under contract
   No. 659 1346.

MULTIPLE EXCITATION IN FREE MOLECULES*

Thomas A. Carlson

Oak Ridge National Laboratory

Oak Ridge, Tennessee 37830, USA

INTRODUCTION

This lecture will be the only one in the book that will deal specifically with free molecules. It thus faces the problem of covering in a relatively short space a subject that is more complex than the corresponding one on atomic systems, and whose literature, in the case of certain experimental investigations, is more extensive. However, by making a limited selection of material, and by covering some material superficially, we should be able to obtain an overview and flavor of multiple electron excitation in molecules.

The subject to be covered in this lecture is at the heart of the interest of the school: viz, multiple excitation in the photo-ionization process. We shall restrict ourselves mainly to studying satellite structure in the photoelectron spectra of core shells. In only a few cases has satellite structure been identified [1] in the photoelectron spectra of the valence shell. This is due in part to the fact that the large number of molecular orbitals closely spaced in energy makes difficult the separation of satellite lines from main lines; and in part to the fact that most studies on the valence shell have been done with low energy radiation, where there is often insufficient energy available to give rise to electron shakeup. Core shell photoelectron spectroscopy is usually carried out with Al or Mg Kα x rays, which usually have sufficient energies (1487 eV and 1254 eV) to produce photoelectrons with velocities far in excess of the bound valence electrons. Thus, our field of interest will be limited to the area of the sudden approximation.

For practical reasons we shall also primarily restrict the discussion to electron shakeup (transition to excited but bound states) rather than electron shakeoff (transition to the continuum). In photoelectron spectra the contribution of electron shakeoff is very hard to extract since the continuum spectrum is spread over a large energy range, whereas the discrete peaks due to electron shake-up are easy to detect. The study of electron shakeoff by means of measuring the ionic charge is hampered by fragmentation of the molecular ions. Study of satellite lines in x-ray fluorescence and Auger processes can, however, yield information about both electron shakeup and shakeoff, and will be mentioned briefly at the end of this lecture. Although the material to be discussed will be somewhat circumscribed, I believe that which is covered will give an idea of the added interest and complexity that molecules bring to the problem of many-electron excitation.

II. GENERAL ASPECTS OF ELECTRON SHAKEUP IN MOLECULES

As with atoms the photoelectric cross section for ejecting an electron from orbital, λ, of a neutral molecule, and ending up in the final state $\phi_{n,\lambda}$ of the ion is given by the expression

$$\sigma(E) = \overbrace{\frac{4\pi\alpha \, a_o^2}{3} \, E(h\nu)\left|<\nu_E|\underline{r}|u_\lambda>\right|^2}^{\text{dipole}} \quad \overbrace{\left|<\phi_{n\lambda}|\phi_o>\right|^2}^{\text{monopole}} \tag{1}$$

Equation (1) was derived [2] within the framework of both the dipole and sudden approximation. If we further assume that at higher kinetic energies of the photoelectron there will be little variation in the dipole contribution with changes in the photoelectron energy because of relatively small energy differences between the ground and excited states of $\phi_{n,\lambda}$, then the probability for monopole excitation is

$$P_{n\lambda} = \left|<\phi_{n\lambda}|\phi_o>\right|^2 \tag{2}$$

where ϕ_o is the state of the neutral molecule in which one electron is missing from orbital, λ; or, in other words, is a collection of frozen orbitals of the singly-charged ion in which one electron has been removed from orbital λ. $\phi_{n\lambda}$ is one of a number of n configurations, for both the ground and excited states, in which electrons from all the other orbitals have relaxed to the hole created in λ. $P_{n\lambda}$ is thus the probability for going to a particular final state of the ion. Because of relaxation, the probability for going to the ground state is no longer unity and transitions to various excited states are also possible. For molecules the basic selection rule arising out of equation (2) is that $\phi_{n\lambda}$ has the same spin and space symmetry as ϕ_o. Within the framework of configuration interaction we may consider all possible configurations in the excited state that will properly mix. Single electron excitation, which we normally associate with electron shakeup, is one of the most important means of excitation but not necessarily the only one.

Table I contrasts the selection rules for monopole excitation between atoms and molecules. The capital letters representing the entire state illustrate the basic selection rules; the small letters representing the individual orbitals give the more restricted selection rules for single electron excitation. Γ for molecules are the irreducible representations or as sometimes called Mulliken symbols. Within the selection rule for single electron excitation the symmetry for an individual orbital is maintained: $\Delta\lambda = 0$. Angular momentum is not a good quantum number for molecules, except for linear molecules where the projection of the angular momentum onto the internuclear axis is quantized, and this projection quantum number, λ_p, follows the selection rule $\Delta\lambda_p = 0$ for monopole excitation. An interesting and important consequence of the localization of a core vacancy should also be mentioned. In the linear homonuclear molecule, e.g., N_2, gerade and ungerade symmetry exists. One would expect in this case that the following monopole selection rules pertain: $u \rightarrow u$; $g \rightarrow g$; $u \not\rightarrow g$. However, when a hole is created in one of the 1s shells of N_2, the molecular orbital selection rules regarding u and g symmetry no longer apply.

It is highly instructive in studying molecules to compare the shakeup structure formed by creating core vacancies in each of the different atoms of a molecule. The core binding energies for various elements are sufficiently different that the photoelectron spectra are easily separated experimentally. If the excitation of the valence shell was identical regardless of the core vacancy formed, the shakeup spectra as represented by the satellite lines would be identical. This is rarely the case [3-5]. For example, figure 1 shows the photoelectron spectra of CO_2 with data for photoejecting electrons from either the C(1s) or O(1s) levels. The details of the shakeup structure are quite different. Since molecular orbitals are rarely equally distributed over the whole molecule, it is not surprising to find the shakeup probability for a given orbital strongly dependent on the location of the core hole. If an orbital

TABLE I.   Selection rules for electron shakeup

Atoms	Molecules
$\Delta S = \Delta L = \Delta J = 0$	$\Delta S = \Delta\Gamma = 0$
$\Delta s = \Delta \ell = \Delta j = 0$	$\Delta s = \Delta\lambda = \Delta\lambda_p = 0$
	but $u \not\rightarrow g$ not applicable to core hole

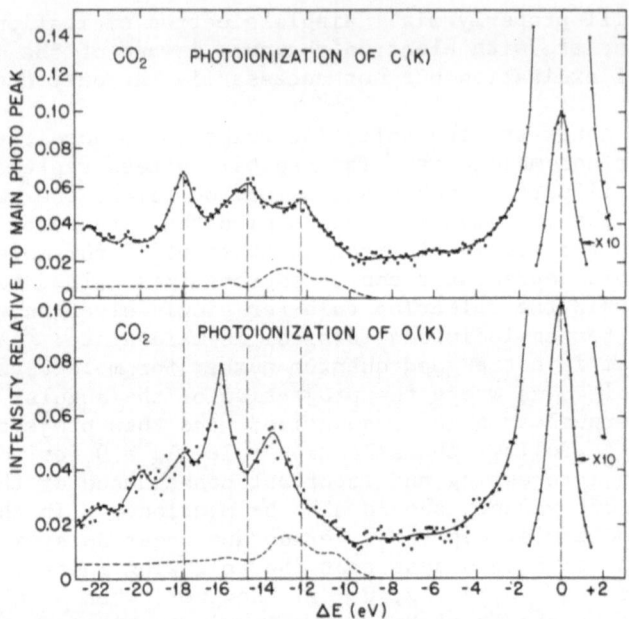

Figure 1.   Comparison of satellite structure found in photoelectron spectrum of C(1s) and O(1s) of CO_2 using Al Kα x rays.   Figure reproduced from Carlson et al., ref. [3].

has a high population density associated with a given atom, a localized hole in that atom will tend to preferentially excite the electron in that orbital.

The energy of excitation is dependent on the location of the core vacancy for two basic reasons. First, the extra atomic relaxation energy can be different according to which element has the core vacancy. This can cause changes in the total energies for both the ground and excited states, the shakeup energy being the difference between the two. Second, there will be a splitting of lines due to the possibility of more than one spin state. When photoionization occurs in a core orbital, an unpaired spin results. If electron shakeup occurs, both the valence shell and excited orbital will also contain unpaired spins. Within the monopole selection rules two states can arise which have the same configuration and state designation, but differ as to which pair of orbitals have parallel spins. The two states are commonly called lower and upper. See figure 2 for a pictorial description. In neon the upper and lower states have been clearly identified [5]. The intensities are roughly equal and the energy separations are about 4 eV. For the other rare gases the splitting between the upper and lower is smaller [5,7]. For molecules the extent of spin coupling is still open to question, but it may be more important than first realized.

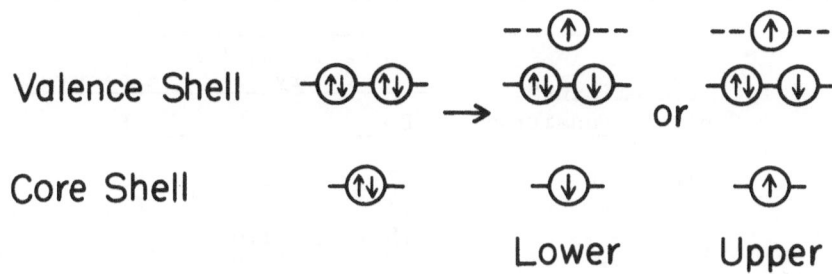

Figure 2. Schematic representation of formation of upper and lower spin states following electron shakeup.

III. COMPARISON OF THEORY AND EXPERIMENT FOR SPECIFIC MOLECULES

The first calculation of electron shakeup for molecules using equation (2) employed [8] semiempirical molecular orbital calculations, such as CNDO [9]. Since the semiempirical calculations do not involve the core electron specifically but treat the core as a fixed potential, it is necessary to use the equivalent charge approximation. That is, the effect of removing a shielding electron from the core is simulated by increasing the nuclear charge by one. This method of course neglects the effect of spin coupling between the partially filled shells. It gives only marginal agreement with experiment, but can be applied to complex systems that are not amenable to detailed calculations.

Recently, improved calculations of electron shakeup for molecules using Hartree-Fock wave functions and configuration interaction have been made [10-13]. The reader is warned tha rarely is a complete basis set used for molecular Hartree-Fock calculations, so the quality of the wave function may change considerably with the choice of the basis set. This uncertainty plus the wide possibilities for choices of configuration interaction result in a large variation in conclusions reached by different calculations.

In Table II are listed results of calculations made by Hillier and Kendrick [11] on nitrogen compared with experiment. The experimental data are from a spectrum reported by Gelius [5]. These data are taken with a monochromatic source of x rays and show a number of peaks that were not possible to resolve previously. Such an experiment requires the use of a powerful x-ray source employing a rotating anode, and a highly efficient spectrometer with a position sensitive detector in order to make up for the loss of intensity caused by the use of a monochromator. The availability of highly resolved spectra will make it possible in the future to test more closely, at least for simple molecules, the validity of theoretical calculations.

TABLE II. Electron shakeup in photoionization of K shell of N_2[a]

Line	"Predominant" Orbital transition	Theory[b] E	Theory[b] I	Experiment[c] E	Experiment[c] I
1	$1\pi \to 2\pi$	7.7	0.2	9.3	2.1
2	$1\pi \to 2\pi$	18.3	5	16.0	24.6
3	$\{\begin{array}{l}5\sigma \to 6\sigma\\5\sigma \to 6\sigma\end{array}$	$\{\begin{array}{l}21.5\\22.3\end{array}$	$\{\begin{array}{l}1.6\\0.2\end{array}$	19.6	0.9
4	$5\sigma \to 7\sigma$	23.0	0.3	20.8	0.3
5	$\{\begin{array}{l}1\pi \to 3\pi\\5\sigma \to 8\sigma\end{array}$	$\{\begin{array}{l}23.7\\24.6\end{array}$	$\{\begin{array}{l}1.6\\0.1\end{array}$	~ 22.1	~ 0.7
6	$5\sigma \to 8\sigma$	25.1	1.2	23.0	1.9
7	$\{\begin{array}{l}4\sigma \to 7\sigma\\5\sigma \to 10\sigma\end{array}$	$\{\begin{array}{l}27.8\\28.0\end{array}$	$\{\begin{array}{l}0.5\\0.1\end{array}$	24.6	3.1
remainder		31–35	4.9	27–29	~ 4

[a] Satellite position (E, eV) and intensity (I, %) are with respect to main peak.

[b] Ref. [11].

[c] Ref. [5].

Hillier and Kendrick [11] have employed configuration interaction in the final state wave function using single-electron excitations which yield configurations of $^2\Sigma^+$ symmetry (the same as
the ground state for a hole in the 1s shell of N_2). Equation (2)
was then used by substituting the wave functions for the various
excited configurations which have been relaxed to the 1s vacancy.
Configuration interaction was not used for the unrelaxed initial
state. The agreement between theory and experiment is only marginal,
particularly with regard to intensity. What is most striking is
the prediction that the lower and upper states due to spin coupling
(e.g., $1\pi \to 2\pi$) can be separated in energy by more than 10 eV and
have intensity differences of more than an order of magnitude. If
the assignment of the peaks are correct these conclusions are verified experimentally, but the assignment could of course be in
error.

Of particular interest are the recent calculations of Martin
et al. [13] on HF. They include configuration interaction in both
the initial and final states and use a good basis set for the
Hartree-Fock calculations. By so doing they have been able to obtain excellent agreement between theory and experiment. (See Table
III). Note that inclusion of configuration interaction in the
initial state raises the transition probability by about a factor

TABLE III. Electron shakeup in HF[a]

Line	"Predominant" Orbital transition	Theory			Experiment	
		I(A)	I(B)	E	I	E
1	$3\sigma \rightarrow 4\sigma$	0	0.1	23.9	-	-
2	$3\sigma \rightarrow 4\sigma$	1.2	2.0	25.9	1.9	22.4
3	$1\pi \rightarrow 2\pi$	1.5	3.0	29.6	3.0	26.5
4	$3\sigma \rightarrow 5\sigma$	0	0	30.9	-	-
5	$1\pi \rightarrow 2\pi$	3.6	6.2	32.4	5.7	29.9
6	$3\sigma \rightarrow 5\sigma$	0.7	1.2	33.3	1.0	30.9
7	$1\pi \rightarrow 4\pi$	2.8	4.1	34.8	3.8	32.7

[a] From ref. [13]. Satellite position (E, eV) and intensity (I, %) are with respect to main peak. Theory A has configuration interaction in the final state only, while theory B employs configuration interaction in both initial and final states.

of 1.4 to 2.0. Thus, configuration interaction in the initial state yields additional pathways to the final excited states. The correlated wave function in the initial state includes contributions of excited configurations that will effectively overlap with some excited configurations in the final state. The calculations suggest that the inclusion of configuration interaction in the initial state will nearly double the total shakeup probability, although the relative shakeup probabilities may not be strongly affected. On viewing Table III it is also interesting to note that the energy splitting for lower and upper states due to spin coupling ranges from 2.0 to 2.8 eV, while the intensity ratio is sometimes greater than a factor of 10.

Though the agreement in Table III between experiment and theory is excellent, the experimental data were not taken with a monochromatic x-ray source and the results of the higher states depend strongly on the goodness of the deconvolution. The authors had to assume a sizeable variation in the width of the different shake-up peaks (from 2.1 to 7.9 eV). In summary, although the agreement between theory and experiment for the sharper and more intense peaks (2, 3 and 5) can be taken seriously, the agreement of the remainder of the spectra should be taken as consistent with theory rather than proven.

To conclude this section, a few words ought to be said on the possible causes for broadening of the shakeup peaks in the photoelectron spectra of molecules. The natural line widths of the Al and Mg x rays are about 1 eV, while a slice of this line, as low as 0.2 eV, can be taken with a monochromator. The lifetime of the shakeup states ought not to be much different from that of the

ground state, and depends primarily on the lifetime of the core
hole. For a 1s vacancy in the first row elements this results in
a natural width of only about 0.1 to 0.2 eV. Excited states which
can lead to autoionization in the valence shell might result in
extra broadening. Large increased widths such as reported in Table
III, may be accounted for by unresolved Rydberg states. With a
monochromatic x-ray source this possibility might be checked. Finally, in molecules broadening can also occur through unresolved
vibration states. One may often expect to find that the shakeup
state has a substantially different molecular potential than the
ground state, even to the point of having Franck-Condon transitions
to a predissociated level. Not only broadening but a shift in the
centroid of the photoelectron peaks of a few volts may occur. This
incidentally means that calculations based on Koopmanns' frozen orbital approximation may be more realistic with regard to comparing
vertical ionization potentials than calculations based on adiabatic
binding energies.

IV. APPROXIMATE METHODS AND COMPLEX MOLECULES

For more complex molecules and even for simple molecules it is
often desirable to use semi-empirical and other approximate methods
for estimating the nature and extent of electron shakeup. The results of satellite structure for a number of gaseous molecules have
been so rationalized [3,4,7]. One method has been to use energy
separations measured in optical data on excited states of the neutral molecules to deduce shakeup energies. The argument for such a
procedure follows from the assumption that the valence electrons
are not strongly perturbed by a core vacancy. This assumption is
rather poor, though data on neutral molecules may give some clues
as to the relative spacing of the various excited states. A sounder
procedure (based on the rationale of equivalent charge) is to use
data on neutral molecules which are equivalent to the creation of
a core vacancy. That is, NO is said to be equivalent to N_2^+ with a
hole in one of the K shells. The limitations of this method are
that the chance of finding data on a neutral molecule in which the
equivalent charge approximation is applicable is rather poor and
that the equivalent charge method neglects the effect of spin
coupling.

Finally, a third approximate method for evaluating electron
shakeup has been to make use of population analysis for a molecular
orbital derived from a MO-LCAO model. That is

$$\phi_j = \sum_{A,\lambda} C_{A\lambda j} \, \phi_{A\lambda} \qquad\qquad (3)$$

where ϕ_j is the molecular orbital, $\phi_{A\lambda}$ the atomic orbital, and
$C_{A\lambda j}$ are the relative atomic densities. When comparing the spectra

made by forming core holes in the different atoms, one anticipates
that the greatest chance for exciting a given molecular orbital
will occur when the vacancy is made in the atom where the orbital
is localized. The approximate methods are highly speculative, but
as better calculations give us more insight into the nature of
electron shakeup, it may be hoped that approximate methods will be
made with more confidence.

Although it might seem that predictions of electron shakeup
which can sometimes be so difficult, even for diatomic molecules,
must be impossible for complex molecular systems. However, in
cases where a special transition has an unusually high probability,
the behavior of this transition can be understood for even very
complex molecular species. For example, in organic compounds a
distinctive satellite is found at low excitation energies (5 to 10
eV) which has been assigned to the π orbital associated with the
carbon-carbon double bond. Clark and his co-workers [14] have
studied this effect for a number of organic solids and polymers.
The intensity of the shakeup peak is found to be dependent on the
relative number of double bonds and the nature of substitutional
groups attached to the carbon. Calculations based on CNDO molecular
orbitals have also been used to help correlate the data. Recently,
these studies have been extended to free molecules in the gas
phase [15].

A large number of studies [16] have been made on the satellite
structure found in the photoionization of the 2p shell of the first
row transition metal compounds. The satellite structure is sometimes very intense, being as large as the main peak. Again the
main satellite lines are found at lower excitation energies (from
4 to 12 eV). The satellite structure has been shown to arise primarily from electron shakeup [17] and to be due to transitions [18]
from ligand orbitals to orbitals made up of the unfilled 3d shell
of the transition metals. The nature of the shakeup structure is
dependent upon both the nature of the metal ion and ligand, and can
be rationalized on the basis of the energy separation between the
ligand and metal orbitals.

V.   USE OF AUGER AND X-RAY SATELLITES

The Auger process and x-ray fluorescence are essentially two
step processes. First, a vacancy is created in the core shell by
photoionization, electron impact, or some other method; and second,
a radiative (x-ray emission) or nonradiative (Auger process) process
fills the hole. In a molecule the hole is usually filled before
molecular decomposition occurs, and the Auger or x-ray spectra will
contain information about the initial ionization process by way of
satellites corresponding to excited configurations in the initial
state. If the most is to be learned about excitation in the molecular

orbital, the x-ray or Auger transition should take place with the valence shell electrons.

Because x-ray emission involves a rearrangement with one electron while Auger processes involve two, the former spectrum ought to be simpler to analyze. However, since Auger transition rates to the valence shell are generally more intense, it is easier to obtain a high resolution Auger spectrum. Recently, high resolution x-ray spectra have been obtained [19], but as yet there have been relatively few attempts to analyze the data in terms of shakeup and shakeoff. I shall thus confine my remarks to Auger processes.

Figure 3 gives schematically the different Auger transitions as a function of initial excitation. When electron shakeup occurs, an Auger process, which involves the excited electron, will gain exactly the same energy that was lost to the photoelectrons, giving rise to the shakeup. For example [20], most of the high energy satellite structure in the KLL Auger of N_2, where the K hole is produced by photon irradiation, is about 16 eV higher than the main normal Auger lines, in good agreement with the excitation energy for the most intense shakeup peak in the photoionization of N_2.

When electron shakeoff occurs, the satellite structure is usually found at energies lower than the normal Auger peaks. For example, satellites observed about 15 eV lower than the main Auger peaks were assigned to electron shakeoff. Auger spectra from proton bombardment [21] were used to confirm this assignment, since at energies below 400 keV, protons give rise to extensive double ionization due to electron pickup and the peaks in the energy region of the Auger spectrum believed to be due to electron shakeoff were greatly enhanced relative to the normal Auger lines.

The use of satellite lines in Auger and x ray spectra for the study of shakeup and shakeoff in molecules has not yet received extensive attention, but the methods have substantial potential.

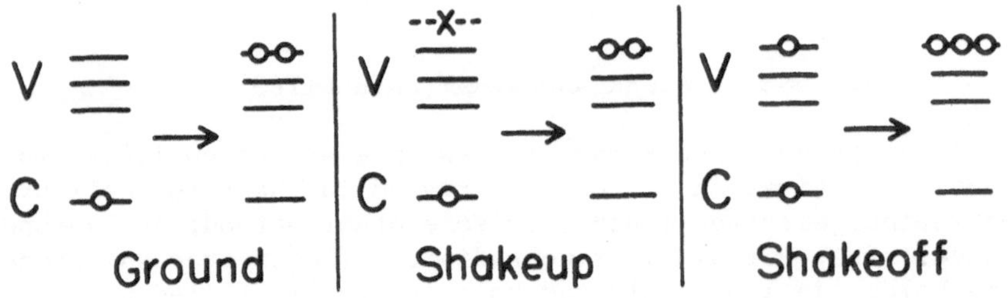

Figure 3. Schematic representation of Auger process as function of initial state excitation. V and C are valence and core shells, O represents a hole and X represents an excited electron.

SUMMARY

Multiple excitation in molecules offers a number of new challenging problems that are not found in atomic systems. In particular, there is the interesting question of the nature of electron shakeup as the result of producing a localized and non-central potential. In fact, are we correct in only assuming that monopole transitions occur? There is a strong need for extending fundamental studies in the general field of molecules, particularly in the area of theory. In addition, the investigation of strong shakeup satellites in complex molecules can be used to shed light on the nature of chemical bonding for these species.

REFERENCES

*Research sponsored by the U. S. Energy Research and Development Administration under contract with Union Carbide Corp.

[1]. A.W.Potts and T.A.Williams, J. Electron Spectrosc. 3, 3 (1974).

[2]. E.g., see H.Basch, Chem. Phys. 10, 157 (1975).

[3]. T.A.Carlson, M.O.Krause and W.E.Moddeman, J. de Phys. (Paris) 32, C4-76 (1971).

[4]. C.J.Allan et al., J. Electron Spectrosc. 1, 131 (1972/3).

[5]. U.Gelius, J. Electron Spectrosc. 5, 985 (1974).

[6]. D.P.Spears, H.J.Fischbeck and T.A.Carlson, J. Electron Spec.-trosc. 6, 411 (1975).

[7]. D.P.Spears, H.J.Fischbeck and T.A.Carlson, Phys. Rev. 9, 1603 (1974).

[8]. L.J.Aarons et al., Mol. Phys. 26, 1247 (1973).

[9]. J.A.Pople and D.L.Beveridge, "Approximate Molecular Orbital Theory," McGraw-Hill, New York (1970).

[10]. I.H.Hillier and J.Kendrick, J.Electron Spectrosc. 6, 325 (1975).

[11]. I.H.Hillier and J.Kendrick, J.Electron Spectrosc. (in press).

[12]. H.Basch, J. Electron Spectrosc. 5, 463 (1974).

[13]. R.L.Martin, B.E.Mills and D.A.Shirley, (to be published).

[14]. D.T.Clark et al., J. Electron Spectrosc. (in press).

[15]. T.Ohta, T.Fujikawa and H.Kuroda, Chem. Phys. Lett. 32, 369 (1975).

[16]. For review, see M.A.Brisk and A.D.Baker, J. Electron Spectrosc. 7, 197 (1975).

[17]. T.A.Carlson, J.C.Carver and G.A.Vernon, J.Chem. Phys. 67, 932 (1975).

[18]. K.S.Kim, J. Electron Spectrosc. 3, 217 (1974).

[19]. L.O.Werme et al., Phys. Rev. Lett. 30, 525 (1973).

[20]. W.E.Moddeman et al., J. Chem. Phys. 55, 2317 (1971).

[21]. M.Stolterfoht, in Proceedings International Conference on Inner Shell Ionization (Atlanta, Georgia, 1972), ed. by R.Fink et al., USAEC Conf-720404 (Washington,D.C., 1973) p. 1043.

HIGH-ENERGY EXCITATIONS IN MOLECULES AND SOLIDS

S. Doniach

Department of Applied Physics and Stanford Synchrotron

Radiation Project, Stanford University, California

1. INTRODUCTION

In these lectures we consider high energy probes of electronic states in molecules and solids as, basically, those involving energies sufficient to reach inner core states of the constituent atoms. From a more fundamental point of view any interaction process involving an energy transfer to the target atoms of $\hbar\omega \gg k_B T$ could be considered as " high energy " ; from an operational point of view, the fact that the binding energies of inner core levels can be specifically attributed to a given element with atomic number Z provides a very important tool for local investigation in a chemically complex material.

The U.V. and X-ray spectroscopy of molecules and solids is a very broad subject and in these lectures a strong selection has been made of a limited number of topics, following the interests of the author. In discussing photoabsorption (and related measurements such as photoemission, X-ray Raman scattering, etc.) in molecules and solids, we can distinguish two classes of physical effects in which the response of the system to the probe differs from that of isolated atoms : one electron effects - multiple scattering, band structure, etc.; and many-electron effects - relaxation, satellites, etc. In both cases we will be interested specifically in the influence of the aggregate of atoms, i.e. in the changes caused by inter-atomic interactions, in contrast to the intra-atomic effects considered in the majority of lectures at the school.

To illustrate these points two specific subjects have been selected for discuss in greater depth. These are :

1) the influence of the fermi gas of conduction electrons in a metal on relaxation effects resulting when an inner core electron is excited ;

2) the large scale modulation of photoabsorption cross sections above an inner shell threshold in molecules and solids - " Kronig structure " or so called EXAFS (Extended X-ray absorption Fine Structure).

As a final topic, we go more deeply into the physics of spectroscopic probes which give more information than simple photoabsorption, namely Compton scattering and X-ray Raman scattering, and discuss how these may be used to obtain more specific information about quantum numbers of high energy excitations in solids.

The printed version of the lectures is in greatly abbreviated form : references are given for more detailed treatment of the subject matter.

2. PHYSICS OF CORE LEVEL RELAXATION IN METALS : THE " X-RAY EDGE SINGULARITY " PHENOMENON

As discussed by other lecturers at the school, the excitation of an electron from a core level in an isolated atom causes a sudden change of charge in the interior of the atom with a resulting severe excitation of other electrons in the atom in the form of a " relaxation " to a new screening configuration. This effect may be observed most straight forwardly by considering the photoemission spectrum resulting from a core level excitation (Figure 1a). The highest energy photoelectron observed from a photon with energy $\hbar\omega$, has energy (setting $\hbar = 1$)

$$\varepsilon = \omega - E_{core} \qquad (2.1)$$

corresponds to a fully relaxed final state, i.e. to a minimum energy transfer to the atom. Lower energy photoelectrons correspond to events in which shake up or shake off occurs, with excitation energy

$$\varepsilon = \omega - E_{core} - \Omega_{\alpha} \qquad (2.2)$$

The lowest such satellite energy would correspond to a monopole shake up of the outer electron of the atom.

If the atom is now a constituent of a metal, then the outer electron orbital is degenerate with that on all the other atoms and broadens out into

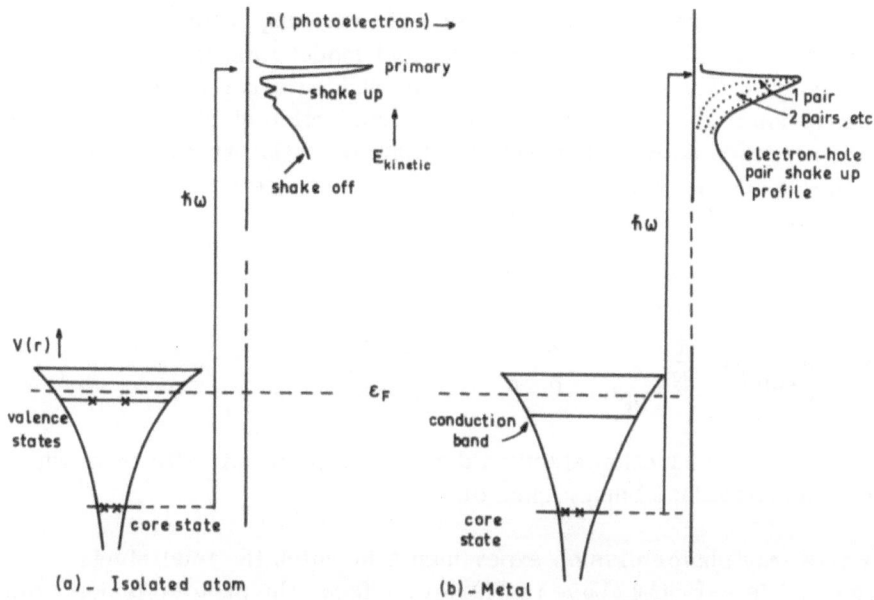

Fig. 1. Comparison of satellite structures : atoms versus solids.

a conduction band of states, occupied up to a fermi level as determined by
the Pauli principle. It is now evident that the discrete shake up energy of
an isolated atom broadens out into a band of possible excitation energies
stretching down to zero. Consequently one expects a band of low energy
satellites as opposed to the discrete shake up states of the isolated atom
(Figure 1b). In fact, it turns out that the possibility of excitation of many
conduction electrons (as opposed to the single shake up electron in the
isolated atom) diverges as $\Omega \longrightarrow 0$. This gives rise to a " singularity " in
the shake up satellite spectral density. In practice this singularity is
broadened out to a finite cross section by core hole lifetime effects.
On the scale of atomic excitation energies (Rydbergs) these satellite
structures may seem to be very small effects (tenths of eV above
threshold), however in terms of the physics of the interaction of an electron gas in a metal with a local perturbation, they do give information of
quite general interest (electron - impurity scattering in metals), besides
providing a challenge to solid state theorists and experimentalists. In
order to provide a detailed treatment of the shake up phenomena in metals,
a simplified model Hamiltonian is required which selects out those degrees
of freedom of the system whose excitation is important near the threshold
for an inner shell absorption process.

[It should be kept in mind that a fairly basic difference in philosophy
distinguishes atomic theory from solid state theory - the qualitatively
different complexity of the solid, from the many body point of view : 10^{22}

electrons as opposed to a finite number of order 10, leads to a considerably greater reliance on semi phenomenological model Hamiltonians in solid state theory.] The principal effect of the sudden creation of a core-hole by absorption of an X-ray photon, on the conduction electrons of the metal, may be represented by a screened Coulomb potential which switches on when the core-hole is present, i.e., as a result of the action of a creation operator a^+ for the core-hole state :

$$H_{int} = v\, a^+ a\, \rho^o_{cond} \tag{2.3}$$

where $\quad \rho^o_{cond} = \dfrac{1}{N} \sum_{pp'} C^+_p C_{p'} \tag{2.4}$

represents the conduction electron density at the atomic site from which the core electron has been kicked out.

For X-ray photoemission experiments in which the final electron energy ε_k is very far above the fermi surface, the photoemission cross section may be written approximately as

$$\frac{d^2\sigma}{d\omega d\varepsilon_k} = 2\pi\,|M_k|^2 \sum_f |<\Psi^o_N\,|\,a^+|\,\Psi^f_{N-1}>|^2\, \delta(\omega - E^o_N + E^f_{N-1} - \varepsilon_k) \tag{2.5}$$

where $|\Psi^o_N>$ is the ground state of the fermi sea (in the absence of the core-hole) and $|\Psi^f_{N-1}>$ represents the manifold of states of the fermi sea in the presence of the core-hole, and M_k is the dipole matrix element for excitation of the core state.

Because the spectrum of final states constitutes a continuum, it is much easier to calculate its properties by Fourier transforming (2.5) to a time representation.

$$\frac{d^2\sigma}{d\Omega d\varepsilon_k} = \int_{-\infty}^{\infty} dt\, e^{i(\omega - \omega_k)t}\,|M_k|^2\, g(t) \tag{2.6}$$

where $\quad g(t) = <\Psi^o_N\,|\,e^{-iH^o t}\, a^+\, e^{iHt}\, a\,|\,\Psi^o_N> \tag{2.7}$

is the " deep hole propagator ". And $H = H^o + H_{int}$ represents the Hamiltonian of the system in the presence of the core hole. In time language equations (2.6, 2.7) express the fact that after the X-ray knocks out the core electron, the metal system then propagates with Hamiltonian H involving the effect of the core-hole charge on the conduction electrons.

This statement, that the effect of the X-ray is to produce a frozen hole state which then propagates in time, is precisely the sudden approximation. The problem of calculation the response of a fermi sea to a sudden impulse was solved by Nozières and de Dominicis (1). For a simplified discussion of the resulting relaxation phenomena, the reader is referred to the book by Doniach and Sondheimer (chapter 10) (2). The general solution is of a singular response

$$g(t) = \frac{1}{(\varepsilon_F t)^\alpha}$$ (2.8)

where α is a " singularity index ", depending on the conduction electron-core hole interaction. From (2.8) one can see that the photoemission profile has the form

$$\frac{d\sigma}{d\omega d\varepsilon_k} = (\omega - E_{hole} - \varepsilon_k)^{-\alpha}$$ (2.9)

when lifetime effects are included, Doniach and Sunjic (3) showed that this becomes an asymmetric Lorentzian line shape.

Recent high resolution XPS experiments (4) have shown that this form works remarkably well, when experimental resolution effects are properly taken into account. For the case of photoabsorption near threshold, as opposed to the simpler case of X-ray photoemission discussed above, the final electron is close to the fermi level and its interaction with the core hole cannot be neglected.

The formula for the absorption cross section may again be written in time-dependent form

$$\frac{d\sigma}{d\omega} \propto \frac{1}{2\pi} \int_{-\infty}^{\infty} dt \, e^{i\omega t} F_L(t)$$ (2.10)

where $F(t)$ is now a propagator representing the evolution of the electron gas in the presence both of the core hole and of the final electron at the fermi surface, represented by the creation operator for an outgoing spherical wave with angular momentum L determined by the dipole selection rule for the core hole in question (L=1 for a K-shell, L=0 or 2 for an L-shell) :

$$C_L^+ = \frac{1}{\sqrt{N}} \sum_p C_p^+ Y_L (\Omega_p)$$ (2.11)

One then has

$$F_L(t) = \langle \Psi_{grd} | \; a^+ C_L^{} e^{-iHt} C_L^+ \, a \; | \Psi_{grd} \rangle \qquad (2.12)$$

where H represents the Hamiltonian of the system in the presence of the core-hole conduction electron interaction. The general solution again gives a singular propagator, which now depends on L, of the form

$$F_L(t) \propto \frac{1}{(\varepsilon_F t)^{\propto L}} \qquad (2.13)$$

Comparison of this form with experiment turns out to be more difficult than in the photoemission case on account of energy dependence of the one-electron matrix elements near threshold which tends to obscure the relaxation effects.

3. X-RAY ABSORPTION SPECTRA IN MOLECULES AND SOLIDS - " EXTENDED FINE STRUCTURE " (EXAFS)

When the absorption spectrum for an inner shell of an atom in a molecule or solid is examined, over the range of several 100's of eV above threshold, a series of oscillations (or "fine structure") with period of order 50 eV is observed which does not occur in the isolated atom. This is well illustrated by a comparison of the K-shell absorption in Kr and in Br_2 gas (Kincaid and Eisenberger) (5) (Figure 2). The origin of the modulations of the atomic absorption was first explained by Kronig (6) (1934). More recently Stern, Sayers and Lytle (7) have successfully applied the Kronig theory to analyse a variety of absorption spectra in solids. The theory is a one-electron calculation of the modulation of the one electron matrix element as a result of interference of the outgoing wave of an ejected electron with electron waves diffracted from neighboring atoms in the molecule or solid. The neglect of many electron relaxation effects may perhaps be justified by the idea that for those absorption events in which some shake up or electron-electron collision occurs, the phase of the diffracted electron wave will vary in a random way so that the interference effect leading to the modulations will tend to get washed out. The conjecture is substantiated by the finding (5) that the calculated amplitudes are of order a factor 2 larger than the measured ones, indicating that about 50% of the absorption process are inelastic (for the case of Br_2). The general 1-electron formula for the absorption process is

$$\frac{d\sigma}{d\omega} \propto |\langle \varphi_{core} | \bar{X} | \Psi_{final} \rangle |^2 N(\varepsilon_k) \qquad (3.1)$$

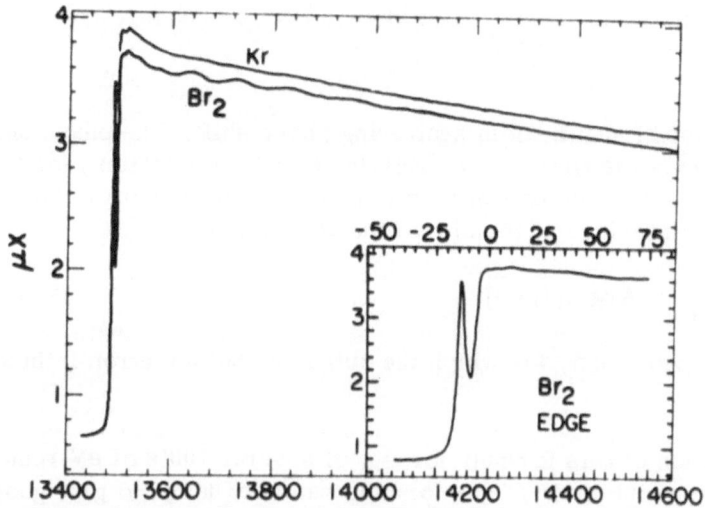

Fig. 2. Photoabsorption above the K-shell of Kr gas and Br_2 gas (Kincaid and Eisenberger, Ref. 5).

where ε_k is the electron energy above threshold

$$\varepsilon_k = \hbar\omega - E_{core} \tag{3.2}$$

If only single scattering with neighbors is included, the electron wave function may be written in terms of the electron-atom scattering amplitude

$$\Psi_{final} = \Psi_{out} + \sum_i \psi^i_{reflected} \tag{3.3}$$

where $\psi^i_{reflected}$ is the amplitude for reflection from an atom at R_i

$$\psi^i_{reflected} = \frac{e^{ik\cdot R_i}}{R_i} f(\theta,k) \frac{e^{ik|r-R_i|}}{|r-R_i|} \tag{3.4}$$

Substituting in (3.1) the absorption cross section may be written in terms of the atomic absorption as

$$\frac{d\sigma}{d\omega} = \frac{d\sigma_{atomic}}{d\omega} \left(1 + \sum_i \frac{|f(\pi,\varepsilon_k)|}{R_i^2 k} \sin(2kR_i + \alpha) \right) \tag{3.5}$$

where $f(\pi,\varepsilon_k)$ is the electron-atom back scattering amplitude

$$f(\pi, \varepsilon_k) = \sum_{\ell=0}^{\infty} (-1)^{\ell} \frac{e^{2\iota\eta_\ell}-1}{2ik} \qquad (3.6)$$

and η_ℓ is the electron-atom scattering phase shift. The phase shift α in (3.5) represents the time delay (relative to a free electron) for the electron to escape from the parent atomic potential, bounce off the neighbor, and return to interfere at the site of the absorption

$$\alpha = 2\eta_1 + \text{Arg } [f(\pi)] \qquad (3.7)$$

for K-shell absorption, for which the outgoing photoelectron is in a p wave ($\ell = 1$).

Application of this formula for ε_k of several 100's of eV requires η_ℓ for ℓ up to about 15 (8). This formula has been found to give good fits to the phases of the observed modulations using phase shifts calculated from self consistent atomic potentials for the case of Br_2 and reasonable fits for the metal Cu by Ashley and Doniach (9). In the case of solids, it is clear from the analogous LEED calculations that multiple scattering of the fast photoelectron needs to be taken into account. This is discussed in more detail by Ashley and Doniach and by Lee and Pendry (10).

The really interesting applications of the EXAFS technique are to complex systems involving many different chemical elements. By tuning to the absorption edge of a specific atomic species, information about the local environment of this element can be sorted out from that of other elements. This technique appears particularly promising for organo-metallic systems where, for example, in Fig.3, spectra of three different porphyrins show clear similarities for the local environment of the metallic atom (11).

4. OTHER PROBES OF EXCITATIONS IN ATOMS, MOLECULES AND SOLIDS : X-RAY COMPTON AND RAMAN SCATTERING

In addition to photoabsorption, which gives a total cross section for the interaction of a photon with matter, more detailed techniques involving photons are gaining interest as more intense synchrotron sources become available.

In this section a brief mention of X-ray Compton and Raman scattering will be made.

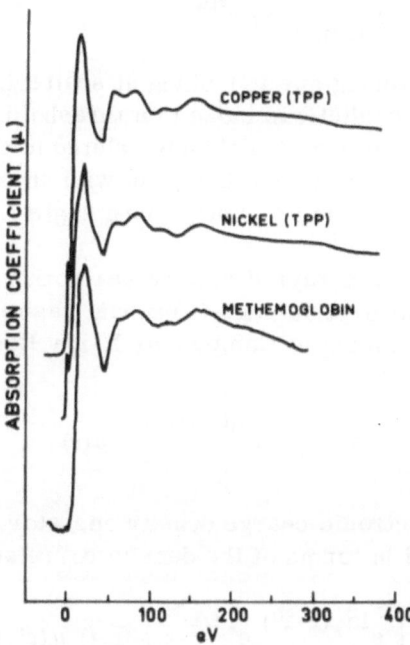

Fig. 3. Photoabsorption in a series of tetraphenyl porphyrins (TPP) -
(Kincaid, Eisenberger, Hodgson and Doniach, Ref. 11).

In those experiments, which are identical from the point of view of
measurement, a photon beam is inelastically scattered from a solid leaving
it in an excited state (energy loss)

$$\hbar\omega_{\underset{\sim}{k}} + M \quad \longrightarrow \quad M^* + \hbar\omega_{\underset{\sim}{k'}}$$

By energy and momentum conservation, the resulting excitations in the
solid carry total momentum and energy :

$$\left.\begin{aligned} \underset{\sim}{q} &= \underset{\sim}{k} - \underset{\sim}{k'} \\ \hbar\Omega &= \hbar\omega_{\underset{\sim}{k}} - \hbar\omega_{\underset{\sim}{k'}} \end{aligned}\right\}$$

Thus the total cross section, $\dfrac{d\sigma}{d\Omega dq}$ contains both momentum and energy
information. The distinction between Compton and X-ray Raman
scattering results from a comparison of the binding energy E_B of electrons
being excited with the excitation energy $\hbar\Omega$.

For Compton scattering $\hbar\Omega \gg E_B$ so that the binding energy may be
neglected relative to the energy of the outgoing electron. The cross section
then measures the form factor of the bound electron (12)

$$f_{\underset{\sim}{p}}^{bound} = \int d^3 r \ \Psi_{bound} (r) \ e^{i\vec{p} \cdot \vec{r}}$$

For X-ray Raman scattering one is looking at scattering from electrons for which $\hbar\Omega \cong E_B$, i.e. excitations close to a threshold for exciting a core electron. In this case, information closely related to that obtained in a photoabsorption experiment is obtained, but with an extra parameter, namely the momentum transfer (or scattering angle).

For the scattering of X-rays it may be shown that provided the incident photon is not near an absorption edge (resonant case), the principle interaction with the photon leading to Compton or X-ray Raman scattering is the Thompson term (12),

$$H_{int} = \int d^3 r \ \sum_{kk'} A_k A_{k'}^* \ e^{i(k-k') \cdot r} \ \rho(r)$$

where $\rho(\underset{\sim}{r})$ is the electronic charge density operator. The cross section may then be expressed in terms of the density correlation function

$$\frac{d\sigma}{d\Omega dq} = \int_{-\infty}^{\infty} dt \int d^3 r \ e^{i\vec{q} \cdot (\vec{r} - \vec{r}')} \ e^{i\Omega t} \ < \rho(\vec{r}, t) \ \rho(r', 0) >$$

In the forward direction ($q \rightarrow 0$) it then follows that

$$\lim_{q \rightarrow 0} \frac{d\sigma}{d\Omega dq} \ \propto \ q^2 \ \epsilon''(\Omega)$$

where $\epsilon(\Omega)$ is the dielectric function of the material, i.e. in the forward direction, X-ray Raman scattering is identical to photoabsorption.

For $q \neq 0$ the X-ray Raman experiment corresponds to a photoabsorption with finite momentum transfer. In general this leads to a breakdown of the dipole selection rule for photoabsorption, so that the X-ray Raman cross section will also lead to information about monopole and quadrupole transitions. This additional information can prove useful in analysing the satellite structure of a given absorbtion spectrum as shown by Doniach, Platzmann and Yue (13).

For the case that the incident photon is near an absorption edge, additional effects occur which lead to a " resonant X-ray Raman " scattering, below threshold and a " resonant fluorescence " above threshold (14, 15). This subject has not yet been explored in detail in the X-ray region (16).

REFERENCES

(1) P. Nozières and C. De Dominicis, Phys. Rev. 178, 1097 (1969).
(see also G. Mahan " Solid State Physics " 29, Academic Press 1974,
p.75.

(2) S. Doniach and E.H. Sondheimer, " Green's functions for solid state
physicists ", Addison Wesley 1974, Chapter X.

(3) S. Doniach and M. Sunjic, J. Phys. C3, 285 (1970).

(4) P. Citrin, G. Wertheim and Y. Baer, Phys. Rev. Lett. 35, 885 (1975).

(5) B. Kincaid and P. Eisenberger, Phys. Rev. Lett. 34, 1361 (1975).

(6) L. Kronig, Z. f.Phys. 70, 317 (1931).

(7) D. Sayers, F. Lytle and E. Stern, " Advances in X-ray analysis " 13,
248 (1970) - Plenum Press.

(8) B. Kincaid, Thesis (unpublished), Stanford University 1975.

(9) C. Ashley and S. Doniach, Phys. Rev. B11, 1279 (1975).

(10) P.A. Lee and J. Pendry, Phys. Rev. B11, 2795 (1975).

(11) B. Kincaid, P. Eisenberger, K.O. Hodgson and S. Doniach, Proc. Nat.
Acad. Sci. (U.S.A.) 72, 23 (1975).

(12) See for instance P. Platzmann and N. Zoan, Phys. Rev. 139, 410
(1965).

(13) S. Doniach, P. Platzmann and J.Yue, Phys. Rev. B4, 3345 (1971).

(14) C. Sparks, Phys. Rev. Lett. 33, 262 (1974).

(15) Y. Bannet and I. Freund, Phys. Rev. Lett. 34, 372 (1975).

(16) P. Nozières and E. Abrahams, Phys. Rev. B10, 3099 (1974).

RELATIVISTIC EFFECTS IN ATOMIC STRUCTURE CALCULATIONS :

AN INTRODUCTION

J. P. Desclaux

Institut Max von Laue - Paul Langevin

B.P. 156   38042 GRENOBLE Cedex   France

I GENERAL CONSIDERATIONS

Let us first consider qualitatively how relativistic effects influence atomic structure calculations. Using atomic units (i.e. $e = \hbar = m_o = 1$), the mean velocity of a 1s electron in a Coulomb field of charge Z is exactly equal to Z. In the same system of units the velocity of light is roughly 137; thus, for a rare earth atom the mean velocity of the 1s electron is already almost half the velocity of light, and its mass is about 1.15 that of the rest mass. We may get a feeling of the main changes induced by this variation in the mass by simply considering the Schrödinger equation for the apparent mass instead of the rest mass. As the energy is directly proportional to the mass ($E = -Z^2 m/2n^2$) we conclude that the binding energy should increase while at the same time the charge density should contract towards the origin (the scaling factor is r/m). If we now consider electrons of higher quantum number n, we will conclude from the preceding argument that they will be essentially unaffected since their velocity is rather small compared to the light velocity. But as their wavefunctions have to be orthogonal to the inner electron ones, they experience an indirect relativistic effect which also results in a contraction of the charge density. For example, consider the 1s orbital for the Z = 10 hydrogenic case and the 10s orbital of the Z = 100 case. Both have the same energy but while the 1s orbital contracts by only about 0.2% the orthogonality constraint causes the 10s orbital to shrink by 6% (the 1s being contracted by 21% in this Z = 100 case).

In illustration, the normalized charge densities for 1s and 2s electrons in a Coulomb field of charge Z = 80 are plotted in Fig. 1. Note that while the non-relativistic density profile vanishes at the node of the 2s radial function, the relativistic one exhibits only a non-zero minimum. This is due to the fact that the nodes of the large and small components of the relativistic wavefunction interlace. In Table 1 some eigen-values and expectation values of r are listed for various s orbitals in the same Z = 80 hydrogenic case. If we measure the relativistic contrac-tion of the s orbitals by the ratio of the relativistic to the non-relativistic values, we notice that the 2s orbital is relatively more contracted than the 1s. This is a consequence of the orthogonality constraint, and this result holds for all 2s shells. For other s orbitals the relative contraction is a decreasing function of the principal quantum number n. For non-zero angular momentum orbitals, the relativistic effects also result in a contraction of the charge density towards the nucleus.

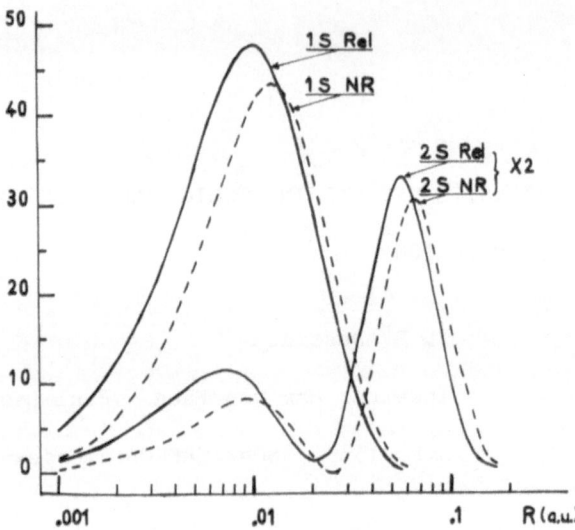

Fig. 1   Normalized charge densities for 1s and 2s electrons in a Coulomb field of charge Z = 80.

Table 1   Relativistic binding energies and expectation values of r in the Z = 80 hydrogenic case. The ratio to the non-relativistic results is also given.

This contraction is more important for the j = ℓ-1/2 subshell than for the j = ℓ+1/2 one. For a given set of values (j = ℓ-1/2 or j = ℓ+1/2) the contrac-tion decreases with ℓ when n is held fixed. A detailed study of the influence of rela-tivistic effects on hydrogenic wavefunc-tions has been given by Burke and Grant (Ref. 1).

	ε	Ratio	<r>	Ratio
1s	3532.18	1.104	0.00164	0.875
2s	904.845	1.131	0.06497	0.866
3s	392.083	1.103	0.15154	0.898
4s	216.424	1.082	0.27570	0.919
5s	136.695	1.068	0.43737	0.933
6s	94.0245	1.058	0.63656	0.943
9s	41.0755	1.040	1.45914	0.961

If we now consider the many-electron problem, it is clear from the results we obtain in the one-electron case that the inner orbitals will contract towards the nucleus and thus screen more effectively the nuclear potential experienced by the outer electrons. Because of this more effective screening, the charge density of these electrons will tend to expand. This effect, an indirect one, or a self-consistency effect, will compete with the direct relativistic effect due to the variation of mass for almost every shell and the net result on the valence orbitals cannot be predicted a priori. But we may expect that for some orbitals, contrary to the one-electron case, a relativistic treatment will result in an expansion of the charge density. Furthermore, even if not "relativistic" by themselves, the valence orbitals may be strongly affected by relativistic effects, as these effects propagate through the self-consistency requirement of the atomic field. Before discussing in more detail the many-electron problem, let us summarize the theory underlying these calculations.

II BACKGROUND THEORY

For one-electron atoms the relativistic counterpart of the Schrödinger hamiltonian is the Dirac hamiltonian (Ref. 2) which in a non-covarient form may be written as:

$$h_D = c\underset{\sim}{\alpha} \cdot \underset{\sim}{p} + c^2\beta + V(r) \tag{1}$$

where c is the velocity of light, p is the momentum operator, $V(r)$ is the potential energy in the field of the nucleus. The operators $\underset{\sim}{\alpha}$ and β are given by:

$$\underset{\sim}{\alpha} = \begin{pmatrix} 0 & \underset{\sim}{\sigma}^P \\ \underset{\sim}{\sigma}^P & 0 \end{pmatrix} \qquad \beta = \begin{pmatrix} 0 & 0 \\ 0 & -2I \end{pmatrix} \tag{2}$$

where the $\underset{\sim}{\sigma}^P$ are the usual second-order Pauli matrices, 0 and I are respectively the second-order zero, and unit matrices. With the definition of β as given in Eq. 2 the rest mass energy of the electron has been subtracted. In Eq. 1 the nucleus is considered as a classical system, i.e. its field is not quantized. Corrections to this approximation are taken into account by quantum electrodynamics and are responsible for the well-known Lamb shift. As the spin-orbit interaction is explicitly included, the orbital (ℓ) and spin (s) angular momenta do not commute separately with $h_{\underset{\sim}{D}}$. Only the total angular momentum of the electron $j = \ell+s$ and the parity operator commutes with the Dirac hamiltonian. It can be shown that the operator:

$$K = \beta(I + \underset{\sim}{\sigma} \cdot \underset{\sim}{\ell}) \qquad \text{where} \qquad \underset{\sim}{\sigma} = \begin{pmatrix} \underset{\sim}{\sigma}^P & 0 \\ 0 & \underset{\sim}{\sigma}^P \end{pmatrix} \tag{3}$$

also commutes with h_D. Its eigenvalues ($k = -(\ell+1)$ if $j = \ell+1/2$ and $k = \ell$ if $j = \ell-1/2$) associated with those of j_z can be used to classify the eigenstates of h_D. The eigenfunctions (four row vectors) can be written as:

$$\phi_{nkm}(r) = \frac{1}{r} \begin{pmatrix} P_{nk}(r) \; X_{km}(\theta,\phi) \\ iQ_{nk}(r) \; X_{-km}(\theta,\phi) \end{pmatrix} \tag{4}$$

where P and Q are respectively the large and small radial functions while the X's are two row spin-orbit eigenfunctions.

If the Dirac theory associated with quantum electrodynamics provides a quite adequate relativistic description of one-electron systems, the extension of the theory to the many-electron case is far from being a trivial problem. In fact no Lorentz invariant hamiltonian exists in a closed form, i.e. the fully relativistic interaction operator between two electrons consists of an infinite series of terms. The only practical formulation for a relativistic Hartree-Fock type of calculation is the approximate Dirac-Breit hamiltonian (Ref. 3):

$$H = H_o + H_B \tag{5}$$

where

$$H_o = \sum_i h_D(i) + \frac{1}{2} \sum_{i,j} \frac{1}{r_{ij}} \tag{6}$$

$$H_B = -\frac{1}{2} \sum_{i,j} \frac{1}{r_{ij}} \left[\underset{\sim}{\alpha}_i \cdot \underset{\sim}{\alpha}_j + \frac{(\underset{\sim}{\alpha}_i \cdot \underset{\sim}{r}_{ij})(\underset{\sim}{\alpha}_j \cdot \underset{\sim}{r}_{ij})}{r_{ij}^2} \right] \tag{7}$$

h_D is the Dirac one-electron hamiltonian as defined by Eq. 1. The second term of Eq. 6 is the classical Coulomb repulsion between the electrons while the Breit operator (Eq. 7) corrects for the relativistic interaction between the electrons up to the order $1/c^2$ compared to Coulomb repulsion. Because of the approximations involved in the derivation of the Breit operator, the wavefunctions have to be calculated as the eigenfunctions of H_o and the Breit interaction can be included only as a first-order perturbation correction to the energy. For a discussion of this point see, for example, Bethe and Salpeter (Ref. 4). Quantum electrodynamic effects appear as higher-order correction terms; for example, the Lamb shift is of order $(Z/c)\log(1/c)$ compared to the Breit interaction. If we assume that the Dirac-Breit hamiltonian provides a good approximation, we can use it to derive Dirac-Fock equations quite analogous to the non-relativistic Hartree Fock ones. The radial functions $P(r)$ and $Q(r)$ are then determined by the usual self-consistent-field (SCF) technique. The relativistic SCF procedure has been described in

great detail by Grant (Ref. 5) and we shall not discuss it here.
Compared to the non-relativistic usual hamiltonian, the Dirac-Breit
formalism includes:

    - the variation of mass and the spin-orbit coupling through
the one electron Dirac hamiltonian

    - the retardation in the Coulomb interaction, the spin-spin,
spin-other-orbit and orbit-orbit interactions through the Breit
operator.

Before concluding theoretical considerations, let us point out
some of the difficulties encountered in going beyond the Hartree-
Fock approximation. As the multiconfiguration technique will be
discussed in a following section, we shall consider only perturbation
theory here. In such theories one starts with a zero order hamiltonian h_o, the wavefunctions of which are known, and treat the
remaining part of the true hamiltonian according to various techniques. But in any case one needs the complete set of eigenfunctions
of h_o. In the non-relativistic case this complete set generally
involves a discrete spectrum and a continuum spectrum while in the
relativistic case we must consider a discrete spectrum and two continuum spectra. The existence of two continua arises from the fact
that the Dirac equation gives two solutions for the energy, one of
the order $+m_o c^2$ and the other $-m_o c^2$. The first set of eigenvalues
is associated, as in the non-relativistic case, both with bound and
continuum states, while the second set presents only continuum states
corresponding to positron states. The existence of the two continua
will of course increase the complexity of the calculation and furthermore it can be shown (Ref. 6) that even when considering Hartree-Fock
solutions as zero-order solutions the Brillouin theorem holds only
for transitions to states of positive energy. Considering the complexity of the problem it may seem attractive to calculate separately
correlation and relativistic effects. The validity of such an
approach has been studied for atomic hyperfine calculations (Ref. 7)
and it was demonstrated that corrections to this additive approach
are of order $(Z/c)^2$ smaller than the hyperfine structure itself.
Calculations of correlation effects have now reached such a level
of sophistication that in a certain number of cases this correction
cannot be neglected.

III SOME APPLICATIONS OF THE DIRAC-FOCK THEORY

III.a Charge Densities

Relativistic calculations have now been performed for all neutral
atoms of the periodic table (and beyond!) and for a variety of ions.
A systematic comparison between Hartree-Fock (HF) and Dirac-Fock (DF)
results for the neutral atoms has been given by the author (Ref. 8)
and we shall summarize here only the main results concerning the

charge densities of the various orbitals:

-s orbitals always contract towards the nucleus. As for the hydrogenic case, the relative contraction of the 2s is always greater than for the 1s. The relative contraction of outer ns shells is modulated essentially by the changes in d and f shells and the behaviour is far from being a smooth function of Z. In some cases the relative contraction of the valence s orbitals is greater than the K contraction (a typical example is for the 6s orbital around Z = 80).

-the $p_{1/2}$ subshells behave very similarly to the s shells (the small component of these orbitals is of the "s" type) while the $p_{3/2}$ subshells are less contracted with even a small expansion as the shell begins to be filled.

-d shells show the same general behaviour as $p_{3/2}$ subshells with a more pronounced expansion and a less important contraction. f shells are always expanded.

The non-smooth behaviour as a function of Z and the expansion of some of the charge densities is the result of the self-consistent effect already mentioned in Section I.

III.b Binding Energies

The neon and fermium 1s binding energies are given in Table 2 to illustrate the order of magnitude of the various contributions. For neon the HF result (calculated as the difference between total energies) is 868.6 eV and it can be seen from the values listed that correlation and relativistic effects are of the same importance, even for such a light element. The agreement with experiment may have been anticipated to be better but the two contributions have been calculated independently and as already mentioned in the previous section, they are not truly additive. For the fermium atom the direct relativistic correction amounts to about 20% (28 KeV!) of the binding energy and is by far the most important correction to be considered. The higher order relativistic corrections have to be carefully examined to obtain good agreement with experiment. Even if the correlation energy is not

Table 2

K shell binding energy (in eV)

	Ne (Z=10)	Fm (Z=100)[+]
Elec.[(a)]	869.78	142 929
Magn.	-0.34	-715
Retard.	<0.01	41
Vac. Flu.	-0.10	-457
Vac. Pol.		155
Correl.	0.81[*]	?
Total	870.15	141 953
Exper.	870.32	141 963

+ see Ref. 10   * see Ref. 9

(a) Unperturbed total energies difference

known, it may be argued that its value is certainly at least an order or magnitude smaller than the Lamb shift contribution. It is interesting to note that for such a high Z element the proton charge can no longer be considered as punctual but must be distributed over the nuclear volume. We noticed (Ref. 10) that a change of 0.1 Fm in the nuclear radius modifies the K binding energy by nearly 8eV. From such considerations it is clear that in calculating heavy elements one must consider with great care every parameter (even the value of the velocity of light!).

We now consider modifications in outer shells induced essentially by the self-consistent effect. In Table 3 we list binding energies for 4f and 5s electrons in mercury. The HF result predicts that the 4f electrons are more bound than the 5s while the contrary is obtained (and in agreement with experiment) with the DF method. As already mentioned, the inner shells have contracted and thus screened more effectively the Coulomb potential experienced by the 4f electrons. The charge density of these electrons expands and consequently screens less effectively the nuclear potential seen by the 5s electrons. This effect, combined with the orthogonality requirement, has pushed the 5s electrons well inside, compared to the non-relativistic result. From experimental data (Ref. 11), the binding energy of the $4f_{5/2}$ electron exceeds that of the 5s for bismuth (Z = 83) and heavier atoms. This result is in agreement only with relativistic calculations (Ref. 8); the non-relativistic calculations (Ref. 12) predict crossover at a lower atomic number (Z = 78).

IV BEYOND THE SINGLE jj CONFIGURATION

Up to now we have systematically avoided mentioning the problem of angular momentum coupling for many-electron systems which, as we will see, introduces further complications into relativistic calculations. As mentioned in Section II, the one-electron wavefunctions ϕ_{nkm} defined by Eq. 4 are only eigenfunctions of the total angular momentum of the electron. Thus the total wavefunction of the system (built up as an antisymmetric product of ϕ_{nkm} functions) is itself only an eigenfunction of the total angular momentum $J = L + S$ of the system and not a simultaneous eigenfunction of the orbital L and spin S angular momenta. Stated in another way, this implies

Table 3

Ionization potentials for mercury (in eV)

		$4f_{5/2}$	$4f_{7/2}$	5s
Koopmans'	H.F.	136.4	136.4	113.8
Theorem	D.F.	121.7	117.3	138.9
Total Energy	D.F.	108.8	104.8	133.7
Experiment *		111.1	107.1	134

* Ref. 13

that for open shell systems we are forced to expand the total wave-function on a basis defined in the (j,j) coupling scheme and not in the (L,S) scheme. If we restricted ourselves to a single jj configuration, such a procedure would be meaningful only when spin-orbit interaction dominates the Coulomb repulsion, which obviously is not true for most atomic systems. It is thus necessary to define a procedure taking into account the whole LS configuration and not to restrict calculations to the lowest jj subconfiguration. To illustrate the problem we are faced with, consider the very simple p^2 configuration. The spin-orbit interaction splits the p shell in the two $p_{1/2}$ and $p_{3/2}$ subshells and consequently gives rise to the three $p_{1/2}^2$, $p_{1/2}p_{3/2}$ and $p_{3/2}^2$ jj subconfigurations. Consider now the ground state of this configuration which in the pure LS limit is the $^3P_0$ term. As two of the three subconfigurations exhibit an eigenfunction of J^2 with J = 0 as eigenvalue, this ground state will involve a strong mixture among these two configurations if Coulomb repulsion dominates spin-orbit interaction. Again in a pure LS limit, the ground state wavefunction is given by (Ref. 14):

$$|^3P_0> = \{|p_{3/2}^2> - \sqrt{2} \ |p_{1/2}^2>\}/\sqrt{3} \tag{8}$$

where for each subconfiguration we consider the J = 0 eigenvector. More generally we can express the wavefunction as:

$$\psi_G = a \ p_{3/2}^2> + b \ p_{1/2}^2> \tag{9}$$

The departure of the coefficients a and b from the values given in Eq. 8 will only be a measure of the breakdown of (L,S) coupling, and corresponds to a mixture with the $^1S_0$ higher state. From this simple example it is obvious that in the relativistic case a physically realistic calculation should take into consideration the various jj subconfigurations arising from the single LS configuration and will be, for open shell systems, more complex than its non-relativistic counterpart. We will now briefly review the various techniques available to tackle the problem.

IV.a   Effective Operators

The effective operator technique first introduced by Sandars and Beck (Ref. 15) has been recently reviewed (Ref. 16) and may be summarized as follows: consider a relativistic state ψ^R which, in the non-relativistic limit, goes into the state ψ. For any relativistic operator 0^R one can define an effective operator 0^E such that:

$$<\psi^R|0^R|\psi^R> = <\psi|0^E|\psi> \tag{10}$$

This makes possible the use of the well-developed LS angular momentum theory, provided that relativistic radial functions are available to compute the required matrix elements. On the other hand, the only way to introduce the breakdown of LS coupling is to rely on experimental results.

This formalism has been used to calculate transition probabilities (Ref. 17) and hyperfine structure constants (Refs. 17e and 18). Good results are obtained if one is clever enough in the choice of the relativistic radial functions used in the calculation of the matrix elements.

IV.b Configuration Interaction and Multiconfiguration

If relativistic radial wavefunctions are available, one can, instead of using the effective operator formalism, determine the ψ^R state by diagonalization of the hamiltonian matrix and then compute the mean value $\langle\psi^R|O^R|\psi^R\rangle$ of the operator under consideration. If less familiar to atomic physicists, the angular momentum machinery in (j,j) coupling is after all no more complicated than in (L,S) coupling and is well-known in nuclear physics. This procedure of expanding a given state on a basis set has been used for some time in non-relativistic atomic structure calculations to introduce correlation effects and is known as configuration interaction (CI). If we restrict our basis to only the jj subconfigurations arising from the single LS configuration, we will not introduce real correlation effects but only use the CI technique as a way to handle the intermediate coupling problem.

In both effective operators and CI methods one needs a set of relativistic radial functions, the determination of which is left arbitrary. It is obvious that the results will be greatly dependent upon the choice made for these wavefunctions. Various approximations have been suggested to calculate the radial functions: local potential of Slater type (Ref. 18), parametric potential (Ref. 17e), average of configurations (Refs. 19 and 20). We will not discuss here the relative merits of these various approximations, but only consider the last one which we have used for some of our calculations, essentially to avoid any "ad hoc" adjustable parameter. As in the non-relativistic case (21), we define for each jj subconfiguration the average energy as the centre of gravity of the energies of the various terms, namely:

$$E_{av.}(j,j) = \frac{\sum_i E(J_i)}{\sum_i (2J_i+1)} \tag{11}$$

Then, to take into account all the subconfigurations, we introduce a weighted sum of all the jj average energies arising from a given LS configuration, i.e.:

$$E_{av.}^{G} (LS) = \sum_{n} w_n E_{av.}^{n} (jj) \tag{12}$$

where the weights w_n are set equal to the degeneracy of each sub-configuration. For the p^2 configuration this generalization of the definition of the average energy gives:

$$E_{av.}^{G} (p^2) = \{E_{av.} (p_{1/2}^2) + 8E_{av.} (p_{1/2}p_{3/2}) + 6E_{av.}(p_{3/2}^2)\}/15 \tag{13}$$

Wavefunctions determined in this way have been used in CI calculations to study the optical spectra of neutral atoms with the p^2 ground state configuration (Ref. 22) and it was shown that, starting from germanium, this method of including intermediate coupling improves the agreement with experiment compared to standard Hartree-Fock methods.

This problem of averaging over the entire LS configuration has been rather widely discussed (Refs. 18 and 23) and we would like to point out only the main difference with single jj configuration calculations (Refs. 24 and 25). In the definition of the generalized average energy each $n\ell$ electron is considered as being shared by the two j subshells proportionally to $2j+1$, i.e. a p electron is considered in its non-relativistic limit as $(1/3)p_{1/2}$ and $(2/3)p_{3/2}$. This implies that for open shell atoms the occupation numbers of the two j subshells are closer than in the case of the single jj subconfiguration. We may thus expect a much similar behaviour of the two $j = \ell \pm 1/2$ subshells, a fact which can be verified for the open f shell of the rare earths by comparing the expectation values of r^n obtained in the single subconfiguration approximation (Ref. 24) and for the generalized average energy approximation (Ref. 8).

Both effective operators and CI methods disconnect the determination of radial function from that of the mixing coefficients between the subconfigurations. The next logical step is, of course, to determine them simultaneously and self-consistently. This can be achieved by carrying out multiconfiguration self-consistent field (MCSCF) calculations (Ref. 26). In such a calculation one starts with a total wavefunction such as that defined by Eq. 9 and makes the total energy stationary both with respect to the radial wave-functions and to the a and b mixing coefficients. The MCSCF method is certainly the most attractive one because of its completely ab initio nature but it is also the most complex and computer time-consuming. Furthermore it may become rather tedious for complicated open shell systems. Few calculations have yet been done, mainly because the relativistic MCSCF computer code of the author was only recently made available, but the method has been quite successful for the interpretation of the photoelectron spectrum of mercury (Ref. 27). The relative spacing between the states matches well with experiment and even the absolute energies were found to agree within about 1eV.

On the other hand, recent results on the relative intensities (Ref. 28) are in strong disagreement with our calculations. This discrepancy may doubtless be attributed to correlation effects which were shown to be of great importance in this mercury case (Ref. 29). These correlation effects may be partially taken into account through the MCSCF method if one does not restrict the total wavefunction to be described by a single LS configuration. The feasibility and limitations of the method have been studied for the calculation of the oscillator strengths for the resonance transitions (6S→6P) of gold and mercury (Ref. 29).

CONCLUSION

In this short review we have tried to give a survey of the "state of the art" in relativistic atomic calculations. If the relativistic Hartree-Fock method has been widely applied, very few attempts have been made to go further. The next progress in this field would certainly be the extension of many-body theories to the relativistic case. Even if such calculations appear as a rather formidable task, these two (relativistic and correlation) effects must be treated together and not separately because of their non-additivity.

Relativistic effects have been shown to largely modify the behaviour of valence shells in heavy atoms, and chemical properties calculated from relativistic theory may be considerably different from those obtained by non-relativistic calculations. In this field again only pioneer work has been done and we may expect much progress in the near future.

REFERENCES

1. V.M. Burke and I.P. Grant, Proc. Phys. Soc. 90, 297 (1967)
2. P.M. Dirac, Proc. Roy. Soc. A117, 610 (1928)
3. G. Breit, Phys. Rev. 34, 553 (1929) and 36, 383 (1930)
4. H. Bethe and E.E. Salpeter "Quantum Mechanics of One and Two-Electron Atoms" Springer-Verlag (Berlin 1957)
5. I.P. Grant, Adv. Phys. 19, 747 (1970)
6. L.N. Labzovskii, Sov. Phys. JETP 32, 94 (1971)
7. S. Feneuille and L. Armstrong, Phys. Rev. A8, 1173 (1973)
8. J.P. Desclaux, Atom. Data Nucl. Data Tables 12, 312 (1973)
9. L. Chase, H.P. Kelly and H.S. Köhler, Phys. Rev. A3, 1550 (1971)
10. B. Fricke, J.P. Desclaux and J.T. Waber, Phys. Rev. Lett. 28, 714 (1971)
11. A.J. Bearden and A.F. Burr, Rev. Mod. Phys. 39, 125 (1967)
12. C. Froese-Fischer, Atom. Data Nucl. Data Tables 12, 87 (1973)
13. K. Siegbahn Uppsala University Report UUIP 880-29-31 (1974)
14. E.U. Condon and G.H. Shortley "The Theory of Atomic Spectra" p. 294 (Cambridge U.P. 1935)

15. P.G.H. Sandars and J. Beck, Proc. Roy. Soc. A289, 97 (1966)
16. L. Armstrong and S. Feneuille in Adv. in Atom. and Mol. Phys. 10, 1 (1974)
17. a) E. Luc-Koenig, J. Phys. B7, 1052 (1974)
     b) E. Luc-Koenig, C. Morrillon and J. Vergès, Physica 70, 175 (1973)
     c) L. Holmgren and S. Garpman, Phys. Scripta 10, 215 (1974).
     d) S. Garpman, L. Holmgren and A. Rosén, Phys. Scripta 10, 221 (1974)
     e) E. Luc-Koenig, Thèse d'Etat Université de Paris 1975
18. I. Lindgren and A. Rosén, Case Studies in Atom. Phys. 4, 93 and 197 (1974)
19. D.F. Mayers, J. de Physique 31, C4-213 (1970).
20. J.P. Desclaux, C.M. Moser and G. Verhaegen, J. Phys. B4, 296 1971
21. J.C. Slater in "Quantum Theory of Atomic Structure", Vol. 2, Chap. 17 (McGraw Hill, N.Y. 1960)
22. J.P. Desclaux, Int. J. Quant. Chem. 6, 25 (1972)
23. J. Andriessen and D. van Ormondt, J. Phys. B8, 1993 (1975)
24. J.B. Mann and J.T. Waber, Atom. Data Tables 5, 201 (1973)
25. M.A. Coulthard, J. Phys. B6, 2224 (1973)
26. J.P. Desclaux, Comp. Phys. Comm. 9, 31 (1975)
27. J. Berkowitz, J.L. Dehmer, Y.K. Kim and J.P. Desclaux, J. Chem. Phys. 61, 2556 (1974)
28. H. Hotop, private communication
29. J.P. Desclaux and Y.K. Kim, J. Phys. B8, 1177 (1975)

A NEW EXPERIMENTAL STUDY OF MULTIPLE IONIZATION IN NOBLE GASES BY ELECTRON AND PHOTON IMPACT *

V. Schmidt

Fakultät für Physik, Universität Freiburg

78 Freiburg, W.-Germany

INTRODUCTION

The main topics of this summer school are effects which are due to electron correlations in atoms. One of the most powerful methods used to study the influence of electron correlations is multiple ionization of rare gases produced by electron or photon impact. This is due to the fact that the operator which causes the ionization process is a one-particle operator. Therefore, a theory that takes into account neither electron correlation nor rearrangement of the electron orbitals after the ionization process will allow only singly charged ions. When secondary processes as for instance the Auger effect are excluded, then the appearance of multiply charged ions is due to electron correlations and rearrangement of the electrons. This clear statement is true only for the photon impact experiment. For electron impact, it is valid only in first Born approximation and even then the large range of momentum transfers in the collision process makes the interpretation more difficult.

In order to avoid the problems connected with an experimental determination of absolute values for the multiple ionization cross sections, in most cases only relative values have been determined, for instance the ratio R of doubly to singly charged ions. However, the value of this ratio R has not yet been well established by previous experiments. In the case of electron impact on He, Ne and Ar, the newest values, those of van der Wiel et al. (1), differ systematically from the older ones. It was believed that the reason for this discrepancy is a charge discrimination in the older experiments. In the case of photon impact on He, Ne and Ar a systematic discrepancy exists between the data of Carlson (2) and van

der Wiel and Wiebes (3). Again charge discrimination could be the
reason for this discrepancy, however, in this case one must also
consider a possible influence of the photon source. In the photon
impact experiment by Carlson, the energy of the photons was not
well defined because the photons were selected from an X-ray continuum by special absorption filters. Especially at low photon
energies the ratio R might be influenced (7). Van der Wiel and
Wiebes applied a special method: in an electron impact ionization
process they selected only those events where the momentum transfer in the collision process was very small and where the high energy electron had a special energy loss ΔE. In the limit of zero momentum transfer, an impact of photons with energy $E = \Delta E$ is then
simulated (quasi-photon impact). In this type of experiment, at
higher energy losses the limits of the method might be reached.
Up to now theoretical calculations for the double photoionization
on He (4,5) and Ne (6,7) favoured the experimental results of
Carlson. But these theories made several approximations, such as
the neglect of electron correlation in the final state (4,5) or the
selection of the velocity form for the dipole matrix element.

Due to the fact that double photoionization in noble gases
gives basic information on electron correlations in atoms, a new
experiment has been set up, to clarify the experimental discrepancies
of the ratio R of doubly to singly charged ions caused by electron
or photon impact and to provide good experimental values for a
comparison with theory. This experiment has two essential features:
1) an extensive test of the apparatus concerning the problems of
charge discrimination of the ions and 2) the use of a photon source
with well defined energy namely synchrotron radiation and a specially designed monochromator which prevents spectral order overlapping.

ION APPARATUS AND INITIAL CHECKS

The apparatus for the determination of the relative amount of
differently charged ions produced by electron or photon impact is
a conventional magnetic field mass analyzer. The electron or photon beam traverses an ionization chamber and ionizes the target
gas which is introduced by a collimated holes-structure. The
product ions are extracted out of the electron beam by a weak
homogeneous electric field and accelerated by a stronger electric
field to a potential of - 2025 V. Both electric fields and the
field-free region are separated by Copper meshes. Optics consisting
of two electrostatic quadrupole lenses produce an astigmatic focus
of the ion beam at the entrance of the 60° sector magnetic field
mass analyzer. Ions with certain e/m are refocused at the image
where the detector, a Bendix 4700 channeltron, is placed. To insure
equal detection efficiencies, the differently charged ions are
postaccelerated in an electric field to the same velocity before
striking the detector. Singly charged ions then have an energy of

7 keV. The output pulses of the channeltron are amplified and counted.

Many experimental checks have been performed to examine the reliability of the apparatus with respect to its handling of differently charged ions. The most important checks concerned the problems of the ionization process (additional ionization by possibly secondary electrons or photoelectrons), of the ion-spectrometer (extraction and transmission of the ions, profile of the ion beam, diaphragms, adjustments of quadrupole optics and magnetic field e/m analyzer), of the channeltron detector (sensitivity of the surface area, absolute detection efficiency, linearity of the detector response) and of the working conditions (voltages for the electric fields, pressure of the target gas, background pressure).

RESULTS AND DISCUSSION FOR ELECTRON IMPACT

Table I gives a compilation of the new results for the ratios of doubly to singly charged ions for He, Ne and Ar caused by the impact of electrons with 2 keV energy. These experiments have been carried out together with W. Sandner and H. Kuntzemüller (8). The relative error of 7 % for the data is an intrinsic error of the ion-apparatus which was determined by numerous investigations. In the case of He, an additional error of about 12 % occurs due to the low counting rate of He^{++} and the amount of the H_2^+ peak whose main part coincides with the He^{++} peak. The results of other authors are also shown. It can be seen that the new data do not confirm those of van der Wiel et al. (1) but in general those of the other authors (9,10). The present results are in very good agreement with those of Schram et al (11-13).

He^{++}/He^+

Schram et al.	(4.2 ± 0.4)	$\times 10^{-3}$
Gaudin and H.	(5.5 ± 1.1)	$\times 10^{-3}$
vd Wiel et al.	5.8	$\times 10^{-3}$
this work	(4.2 ± 0.8)	$\times 10^{-3}$

Ar^{++}/Ar^+

Schram et al.	(5.3 ± 0.5)	$\times 10^{-2}$
Gaudin and H.	(5.7 ± 1.1)	$\times 10^{-2}$
vd Wiel et al.	8.8	$\times 10^{-2}$
this work	(5.0 ± 0.4)	$\times 10^{-2}$

Ne^{++}/Ne^+

Ziesel	(4.4 ± 0.2)	$\times 10^{-2}$
Schram et al.	(3.4 ± 0.3)	$\times 10^{-2}$
Schram et al.	(4.1 ± 0.4)	$\times 10^{-2}$
Gaudin and H.	(4.4 ± 0.9)	$\times 10^{-2}$
vd Wiel et al.	6.0	$\times 10^{-2}$
this work	(3.7 ± 0.3)	$\times 10^{-2}$

Table I. Ratios R of doubly to singly charged ions caused by the impact of 2 keV electrons.

EXPERIMENTAL DETAILS FOR THE PHOTON IMPACT EXPERIMENT

When the measurements with the electron gun were finished, the ion-apparatus was transferred to the electron storage ring ACO (Anneau de Collisions d'Orsay) at the University Paris-Sud and connected to a specially designed monochromator of the laboratory L.U.R.E. (Laboratoire pour l'Utilisation du Rayonnement Electro-magnétique). The laboratory L.U.R.E. provided optimum conditions for the photon impact experiment because the storage ring ACO gives a high photon flux at the location of the apparatus (for 540 MeV energy and 100 mA current of the electrons in the ring), the life time is of the order of 20 hours, the stability of the orbit is within 20 microns and the specially designed monochromator delivers a photon flux free of higher order components at the exit slit of the monochromator. This monochromator is the Jaeglé type (14) and it has been adapted to the conditions at the electron storage ring by Dhez et al. (15). It is a grazing incidence instrument in Rowland's circle mounting which covers the wave length range between 40 and 500 Å (between about 300 and 25 eV).

The apparatus for the determination of the relative amount of differently charged ions has been connected to this monochromator. At the entrance to the ionization chamber, a photon beam defining diaphragm has been inserted. In the middle of the ionization chamber, the size of the photon beam is about 2 mm high and about 4 mm wide. The dimensions were determined by photographs taken with Kodak SC5 film. The photon beam is detected by a photomultiplier sensitized with sodium salicylate. The penetration of photoelectrons into the ionization chamber is prevented by several diaphragms with large openings and appropriate potentials.

	ENERGY [eV]	BANDPASS [eV]	RATIO [×100]		ENERGY [eV]	BANDPASS [eV]	RATIO [×100]
He^{++}/He^{+}	100	6	2.4 ± 0.2	Ar^{++}/Ar^{+}	73	2	18.1 ± 2.0
	125	10	3.7 ± 0.3		115	4	18.0 ± 2.3
	150	14	4.1 ± 0.4		125	5	21.2 ± 1.8
					150	7	19.0 ± 1.6
Ne^{++}/Ne^{+}	73	2	1.5 ± 0.2		175	9	20.7 ± 1.9
	100	3	5.4 ± 0.5		200	12	19.1 ± 1.7
	125	5	8.9 ± 0.8				
	150	4	11.2 ± 1.0				
	175	6	12.6 ± 1.1				
	200	12	11.5 ± 1.1				

Table II. Ratios R of doubly to singly charged ions caused by the impact of photons.

RESULTS AND DISCUSSION FOR PHOTON IMPACT

Table II gives a compilation of the experimental results for the ratios of doubly to singly charged ions for He, Ne and Ar caused by photon impact. These experiments have been carried out together with N. Sandner, H. Kuntzemüller, P. Dhez, F. Wuilleumier and E. Källne (16). The errors for the ratio R are due to the statistical error and the intrinsic error of 7 % of the ion-apparatus. This intrinsic error has been added to the statistical one.

In Fig. 1 and 2 the experimental results for He and Ne are plotted together with the data of other authors. The new data for He, Ne and Ar do not confirm the values of van der Wiel and Wiebes (3), they favour in general the results of Carlson (2) for photon energies above 100 eV (for He the new data are between those of Carlson and van der Wiel and Wiebes) and they are close to the data of Samson and Haddad (17).

Now I am very glad to present more information than our new set of data. During the initial phases of this work with synchrotron radiation, we had several discussions with Dr. M. van der Wiel concerning the topic of multiple ionization and, especially, the discrepancy between his and our results for the ratio R in the case of electron impact. In the course of our experiments using the synchrotron radiation, Dr. M. van der Wiel repeated his quasi-photon measurements in a different way and sent us his latest, revised values prior to publication. There exists now gratifying agreement between his new results and our values for He and Ar over our whole energy range and for Ne up to 175 eV. As an example, these new data are included in Fig. 3 which shows the ratio R for Ar (18).

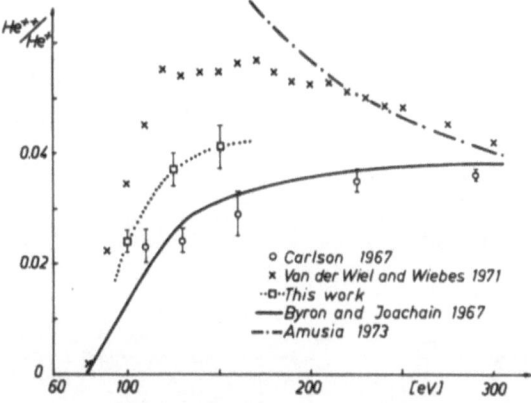

Fig. 1. The ratio of doubly to singly charged He ions as a function of the photon energy $h\nu$ (Brown's curve (5) is rather close to this of Byron and Joachain (4)).

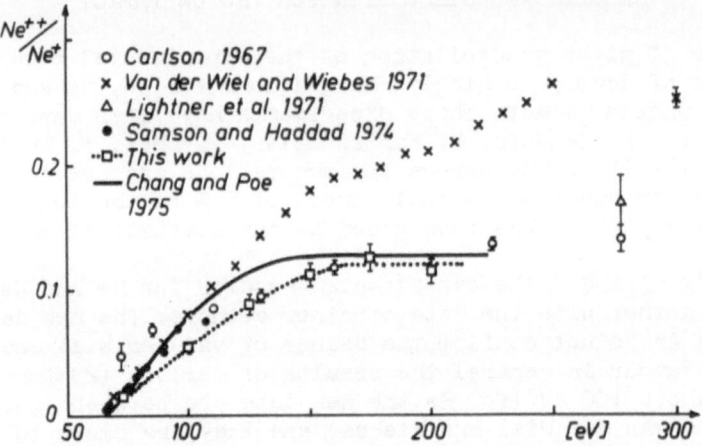

Fig. 2. The ratio of doubly to singly charged Ne ions as a function
          of the photon energy hν (Lightner et al. from Ref. (21)).

        It is interesting to compare the new experimental data with
the theory. For He (see Fig. 1) the experimental values for the
rise above the threshold are larger than given by the calculation
of Byron and Joachain (4) and Brown (5) (whose curve is rather
close to this of Byron and Joachain in our energy region) when the
momentum form for the dipole matrix element is used as anticipated
by the authors. The curve calculated by Amusia (19) is valid only
for high photon energies. For Ne (see Fig. 2) the rise for the experimental data is smaller than those calculated by Chang and Poe (7)

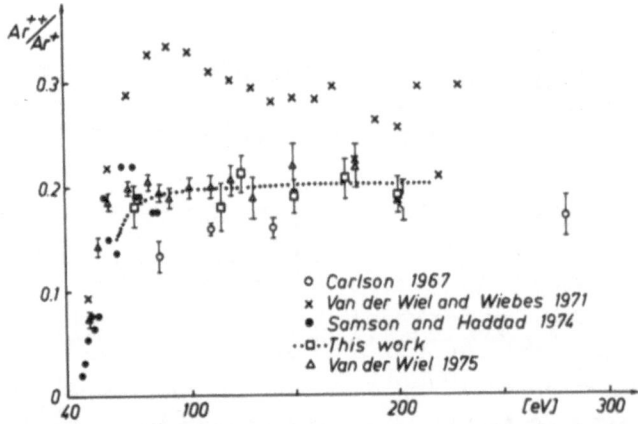

Fig. 3. The ratio of doubly to singly charged Ar ions as a function
          of the photon energy hν.

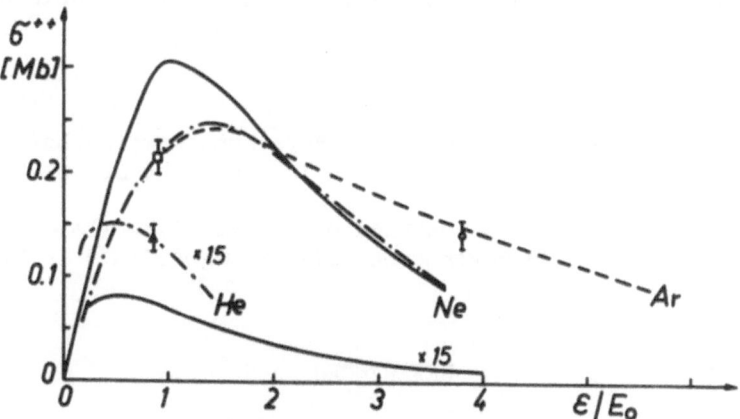

Fig. 4. Absolute values of the double photoionization cross section σ^{++} in He, Ne and Ar as function of the relative ionization energy ϵ/E_o where ϵ = photon energy minus threshold energy for double ionization, E_o = energy necessary to remove a second outer electron in the presence of one hole in the outer shell. Dashed lines = experimental curves; the magnitude of the error is indicated on each curve, the error bar does not include the error on the total photoabsorption cross section. Full lines = theoretical curves for He (4) and Ne (7), in both cases velocity form for the matrix element.

(also the momentum form for the matrix element has been employed). Above the rise there is good agreement between experiment and theory.

In order to have a direct criterion for the importance of electron correlations (and rearrangement of the atomic orbitals) the absolute value of the double photoionization cross section σ^{++} has to be used. Neglecting the triple photoionization, the value σ^{++} has been calculated using our values for the ratio R and the total photoabsorption cross sections as determined by West and Marr (20). Fig. 4 shows the results for σ^{++} as a function of the photon energy (in a normalized energy scale) and also the theoretical curves for He and Ne. From this figure it can be seen that the correlations for 1s electrons in He are much weaker than those for the 2p (2s) electrons in Ne or the 3p (3s) electrons in Ar. In this <u>absolute</u> scale the experimental and theoretical curves show the same deviations as previously discussed for the plot of the ratio R. Especially for Ne, there is good agreement at higher energies and a small discrepancy at lower energies which might be due to the special choice of the wave functions used by Chang and Poe (7). Therefore, the many body perturbation theory approach of Chang and Poe seems

to correctly describe the absolute magnitude of the double photoionization cross section.

Research supported by the Deutsche Forschungsgemeinschaft, Ger-
many, and by the Centre National de la Recherche Scientifique,
France.
Work of photon impact was carried out at L.U.R.E. (Laboratoire
pour l'Utilisation du Rayonnement Electromagnétique) in Orsay/
Paris.

REFERENCES

1) M.J. van der Wiel, Th.M. El-Sherbini and L. Vriens, Physica 42
   (1969) 411.
2) T.A. Carlson, Phys. Rev. 156 (1967) 142.
3) M.J. van der Wiel and G. Wiebes, Physica 53 (1971) 225 and
   Physica 54 (1971) 411.
4) F.W. Byron and C.J. Joachain, Phys. Rev. 164 (1967) 1.
5) R.L. Brown, Phys. Rev. A1 (1970) 586.
6) T.N. Chang, T. Ishihara and R.T. Poe, Phys. Rev. Lett. 27 (1971)
   838.
7) T.N. Chang and R.T. Poe, Phys. Rev. A12, 1432 (1975).
8) V. Schmidt, N. Sandner and H. Kuntzemüller, submitted to Phys.
   Rev. A.
9) J.P. Ziesel, J. Chim. phys. 62 (1965) 328.
10) A. Gaudin et R. Hagemann, J. Chim. Phys. 64 (1967) 1209.
11) B.L. Schram, A.J.H. Boerboom and J. Kistemaker, Physica 32
    (1966) 185.
12) B.L. Schram, Physica 32 (1966) 197.
13) B. Adamczyk, A.J.H. Boerboom, B.L. Schram and J. Kistemaker,
    J. Chem. Phys. 44 (1966) 4640.
14) P. Jaeglé, J. Physique 24 (1963) 179 and Comptes Rendus 259
    (1964) 533 and 4556.
15) P. Dhez, P. Jaeglé and F. Wuilleumier, in Vacuum UV Radiation
    Physics, ed. by E.E. Koch, R. Haensel and C. Kunz, Pergamon
    Vieweg (1974) 788.
16) V. Schmidt, N. Sandner, H. Kuntzemüller, P. Dhez, F. Wuilleumier
    and E. Källne, submitted to Phys. Rev. A.
17) J.A.R. Samson and G.N. Haddad, Phys. Rev. Lett. 33 (1974) 875.
18) I thank Dr. M. van der Wiel very much for the early communica-
    tion of his new results.
19) M.Ya. Amusia, VIII ICPEAC, ed. by B.C. Cobic and M.V. Kurepa,
    Inst. of Phys., Beograd, Yugoslavia (1973) 171.
20) W.D. West and G.V. Marr, to be published in Proc. Roy. Soc.
21) G.S. Lightner, R.J. van Brunt and D. Whitehead, Phys. Rev. A4
    (1971) 602.

A PRIMER ON PHOTOELECTRON ANGULAR DISTRIBUTIONS

Dan Dill[*]

Department of Chemistry, Boston University

Boston, Massachusetts   02215

1. INTRODUCTION

We consider here the angular distribution of electrons ejected by electric dipole interaction from randomly oriented atomic or molecular targets. This corresponds to the usual gas phase photo-ionization experiment. The angular distribution is restricted by Yang's theorem [1] to the form

$$\frac{d\sigma}{d\Omega} = \frac{\sigma}{4\pi} \{1 + \beta P_2(\cos\Theta)\}. \tag{1}$$

Here σ is the integrated cross section and Θ is the ejection angle measured from the electric vector of the light. The angular distribution is characterized by the asymmetry parameter β, which is restricted to the range $-1 \leqslant \beta \leqslant 2$ by the requirement that $d\sigma/d\Omega$ be non-negative. In particular $\beta = 2$ corresponds to a $\cos^2\Theta$ distribution, peaked along the electric vector of the light, $\beta = 0$ corresponds to an isotropic distribution, and $\beta = -1$ corresponds to a $\sin^2\Theta$ distribution, peaked perpendicular to the electric vector of the light.

More generally, ionization by light with arbitrary polarization gives [1]

$$\frac{d\sigma}{d\Omega'} = a + b \cos^2\Theta', \tag{2}$$

---

[*]Alfred P. Sloan Research Fellow

where the ejection angle Θ' is measured from the propagation direction of the light and the parameters a and b are known functions of σ, β and the polarization [2]. For example, with unpolarized light one obtains

$$\frac{d\sigma}{d\Omega'} = \frac{\sigma}{4\pi} \{1 - (1/2)\beta P_2(\cos\Theta')\}. \tag{3}$$

Maximum anisotropy results when linearly polarized light is used.

The goal of angular distribution studies is to utilize β, obtained either by measurement or by calculation, as a probe of both static (e.g., symmetry) information on initial and final target states, and dynamic (e.g., electron correlation) information on the photoionization process itself. To extract such information essentially involves unravelling geometry from dynamics in the theoretical analysis of β, through identification of universal geometrical elements common to all processes (and therefore not of primary specific interest), due to conservation of total angular momentum and parity.

2. ANGULAR-MOMENTUM-TRANSFER DECOMPOSITION

This unravelling has yielded a general formulation of angular distribution theory in terms of the angular momentum transferred to the target in the photoionization process [3]. The formulation provides a general framework for calculation and analysis of distributions. Further, it reduces the calculational problem to the evaluation of a small number of dynamical, rotationally invariant parameters which characterize the unique properties of any particular system; geometrical parameters common to all processes have been evaluated once and for all. Finally, it provides new insight into dynamical information uniquely probed by angular distribution studies, notably, anisotropic interactions experienced by the electron as it leaves the target.

The generic electric dipole photoionization process can be expressed as

$$T(J_0\pi_0) + \gamma(j_\gamma=1, \pi_\gamma=-1) \longrightarrow T^+(J_c\pi_c) + e^-(\ell s \pi_e). \tag{4}$$

The differential cross section for this process can be decomposed into contributions characterized by the angular momentum transfer

$$\vec{j}_t \equiv \vec{j}_\gamma - \vec{\ell} = \vec{j}_c + \vec{s} - \vec{j}_0 , \tag{5}$$

provided no measurement is made of either the photoelectron spin or the orientation of the residual ion. The vector $\vec{j}_t$ is the angular momentum exchanged between the unobserved initial and final angular momenta in the reaction (4), i.e., between the total angular momentum

$\vec{J}_0$ of the target T and the combined angular momentum of the residual ion T^+ and the photoelectron spin, which is denoted $\vec{J}_{cs} \equiv \vec{J}_c + \vec{s}$. Allowed values of j_t are determined by conservation of angular momentum J and parity π in the reaction (4):

$$\vec{J} = \vec{J}_0 + \vec{J}_\gamma = \vec{J}_c + \vec{s} + \vec{\ell} \tag{6}$$

$$\pi = -\pi_0 = \pi_c(-1)^\ell, \tag{7}$$

where $(-1)^\ell = \pi_e$.

The angular-momentum-transfer decomposition has four key elements: (1) Photoionization amplitude can be characterized by a particular value j_t of the angular momentum transfer. (2) Amplitudes for different j_t superimpose <u>incoherently</u> in both the integrated cross section and the angular distribution,

$$\sigma = \sum_{j_t} \sigma(j_t), \tag{8}$$

$$\frac{d\sigma}{d\Omega} = \sum_{j_t} \frac{d\sigma(j_t)}{d\Omega} = \sum_{j_t} \frac{\sigma(j_t)}{4\pi} \{1 + \beta(j_t)P_2(\cos\theta)\}. \tag{9}$$

(3) Photoelectric current corresponding to a given j_t has a characteristic <u>geometry</u>, according to the parity change $\pi_0\pi_c$. For $\pi_0\pi_c = +(-1)^{j_t}$ (parity favored j_t),

$$\sigma(j_t) = \pi\lambda^2 \frac{2j_t+1}{2J_0+1} \{|S_+(j_t)|^2 + |S_-(_t|^2\}, \tag{10}$$

$$\sigma(j_t)\beta(j_t)\{\pi\lambda^2(2J_0+1)^{-1}\}^{-1}$$

$$= (j_t+2)|S_+(j_t)|^2 + (j_t-1)|S_-(j_t)|^2$$

$$-3\{j_t(j_t+1)\}^{1/2}\{S_+(j_t)S_-^*(j_t) + c.c.\} \tag{11}$$

where λ is the wavelength of the light divided by 2π, c.c. denotes complex conjugate, and the subscripts $\pm$ denote the value of the photoelectron's orbital momentum, $\ell = j_t \pm 1$. For $\pi_0\pi_c = -(-1)^{j_t}$ (parity unfavored j_t),

$$\sigma(j_t) = \pi\lambda^2 \frac{2j_t+1}{2J_0+1} |S_0(j_t)|^2, \tag{12}$$

$$\beta(j_t) = -1, \tag{13}$$

where the subscript 0 denotes $\ell = j_t$. (4) The <u>dynamics</u> of a particular process (4) is contained entirely within the rotationally invariant matrix elements $S_\ell(j_t)$, which should therefore be the starting point of specific calculations.

3. EXAMPLES

Let us now take up three examples, of increasing complexity and richness. These illustrate how the angular-momentum-transfer analysis proceeds.

The first example is the photoionization of atomic hydrogen,

$$H(1s\ ^2S_{1/2}) + \gamma \longrightarrow H^+(^1S_0) + e^-. \tag{14}$$

This is a bit like "shooting with cannon at a fly" [4], but it is a good place to start. The geometry of the process is

$$J_0 = s = 1/2, \quad j_\gamma = 1, \quad J_c = 0, \quad \ell \leqslant 2, \tag{15a}$$

$$\pi_0 = +1, \quad\quad \pi_\gamma = -1, \quad \pi_c = +1, \quad \pi_e = -1,$$

where in fact $\ell = 1$ since $\pi_e = -1$. The angular momentum transfer is

$$\vec{j}_t = \vec{J}_c + \vec{s} - \vec{J}_0 = \vec{s} - \vec{s} = 0, \tag{16}$$

where the last equality holds because the spin orientation is unaffected in the photoionization process. Thus we have the single value $j_t = 0$ and since $\pi_0\pi_c = +1$ we use Eqs. (10) and (11) to obtain $\beta = \beta(j_t=0) = 2$, i.e., a pure $\cos^2\theta$ distribution.

The next example is the photoionization of atomic mercury,

$$Hg(5d^{10}6s^2\ ^1S_0) + \gamma \longrightarrow Hg^+(5d^{10}6s\ ^2S_{1/2}) + e^-. \tag{17}$$

The geometry of the process is

$$J_0 = 0, \quad j_\gamma = 1, \quad J_c = 1/2, \quad \ell \leqslant 2, \tag{18a}$$

$$\pi_0 = +1, \quad \pi_\gamma = -1, \quad \pi_c = +1, \quad \pi_e = -1, \tag{18b}$$

where again parity conservation restricts ℓ to the single value $\ell = 1$. The angular momentum transfer is

$$\vec{j}_t = \vec{J}_c + \vec{s} - \vec{J}_0 = \vec{S}_c + \vec{s}. \tag{19}$$

Since $S_c = s = 1/2$, two values of $\vec{j}_t$ can occur. Before ionization $\vec{S}$ and $\vec{s}$ are antiparallel, coupled into a singlet. If the spin orientation is undisturbed in the ionization, then $\vec{S}_c$ and $\vec{s}$ will remain antiparallel and $j_t = 0$ (parity favored). If, however, the orientation of $\vec{s}$ can change, through spin-orbit coupling, then $j_t = 1$ (parity unfavored) can also occur. Thus, in general the asymmetry parameter is given by

$$\beta = \{2\sigma(0) - \sigma(1)\}/\{\sigma(0) + \sigma(1)\}, \tag{20}$$

with deviations from $\beta = 2$ arising from the spin–orbit interaction [5]. Niehaus and Ruf [6] have measured β at 11.63 eV and 11.83 eV and found $\beta = 1.2$ and $\beta = 2$, respectively. In fact the lower energy coincides with the resonant state $5d^9(^2D_{3/2})6s^26p_{3/2}$ ($J = 1$, $\pi = -1$) calculated [7] to have 93% <u>triplet</u> character (30% $^3P_1^o$ and 63% $^3D_1^o$).

The last example is photoionization of atomic xenon between the fine structure thresholds of the ground state ion,

$$Xe(5p^6\ {}^1S_0) + \gamma \longrightarrow Xe^+(5p^5\ {}^2P_{3/2}^o) + e^-. \tag{21}$$

The photoionization geometry is

$$J_0 = 0, \qquad j_\gamma = 1, \qquad J_c = 3/2, \qquad \ell \leqslant 3, \tag{22a}$$

$$\pi_0 = +1, \qquad \pi_\gamma = -1, \qquad \pi_c = -1, \qquad \pi_\ell = +1, \tag{22b}$$

where parity conservation requires $\ell = 0,2$. The angular momentum transfer is

$$\vec{j}_t = \vec{J}_c + \vec{s} - \vec{J}_0 = \vec{L}_c + \vec{S}_c + \vec{s}. \tag{23}$$

Again, there are two possible values of $\vec{j}_t$. If the spins remain coupled into a singlet then $\vec{S}_c + \vec{s} = 0$ and $j_t = L_c = 1$ (parity favored). If the spin orientation is disturbed, then triplet coupling can occur, $S_c + s = 1$ and $j_t = 2$ (parity unfavored) is also possible; $j_t = 0$ is excluded by the angular momentum balance $\vec{j}_t = \vec{j}_\gamma - \vec{\ell}$. Thus, we have the asymmetry parameter

$$\beta = \{\sigma(1)\beta(1) - \sigma(2)\}/\{\sigma(1) + \sigma(2)\},$$

$$= \frac{|S_d(1)|^2 - \frac{5}{3}|S_d(2)|^2 - \sqrt{2}\{S_d(1)S_s^*(1) + c.c.\}}{|S_d(1)|^2 + \frac{5}{3}|S_d(2)|^2 + |S_s(1)|^2}. \tag{24}$$

The transition amplitudes $S_\ell(j_t)$ exhibit a pronounced spectral variation in the 1.3 eV region between the thresholds for ionization to the fine structure levels $^2P_{3/2}^o$ and $^2P_{1/2}^o$ of the ion, owing to quasi-discrete levels converging to the $^2P_{1/2}^o$ threshold. The asymmetry parameter has a corresponding resonant variation, observed by Samson and Gardner [8]. Such resonant variation of the asymmetry parameters is expected to be a general feature of autoionization resonances [3].

4. PHYSICAL SIGNIFICANCE OF ANGULAR MOMENTUM TRANSFER

In essence angular momentum transfer is a probe of anisotropic electron-target interactions [3,9]. To see this, it is convenient to consider the photoionization process in two stages, namely, an initial stage A of photoabsorption proper and a subsequent stage B of escape of the photoelectron from the rest of the target   The key point is the allowed values of $\vec{j}_t$ are different in the two stages of the photoionization process as a result of anisotropic interactions.

In the initial stage A, illustrated in Fig. 1a, the photoabsorption imparts $j_\gamma = 1$ units of orbital momentum to the photoelectron, which has orbital momentum $\vec{\ell}_o$ (say), yielding a final orbital momentum $\vec{\ell}' = \vec{\ell}_o + \vec{j}_\gamma$. Therefore, in stage A the angular momentum transferred to the target is

$$\vec{j}_t' \equiv \vec{j}_\gamma - \vec{\ell}' = -\vec{\ell}_o, \qquad \text{(Stage A)} \qquad (25)$$

with the single value $j_t' = \ell_o$. Furthermore, owing to parity conservation, $\ell' = \ell_o \pm 1$, and hence $j_t' = \ell_o$ is a parity favored angular momentum transfer.

During the subsequent escape of the photoelectron in stage B, illustrated in Fig. 1b, additional angular momentum transfers can occur, within the allowed range determined by Eq. (5), from anisotropic interactions of the photoelectron with the rest of the target. Owing to the resulting angular momentum exchanges $\vec{k}$ between electron and target, the orientation and the magnitude of the electronic orbital momentum can change from $\vec{\ell}'$ to $\vec{\ell}$. As illustrated in Fig. 1c, the angular momentum transfer is ultimately no longer restricted to the single value $j_t = \ell_o$ but rather

$$\vec{j}_t \equiv \vec{j}_\gamma - \vec{\ell} = \vec{j}_t' - \vec{k}. \qquad \text{(Stage B)} \qquad (26)$$

A widely used treatment of atomic photoelectron angular distributions is the Cooper-Zare independent-particle model [10]. Its connection with the angular-momentum-transfer analysis emerges at this point. The Cooper-Zare model treats the residual ion-core as a spectator to the photoionization process. That is, stage B is ignored altogether, in which case only the single, parity favored angular momentum transfer $j_t = \ell_o$ arises. Further the amplitudes $S_\pm(j_t)$ then assume the limiting form

$$S(j_t = \ell_o) = i^{-\ell} e^{i\delta\ell} (\ell || C^{[1]} || \ell_o) \, R(\ell, \ell_o), \qquad (27)$$

where R is a reduced radial dipole matrix element and the other symbols have their usual meaning. Substitution of Eq. (27) in Eqs. (10) and (11), gives the Cooper-Zare formula for the asymmetry parameter.

It is emphasized that a variety of dynamical interactions can result in multiple angular momentum transfers; forces are not restricted to those of magnetic origin. In the photoionization of atoms, the electrostatic (exchange) interaction which separates different LS terms of the electron-ion complex results in marked deviations from predictions based on the Cooper-Zare model [9]. In the photoionization of molecules, the torque due to the anisotropic electric field of the molecular ion can lead to many angular momentum transfers

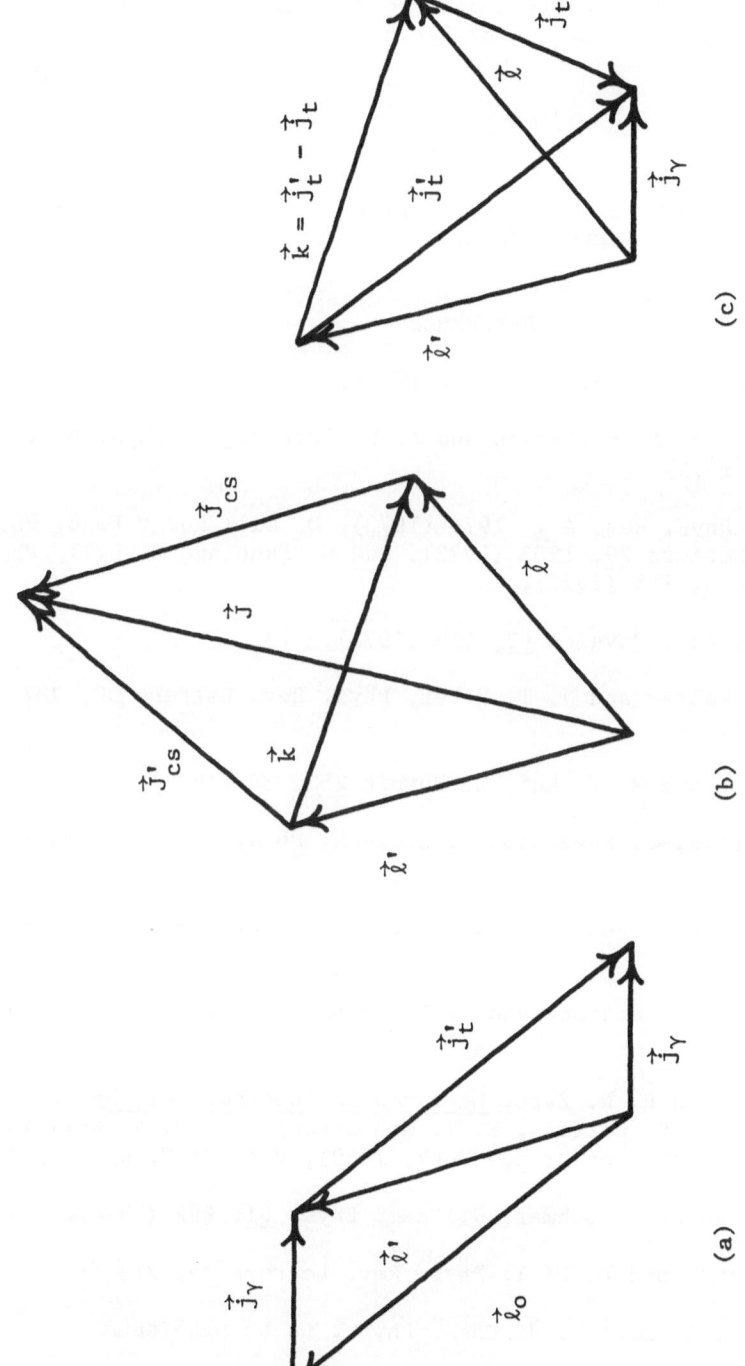

Fig. 1. Schematic vector model of angular momentum exchanges in the stages of the photoionization process: (a) Electric dipole ($j_\gamma = 1$) interaction "within" the target; (b) exchange of angular momentum $\vec{k}$ between the target and the "escaping" electron; (c) net angular momentum transfer $\vec{j}_t$ "outside" the target.

[11,12]. In fact it is anticipated that analysis in the spirit described here [13] will greatly enhance our understanding of the
uniquely molecular information probed by photoelectron angular distributions.

ACKNOWLEDGEMENTS

Critical reading of the manuscript by Scott Wallace and Jon
Siegel is gratefully acknowledged.

REFERENCES

1. C. N. Yang, Phys. Rev. $\underline{74}$, 764 (1948).

2. See, e.g., J. A. R. Samson and A. F. Starace, J. Phys. B $\underline{8}$,
    1806 (1975)

3. D. Dill, Phys. Rev. A $\underline{7}$, 1976 (1973), D. Dill and U Fano, Phys.
    Rev. Letters $\underline{29}$, 1203 (1972), and U. Fano and D. Dill, Phys.
    Rev. A $\underline{6}$, 185 (1972).

4. M. Danos, Ann. Physics $\underline{63}$, 319 (1971).

5. T. E. H. Walker and J. T. Waber, Phys. Rev. Letters $\underline{30}$, 307
    (1973).

6. A. Niehaus and M. W. Ruf, Z. Physik $\underline{252}$, 84 (1972).

7. W. C. Martin, J. Sugar and J. L. Tech, Phys. Rev. A $\underline{6}$, 2022
    (1972).

8. J. A. R. Samson and J. L. Gardner, Phys. Rev. Letters $\underline{31}$, 1327
    (1973).

9. D. Dill, A. F. Starace and S. T. Manson, Phys. Rev. A $\underline{11}$, 1596
    (1975).

10. J. Cooper and R. N. Zare, <u>Lectures in Theorectical Physics</u>,
     edited by S. Geltman, K. T. Mahanthappa and W. E. Brittin
     (Gordon and Breach, New York, 1969), Vol. XI-C, pp. 317-37.

11. D. Dill and J. L. Dehmer, J. Chem. Phys. $\underline{61}$, 692 (1974).

12. J. L. Dehmer and D. Dill, Phys. Rev. Letters $\underline{35}$, 213 (1975).

13. See, e.g., S. Druger, J. Chem. Phys., to be published.

THE QUANTUM DEFECT THEORY APPROACH

Anthony F. Starace*

Behlen Laboratory of Physics

The University of Nebraska, Lincoln, NE 68588, U.S.A.

The Quantum Defect Theory (QDT) is a method of using the analytically known properties of excited electrons moving in a pure Coulomb field to describe atomic photoabsorption and electron-ion scattering processes in terms of a few parameters. These parameters may be determined either from experimental data or from ab initio theoretical calculations. In addition, they are usually nearly independent of energy in the threshold energy region (i.e., within a few eV of the atomic ionization threshold). Thus the determination of these parameters at any <u>single</u> energy suffices to predict the <u>variation with energy</u> of numerous atomic properties in the threshold energy region such as total and partial photoionization or scattering cross sections, photoelectron asymmetry parameters, discrete line strengths, autoionization profiles, etc. These properties are often very strongly energy-dependent and difficult to measure or to calculate by other methods. Yet all these phenomena, according to the QDT, depend on only a few essential parameters which represent the proper interface between theory and experiment. The determination of these parameters should thus be the goal of both theory and experiment rather than the calculation or measurement of the various phenomena dependent on these parameters.

This article aims to describe the essence of the QDT for the non-specialist. More extensive surveys of the theory for the non-specialist by Seaton[1] and Fano[2] should be consulted for more complete references to the original literature. Here we shall first discuss the one channel theory since it embodies the main content of the general multichannel theory. Then we shall sketch the multichannel treatment of Lu and Fano and its applications.

*Alfred P. Sloan Foundation Fellow

In common with the closely related R-matrix theory, the QDT assumes that the configuration space for an excited atomic electron can be divided into two regions: an inner region, $0 \leq r \leq r_0$, where electron correlations are strong and difficult to treat; and an outer region, $r_0 \leq r \leq \infty$, where the electron-ion interaction potential is assumed to be purely Coulombic and where the form of the electron wavefunction is known analytically. The boundary radius r_0 between the two regions is typically of the order of the atomic radius: one wants r_0 as small as possible so that the electron wavefunction can be known exactly over as great a range of $r \geq r_0$ as possible and yet one wants r_0 large enough so that the approximation of a pure Coulomb field for $r \geq r_0$ makes sense. (The inclusion of r^{-2} long range potentials for $r \geq r_0$ has been treated by Bely.[3])

Consider now the one-channel problem of an excited electron in an alkali atom: the electron sees a Coulomb field for $r \geq r_0$, where r_0 is roughly the ionic radius. We measure the energy ε of the excited electron relative to the ionization threshold and make the change of variables $\varepsilon = -0.5 \nu^{-2}$. The parameter ν will thus be our measure of energy. The Schrödinger equation for $r \geq r_0$ has two solutions, one regular and one irregular for small values of r:

$$f(\nu,r) \sim r^{\ell+1} \quad \text{as } r \to 0 \tag{1a}$$

$$g(\nu,r) \sim r^{-\ell} \quad \text{as } r \to 0 \tag{1b}$$

A general solution of the Schrödinger equation for $r \geq r_0$ is a linear combination of $f(\nu,r)$ and $g(\nu,r)$ with coefficients to be determined by application of boundary conditions at infinity and at r_0. The form of this general solution turns out to be

$$\psi(\nu,r) = N_\nu \{f(\nu,r)\cos\pi\mu - g(\nu,r)\sin\pi\mu\} \quad \text{for } r \geq r_0 \tag{2}$$

where N_ν is a normalization factor which is determined by the behavior of $\psi(\nu,r)$ at large r. μ, on the other hand, is the relative phase with which the regular and irregular solutions are superposed. Its value is determined by the behavior of $\psi(\nu,r)$ in the core region, $0 \leq r \leq r_0$, where the effective potential is non-Coulombic: i.e., μ has that value which allows the analytically determined $\psi(\nu,r)$ given by Eq. (2) for $r \geq r_0$ to be joined smoothly at $r = r_0$ onto the numerically determined portion of $\psi(\nu,r)$ that obtains in the inner core region, $0 \leq r \leq r_0$, which we shall call $\psi_c(\nu,r)$. (Thus $\psi(\nu,r)$ without other specification is assumed to be the electron wavefunction over all space; $\psi(\nu,r)$ is given by Eq. (2) for $r \geq r_0$ and by $\psi_c(\nu,r)$ for $0 \leq r \leq r_0$.)

Ab initio determination of the parameter μ at any energy requires numerical calculation of the inner core wavefunction $\psi_c(\nu,r)$. The R-matrix theory[4,5] is designed for this. The procedure is to compute the complete set of <u>discrete</u> eigenfunctions $u_\lambda(r)$ for the spherical "box" $0 \leq r \leq r_0$. The inner core wavefunction for energy ν is then represented as a linear combination of this complete set of eigenfunctions:

$$\psi_c(\nu,r) = \sum_\lambda c_\lambda(\nu)u_\lambda(r). \tag{3}$$

The coefficients $c_\lambda(\nu)$ are determined by the R-matrix theory to depend on energy ν and on values of $\psi_c(\nu,r)$ and $u_\lambda(r)$ and their derivatives at $r = r_0$. Requiring the logarithmic derivatives of Eqs. (2) and (3) to be equal at $r = r_0$ permits the determination of μ.[4-6] Other more approximate methods may also be used to calculate μ: e.**g**., if a model potential $V(r)$ is used to describe the electron's motion in the region $0 \leq r \leq r_0$, then μ may be calculated by the Phase-Amplitude Method.[7]

Alternatively, μ may be determined semi-empirically and, provided sufficient empirical data are available, there may be no need to know the inner core wavefunction in order to predict the variation with energy of various atomic properties in the threshold energy region. Specifically, consider the asymptotic behavior of $\psi(\nu,r)$ in the case of excited electron energies below threshold, i.e., $\psi(\nu,r)$ must tend toward zero. The asymptotic forms of the regular and irregular Coulomb functions are:[8]

$$f(\nu,r) \to u(\nu,r)\sin\pi\nu - v(\nu,r)\exp i\pi\nu \quad \text{as } r \to \infty \tag{4a}$$

$$g(\nu,r) \to -u(\nu,r)\cos\pi\nu + v(\nu,r)\exp i\pi(\nu+\tfrac{1}{2}) \text{ as } r \to \infty \tag{4b}$$

where $u(\nu,r)$ is an exponentially increasing function of r and $v(\nu,r)$ is an exponentially decreasing function of r. Substituting Eq. (4) in Eq. (2) gives:

$$\psi(\nu,r) \to N_\nu\{u(\nu,r)\sin\pi(\nu+\mu) - v(\nu,r)\exp i\pi(\nu+\mu)\} \text{ as } r \to \infty \tag{5}$$

In order that $\psi(\nu,r)$ tend toward zero the coefficient of $u(\nu,r)$ must be zero; i.e. $\sin\pi(\nu+\mu) = 0$ or $\nu+\mu = n$, where n is an integer. Substituting $\nu = n-\mu$ in the expression for the electron's energy gives

$$\varepsilon = -\frac{1}{2\nu^2} = -\frac{1}{2(n-\mu)^2} \tag{6}$$

μ is thus the quantum defect of spectroscopy and may be determined directly from Rydberg energy level data for the alkalis.

For positive excited electron energies, on the other hand, $\nu = i/k$, where k is the electron momentum. The asymptotic forms of the regular and irregular Coulomb functions are:[8]

$$f(\nu,r) \rightarrow (\frac{2}{\pi k})^{\frac{1}{2}} \sin(kr+\theta) \text{ as } r\rightarrow\infty \qquad\qquad (7a)$$

$$g(\nu,r) \rightarrow -(\frac{2}{\pi k})^{\frac{1}{2}} \cos(kr+\theta) \text{ as } r\rightarrow\infty \qquad\qquad (7b)$$

where

$$\theta = -\tfrac{1}{2}\ell\pi + \frac{1}{k} \ell n(2kr) + \arg\Gamma(\ell+1-i/k). \qquad\qquad (7c)$$

Substituting Eq. (7) in Eq. (2) gives

$$\psi(k,r) \rightarrow N_k (\frac{2}{\pi k})^{\frac{1}{2}} \sin(kr+\theta+\pi\mu) \text{ as } r\rightarrow\infty \qquad\qquad (8)$$

Eq. (8) implies that $\pi\mu$ is the scattering phase shift for the continuum electron at energies near threshold.

Seaton[9,10] first showed this connection, via μ, between
discrete energy level data and scattering phase shift data. Subsequent work by Seaton and collaborators[1] used empirical energy
level data to obtain electron-ion scattering phase shifts, and,
conversely, ab initio calculated phase shifts to obtain the quantum
defects. Lu, Fano, and collaborators[2] have developed alternative
methods, described below, for obtaining QDT parameters such as μ
from empirical energy level data, and have begun the ab initio
calculation of these parameters by the R-matrix theory.[6]

The theory described above would be in vain if μ were a
rapidly varying function of energy since then separate calculations
would be needed at each energy. Fortunately μ is a slowly varying
function of energy since it is determined from the inner core wavefunction $\psi_c(\nu,r)$, which except for a normalization factor is insensitive to small changes in the electron's energy. This insensitivity
is due to the electron's large kinetic energy in the inner core
region. From Eq. (2) we can see directly that for small $r \gtrsim r_0$,
$\psi(\nu,r)$ depends on energy mainly through the normalization factor
N_ν since μ is weakly energy-dependent and so are $f(\nu,r)$ and $g(\nu,r)$
at small r (cf. Eq. (1)). The normalization factor N_ν is determined
by the asymptotic behavior of $\psi(\nu,r)$ and may be very energy-dependent. The point of this discussion thus is that at small radii
$\psi(\nu,r)/N_\nu$ is likely to be quite insensitive to energy in the threshold energy region and this is one reason why the predictions of
the QDT are often so uncannily good.

To show how energy-independent is the form of $\psi(\nu,r)$ for small
r, consider Table I, which presents data for the ns($1\leq n\leq 7$) bound
wavefunctions of atomic uranium.[11] Notice that as the orbital
energies increase there is a remarkable convergence of the positions
of the first maxima, first minima, and the first two nodes of the

radial wavefunctions to energy-independent values. We have also arbitrarily re-normalized each one of the seven radial wavefunctions to unity at the position of its first maximum by dividing $P_{ns}(r)$ by $P_{ns}(r_{MAX1})$, where r_{MAX1} is the position of the first maximum. This re-normalization should eliminate any energy dependence of the wavefunction amplitude arising from the normalization factor. The last column of Table I shows the re-normalized amplitudes at the positions of the first minima of the wavefunctions. Once again, as the orbital energies increase there is a convergence of the re-normalized amplitudes to an energy-independent value. Note that Table I includes only unexcited electron orbitals since these are tabulated in readily available references.[11] QDT, however, is concerned with excited electron orbitals having energies within a few eV of threshold. Table I shows clearly (cf. the 6s and 7s orbitals) that over a range of a few eV the <u>form</u> of $\psi(\nu,r)$ for small r is to an excellent approximation energy-independent or at most very weakly dependent on energy.

Direct applications of the insensitivity of $\psi(\nu,r)/N_\nu$ upon energy are readily made and serve to indicate the power of the QDT. Consider the calculation of discrete oscillator strengths for transitions to members of a one-channel series of Rydberg levels. The oscillator strength f_n is proportional to the square of the electric dipole matrix element, whose radial part is:

$$R_\nu = \int_0^\infty r^2 dr \psi_0(r) r \psi(\nu,r), \tag{9}$$

where $\nu=(n-\mu)$ and $\psi_0(r)$ is the electron's initial state wavefunction, whose range is comparable to r_0. Over the effective range of inte-

TABLE I: Behavior of Uranium ns Radial Wavefunctions $P_{ns}(r)/r$ for Small Radial Distances r.

Positions (in Bohr) of the first maximum (r_{MAX1}), first minimum (r_{MIN1}), and first two nodes ($r_{NODE1,2}$) of $P_{ns}(r)$:

n	Energy (Ry)	r_{MAX1}	r_{NODE1}	r_{MIN1}	r_{NODE2}	$\dfrac{P(r_{MIN1})}{P(r_{MAX1})}$
1s	-7409.2	0.01097	--------	--------	--------	--------
2s	-1289.2	0.00844	0.02199	0.05922	--------	-1.9589
3s	- 326.64	0.00823	0.02110	0.04771	0.08230	-1.6643
4s	- 83.16	0.00817	0.02089	0.04620	0.07812	-1.6223
5s	- 18.49	0.00816	0.02083	0.04583	0.07714	-1.6128
6s	- 2.96	0.00815	0.02082	0.04572	0.07691	-1.6095
7s	- 0.35	0.00815	0.02082	0.04581	0.07686	-1.6093

gration, however, $\psi(\nu,r)$ depends on energy mainly through N_ν.
Hence R_ν/N_ν is only weakly energy-dependent, or, equivalently,
f_n/N_ν^2 is only weakly energy-dependent.

It can be shown[7] that N_ν is proportional to $\nu^{-3/2}=(n-\mu)^{-3/2}$
for discrete (i.e., negative) electron energies and that N_ν is
independent of energy for positive electron energies (assuming
normalization per unit energy). This implies that multipli-
cation of the discrete oscillator strength f_n by $(n-\mu)^3$ will
produce a spectrum of oscillator strengths that (1) varies slowly
from one discrete level to another and (2) joins smoothly onto the
continuous spectrum of oscillator strength. (This "renormalization"
serves to give the discrete final state wavefunctions continuum-
type normalizations so that the oscillator strength is then continuous across the ionization threshold.[12])

Fig. 1 illustrates this renormalization procedure for H and
Li discrete and continuous oscillator strengths. The area of each
rectangle corresponds to the value of the discrete oscillator
strength f_n (n is labeled s in the figure). The height of each
rectangle equals $(n-\mu)^3 f_n$. It is seen that when plotted in this
way the discrete oscillator strength joins smoothly onto the continuous oscillator strength at threshold. The deviation from constancy near threshold is due to the residual, weak energy dependence

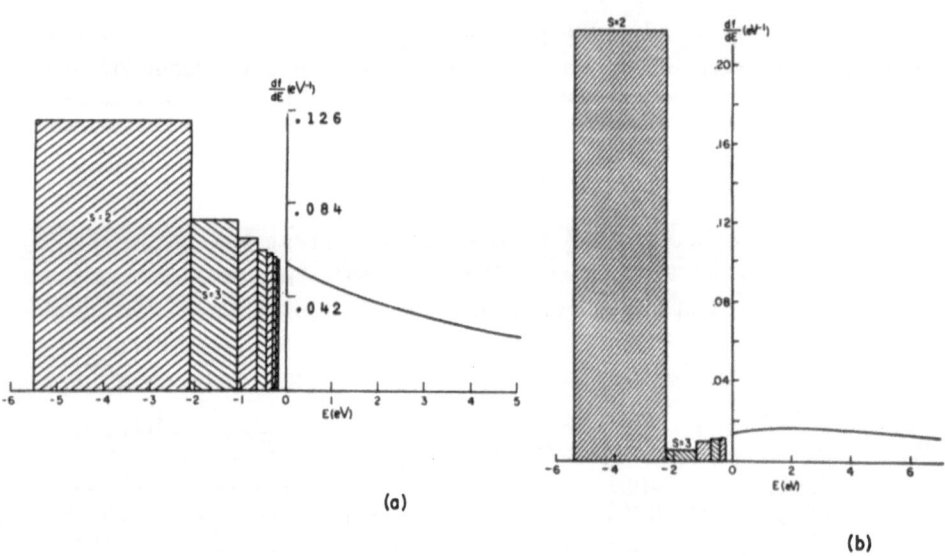

(a)

(b)

Fig. 1. Oscillator strength distribution in the discrete and part
of the continuous spectra of (a) H (theory) (b) Li (experiment).
(From Ref. 12)

of the radial dipole matrix elements and possibly, in the case of Li, of the quantum defect μ. The QDT often takes account of the weak energy dependence of these parameters by expanding them in a power series in energy about their value at threshold and keeping only the first two or three terms.[13,14]

The QDT for the one-channel (alkali-like) spectra thus describes the discrete energy spectrum by the quantum defect μ, which is determined by electron-ion interactions in the inner core region, and by the position of the ionization threshold. The electron-ion scattering phase shift near threshold is equal to $\pi\mu$. In addition, knowledge of the oscillator strength at a single energy near threshold permits the prediction of all discrete and continuous oscillator strengths in the threshold energy region.

Consider now the multi-channel case of rare gas photoabsorption spectra. There are five channels, which in jj-coupling are specified as follows:

$$
\begin{aligned}
p^6(^1S_0) + h\nu &\rightarrow p^5(^2P_{3/2})\varepsilon d_{5/2} \quad J=1 \\
&\rightarrow p^5(^2P_{3/2})\varepsilon d_{3/2} \quad J=1 \\
&\rightarrow p^5(^2P_{3/2})\varepsilon s_{1/2} \quad J=1 \\
&\rightarrow p^5(^2P_{1/2})\varepsilon d_{3/2} \quad J=1 \\
&\rightarrow p^5(^2P_{1/2})\varepsilon s_{1/2} \quad J=1
\end{aligned} \tag{10}
$$

In Eq. (10) ε indicates the excited electron's energy, which may be positive or negative (i.e., discrete). jj-coupling is appropriate at large r, when the ion core and the excited electron are far apart. At such large distances the difference between the two ionization thresholds $I_{3/2}$ and $I_{1/2}$ corresponding to the two levels of the ion dominates the electron-ion core interaction. If these five channels were not interacting one would expect each one to have a characteristic energy-independent quantum defect, just as though each channel could be considered a one-channel case. In fact, at short range the electrostatic interaction between the excited electron and the ion core is dominant and the difference in ionic thresholds pales in significance compared to the excited electron's large kinetic energy. This short range interaction results in what spectroscopists call series perturbations, i.e., quantum defects that vary dramatically along a Rydberg series due to the interaction between nearby levels of different Rydberg series.

The multi-channel QDT treats the electron-ion interaction in the rare gases thusly:[14,15] at large distances r the five excited electron channels are described by the asymptotic coupling appro-

priate to the interactions remaining at large distances, i.e.,
jj-coupling as in Eq. (10). These asymptotic channel states are not
eigenstates of the electron-ion interaction at small distances, but
instead interact. To take a scattering theory point of view, one
says that an asymptotic channel state is scattered, by interactions
predominant in the inner core region $0 \leq r \leq r_0$, into other asymptotic
channel states. If we label two such asymptotic states by i and j,
then S_{ij} indicates the appropriate scattering matrix element arising
from short range interactions. This matrix element may be cast in
terms of the diagonal representation of S as follows:

$$S_{ij}(J=1) = \sum_{\alpha=1}^{5} U_{i\alpha} e^{2i\pi\mu_\alpha} U_{\alpha j}^{\dagger} \quad (i, j=1,5) \quad (11)$$

Here α labels the five scattering eigenstates of the short range
electron-ion interaction. In the rare gases these eigenstates have
been found to be fairly close to the LS-coupled states of the ion
and the excited electron due to the dominance of electrostatic
interactions at short range. The unitary matrix $U_{i\alpha}$ is the matrix
that diagonalizes $S_{ij}(J=1)$. It represents the transformation from
the scattering eigenstates α to the asymptotic, experimentally
observed states i. $U_{i\alpha}$ is thus called the "frame transformation
matrix," because it transforms from the strong interaction frame
appropriate to the inner core region to the weak interaction frame
appropriate to the asymptotic region. The parameters μ_α are the
eigenphase shifts of the scattering eigenstates α: i.e., μ_α is the
phase by which the eigenstate α is shifted in traversing the inner
core region.

In the multi-channel QDT for the rare gases the final state of
the excited electron-ion core system is characterized by the eigenphase shifts μ_α, the ionization thresholds $I_{3/2}$ and $I_{1/2}$, and the
frame transformation matrix $U_{i\alpha}$. It is possible, as in the one-
channel case, to obtain these parameters by ab initio theoretical
calculations, as for example by use of the multi-channel R-matrix
theory.[6] It remains for us in this article, however, to show
how these parameters are related to experimental energy level data
and thus how these parameters may be determined semi-empirically.

As in the one-channel case, the final state wavefunction for
the excited rare gas system may be written analytically in the
region $r \geq r_0$. Since there are two ionic energy levels (i.e., $^2P_{3/2}$
and $^2P_{1/2}$)^o we must include the ionic wavefunction explicitly because
for a given excitation energy E there correspond two excited electron
energies ε_i:

$$E = I_i + \varepsilon_i = I_i - (2\nu_i^2)^{-1} \quad (i=3/2, 1/2) \quad (12)$$

Superposing the regular and irregular Coulomb functions in each channel and summing over the allowed channels gives for the final state wavefunction:

$$\psi = \sum_{\alpha} \psi_{\alpha} A_{\alpha} = \tag{13}$$

$$= \sum_{i} \Phi_i \{ f(\nu_i, \ell_i; r) \sum_{\alpha} U_{i\alpha} \cos\pi\mu_{\alpha} A_{\alpha} - g(\nu_i, \ell_i; r) \sum_{\alpha} U_{i\alpha} \sin\pi\mu_{\alpha} A_{\alpha} \}$$

In Eq. (13) ℓ_i denotes the electron's orbital momentum in channel i, Φ_i represents the ionic wavefunction as well as the angular and spin variables of the excited electron, and the co-efficients A_{α} represent the weights with which the scattering eigen-states ψ_{α} are superposed. The coefficients A_{α} are determined by application of one of the three types of boundary condition at infinity appropriate for ψ, corresponding to each of the following three energy regions: $E \leq I_{3/2}$ (the discrete region), $I_{3/2} \leq E \leq I_{1/2}$ (the autoionizing region), $I_{1/2} \leq E$ (the photoionization region). For example, in the discrete energy region ψ must tend toward zero at large r. Substituting the asymptotic forms for f and g given by Eq. (4) into Eq. (13) and requiring the coefficients of the ex-ponentially increasing functions $u(\nu_i, r)$ to be zero gives

$$\sum_{\alpha} U_{i\alpha} \sin\pi(\nu_i + \mu_{\alpha}) A_{\alpha} = 0, \tag{14}$$

which in turn implies that non-trivial solutions for A_{α} exist only if

$$F(\nu_{3/2}, \nu_{1/2}) \equiv \det |U_{i\alpha} \sin\pi(\nu_i + \mu_{\alpha})| = 0. \tag{15}$$

Lu and Fano[8,15,16] have developed the requisite procedures for determining the eigenphases μ_{α} and portions of the frame transfor-mation matrix $U_{i\alpha}$ by fitting Eq. (15) to experimental energy level data. In the present case of two ionic thresholds, this fitting procedure is best considered with reference to the two-dimensional Lu-Fano plot,[16] although the procedure is no means limited to two threshold problems. A typical Lu-Fano plot is given in Fig. 2 for Ar Rydberg energy levels belonging to the five series of Eq. (10). For each experimental term level E, regardless of the spectroscopic classification of the level, the effective quantum numbers $\nu_{3/2}$ and $\nu_{1/2}$ are calculated from Eq. (12). Since the function in Eq. (15) is invariant to integer changes in the values of ν_i (i=3/2, 1/2), these are only calculated modulo 1. (In treating the energy-dependence of the parameters μ_{α} and $U_{i\alpha}$, however, the absolute values of ν_i must be used[13,14] -- we ignore this complication here and consider only levels close to threshold, where the parameters are energy-independent.) The Lu-Fano plot is constructed by plotting $-\nu_{3/2}$ (mod. 1) vs. $\nu_{1/2}$ (mod. 1) as shown in Fig. 2.

Except for the lowest levels with $n \lesssim 6$, the resulting plot of

levels falls on the solid curves, which are given by the function
in Eq. (15) with appropriately fitted parameters μ_α and $U_{i\alpha}$. This
function is multivalued, having three horizontal branches, corresponding to the three series in Eq. (10) belonging to the $^2P_{3/2}$
ionic threshold, and two vertical portions corresponding to the two
series in Eq. (10) belonging to the $^2P_{1/2}$ ionic threshold. Each
branch of this function is seen to be monotonically increasing and,
in fact, the entire function is continuous in the sense that at any
edge of the unit square it is reflected to the opposite edge with
the same value. It can also be shown that:[15,16] (1) The five intersections of the solid line with the diagonal line partially
drawn in Fig. 2 occur at the five values of the scattering eigenphases μ_α; (2) At these same five intersections, the slopes of the
solid curve give information on the frame transformation matrix
$U_{i\alpha}$. Fig. 2 demonstrates quite well the theoretical correlation
of very many experimental energy level data by means of a very
few parameters (μ_α and $U_{i\alpha}$).

In a similar manner, the intensities of the discrete absorption
lines, the autoionizing level profiles, and the photoionization
cross section at threshold may be determined in the multi-channel
QDT by means of the μ_α, the $U_{i\alpha}$, and five parameters D_α, $1 \leq \alpha \leq 5$,
corresponding to the radial electric dipole matrix elements for
transitions from the ground state to the five eigenstates ψ_α. The
parameters D_α and portions of the $U_{i\alpha}$ matrix are determined by
fitting analytic equations for intensities to available experimental
data. In the paper by Lee and Lu,[14] for example, these parameters
were determined by fitting an experimental autoionization line
profile. The parameters thus determined were then used to predict
discrete oscillator strengths.

It is in this way that the dependence of disparate experimental
data, occuring in different energy regions, on a few common, nearly
energy-independent parameters is exploited by the QDT. An essential
conceptual aspect of the multichannel theory is the connection of
the asymptotic (observable) states to the strong-interaction
scattering eigenstates by means of a frame transformation matrix
$U_{i\alpha}$. In the case of molecules,[8] this transformation is from
the body frame of the molecule, appropriate at short range (where
the excited electron closely follows the motion of the molecule),
to the laboratory frame, appropriate asymptotically (where the
rotational motion of the molecular ion only effects the excited
electron's motion through their mutual energy sharing, which depends on the differing rotational energy levels of the molecular
ion). In a similar way, negative ion photodetachment has been
treated by multichannel QDT.[17]

So far all but one[6] of the applications of the QDT by Lu,
Fano, and collaborators[2] and all applications for complex systems

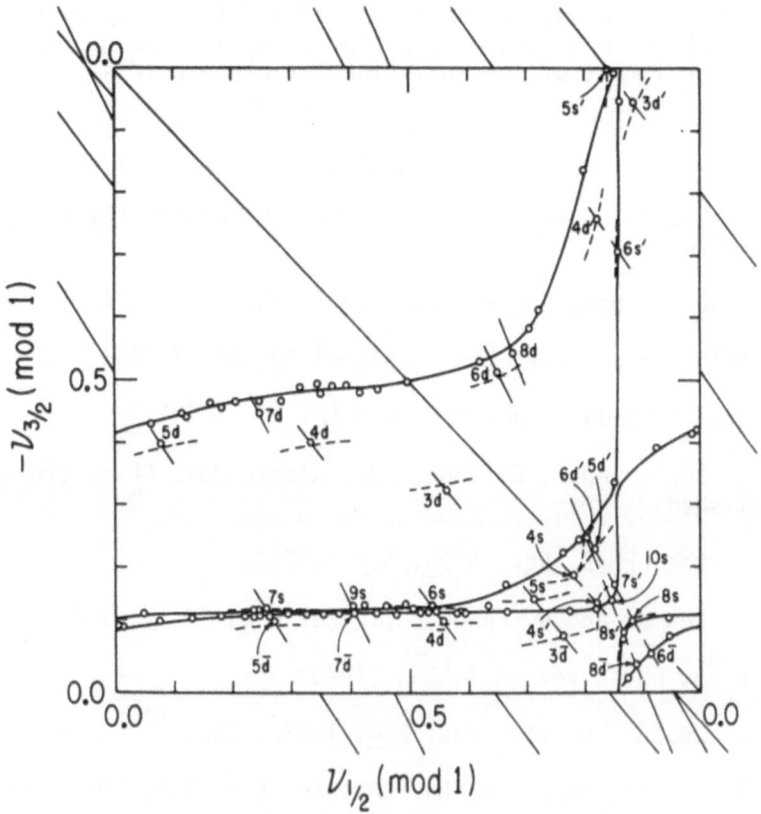

Fig. 2. $-\nu_{3/2}$ (mod. 1) vs. $\nu_{1/2}$(mod. 1):   open circles represent experimental Ar energy levels; solid curves represent Eq. (15) with parameters μ_α and $U_{i\alpha}$ determined by fitting energy level data near the $I_{3/2}$ threshold. (From Ref. 14)

by Seaton and collaborators[1] have depended on the availability of reliable experimental data in order to obtain the necessary parameters in the theory. Seaton and collaborators[1] have also determined the parameters by ab initio calculations, using close-coupling methods, for light atomic systems. Only recently, how-ever, with the development of the R-matrix theory, is it practical to obtain the necessary parameters for heavy atoms by ab initio calculations.[6] Much development of the theory still needs doing. It is worth doing because the focusing of theorists and experimentalists on obtaining the few scattering parameters needed for any atomic or molecular system can only lead eventually to greater unity and coherence in low-energy atomic physics and hopefully permit the treatment of presently intractable problems.

References

1. M. J. Seaton, Comments on Atomic and Molecular Physics II, 37 (1970).

2. U. Fano, J. Opt. Soc. Am 65, 979 (1975).

3. O. Bely, Proc. Phys. Soc. (London) 88, 833 (1966).

4. G. Breit, Handbuch der Physik XLI/1, 1 (1959).

5. P. G. Burke and W. D. Robb, Adv. Atomic Mol. Phys. (to be published).

6. C. M. Lee, Phys. Rev. A 10, 584 (1974).

7. J. L. Dehmer and U. Fano, Phys. Rev A 2, 304 (1970).

8. U. Fano, Phys. Rev. A 2, 353 (1970).

9. M. J. Seaton, Compt. Rend. 240, 1317 (1955).

10. M. J. Seaton, Mon. Not. Roy. Astron. Soc. 118, 504 (1958).

11. The data presented in Table I were obtained by interpolating the values tabulated by F. Herman and S. Skillman, Atomic Structure Calculations (Prentice-Hall, Englewood Cliffs, New Jersey, 1963).

12. U. Fano and J. W. Cooper, Rev. Mod. Phys. 40, 441 (1968), §2.4.

13. A. F. Starace, J. Phys. B 6, 76 (1973).

14. C. M. Lee and K. T. Lu, Phys. Rev. A 8, 1241 (1973).

15. K. T. Lu, Phys. Rev. A 4, 579 (1971).

16. K. T. Lu and U. Fano, Phys. Rev. A 2, 81 (1970).

17. A. R. P. Rau and U. Fano, Phys. Rev. A 4, 1751 (1971).

THE CLOSE-COUPLING METHODS APPLIED TO PHOTOIONISATION

OF NEUTRAL ATOMS AND POSITIVE IONS

Françoise COMBET FARNOUX

E R "Spectroscopie Atomique et Ionique"

Bât.350 - Campus d'Orsay : 91405 Orsay (France)

As Professor Fano and Dr. Inokuti mentioned it in their lectures, close-coupling functions are usually used to describe the electron scattering process. But such continuum wave functions can describe both an ion scattering an electron (calculations of electron impact cross sections) and the final states appropriate to photoionisation. However, close-coupling functions have only been introduced recently in the calculation of photoionisation cross sections at low energies : it was only in 1967 that Henry and Lipsky[1] provided an exact formulation of the photoionisation process within the framework of the dipole approximation, in LS coupling, when the coupling between the final state channels was introduced. In this talk, I will essentially develop three points. First, I shall recall briefly how the close-coupling equations are derived in the scattering process formulation ; second, I shall show how the close-coupling continuum wave functions of the scattering process can be introduced in multichannel photoionisation. Then, I shall emphasize those methods of solving the close-coupling equations which have been used to write the recently available computer codes for the calculation of electron scattering and photoionisation cross sections of atomic systems. Some results concerning photoionisation of Ar isoelectronic series and recently obtained with these various programs will be presented and compared with other ones.

BASIC SCATTERING THEORY

The overall wave function of the projectile + target system in the scattering process by an atom (N electrons) can be expanded in a series of the form :

$$\Psi_j = A \sum_i \Phi_i \varphi_{ij} \tag{1}$$

where Φ_i are target eigenfunctions. φ_{ij} describes the scattered electron and A is the antisymmetrizing operator. $\sum_i$ is taken over all eigenstates of the target which form a complete set of states. Total L,S and their z components are separately conserved in the collision. In the representation diagonal in these quantum numbers, a channel is defined by $\alpha_i L_i S_i l_i LS \; M_L M_S \pi$. $\alpha_i L_i S_i$ represent the target state, while l_i is the orbital angular momentum of the scattered electron and π is the parity. Two conditions restrict the ranges of the quantum numbers :

$$|L_i - l_i| \leqslant L \leqslant L_i + l_i \quad \text{and} \quad S_i - 1/2 \leqslant S \leqslant S_i + 1/2 \tag{2}$$

if we put :
$$\varphi_{ij} = 1/r \cdot F_{ij}(r) \; Y_{l_i}^m i \; (\Omega) \; \delta(\sigma, \mu_i) \tag{3}$$

$F_{ij}(r)$, with $F_{ij}(0) = 0$ describes the radial motion of the scattered electron and satisfies the asymptotic condition :

$$k_i^2 > 0 \quad F_{ij}(r) \longrightarrow k_i^{-1/2} (\sin \theta_i \cdot \delta_{ij} + \cos \theta_i \cdot K_{ij}) \tag{4}$$

$$k_i^2 < 0 \quad F_{ij}(r) \longrightarrow N_{ij} \exp(- |k_i| r) \tag{5}$$

where k_i^2 is the energy of the scattered electron in the i-th channel, expressed in Rydbergs   and

$$\theta_i = k_i r - 1/2 \; l_i + z/k_i \; \ln(2 k_i r) + \arg \Gamma(l_i + 1 - iz/k_i) \tag{6}$$

z is the asymptotic residual charge : z=Z – N. The real and symmetric K_{ij} matrix (reactance matrix) contains all information about the scattering process. A variational principle for the K_{ij} matrix can be derived by developing the integral :

$$I_{ij} = \langle \Psi_i | H - E | \Psi_j \rangle \tag{7}$$

$$I_{ij} = \sum_{kl} \int_0^\infty F_{ki}(r) \left[-(d^2/dr^2 - l_k(l_k+1)/r^2 + k_k^2) F_{kj} \delta_{kl} \right.$$
$$+ 2 V_{kl}(r) F_{1j}(r) + 2 \int_0^\infty W_{kl}(r,s) F_{1j}(s) ds \left. \right] dr \tag{8}$$

V(r) and W(r,s) denote respectively direct and exchange potential terms. It follows that the Kohn variational principle

$$\delta(I - 1/2 \; \delta K) = 0 \tag{9}$$

is satisfied for arbitrary variation δF only if the following equation is satisfied:

$$(d^2/dr^2 - l_k(l_k+1)/r^2 + k_k^2) F_{kj}(r) = 2 \sum_l (V_{kl}(r) F_{1j}(r)$$
$$+ \int_0^\infty W_{kl}(r,s) F_{1j}(s) ds) \tag{10}$$

Let us recall that the K matrix is connected to the scattering matrix S by the relation : $K_{ij} = i \left(1 - S/1 + S\right)_{ij}$

To make solvable the infinite set of coupled integro-differential equations (10) for the scattering functions F_{ij} subject to the boundary conditions (4) and (5), the close-coupling approximation is introduced : it consists in retaining only some states in (1), thus only some equations in (10). The choice of equations retained is dictated by the physics without any general rules.

In other words, in the close-coupling approximation, we suppose that the bound orbitals are known and the task of this approximation is to compute the free orbitals $F_{ij}(r)$ which describe the radial motion of both the incident electron relative to the target (in scattering) and the ejected electron relative to the residual ion (in photoionisation). The number of independent solutions is equal to the number of open channels ($k_i^2 > 0$) at the energy considered, and ψ_{ij} denotes the j-th solution for the electron in the i-th channel.

The close-coupling method for continuum states when Hartree Fock solutions are used to represent the core wave functions has been outlined by Seaton[2] as soon as 1953, but the explicit formulation applicable to final states where the core only consists of closed subshells and a single (closed or open) p electrons subshell has been given by Smith, Henry and Burke[3] in 1966.

THE MULTICHANNEL PHOTOIONISATION CROSS SECTIONS

To resume the same notations as above and keep the similarity between scattering and photoionisation processes, we shall deal with electronic systems (atoms or ions) with charge Z and N+1 electrons. In the dipole-length formulation, the cross section averaged over initial states and summed over final states, for photoionization to the j-th channel was given by Burgess and Seaton[4] :

$$j\sigma_L = -\frac{4\pi \alpha\, a_0^2}{3\,\omega} (I + k^2) \sum_{M_o \mu_o M_j \mu_j} \left| \left\langle \psi_o(x_1 \cdots x_{N+1}) \sum_{t=1}^{N+1} r_t \right. \right.$$

$$\left. \left. \psi_j(x_1 \cdots x_{N+1}) \right\rangle \right|^2 \qquad (11)$$

$\omega = (2S_o+1)(2L_o+1)$ is the statistical weight of the initial term and $I+k^2$ the photon energy in Rydbergs. M_o M_j μ_o and μ_j represent the z component of the orbital and spin angular momentum relative to the initial and final states respectively described by ψ_o and ψ_j subject to the normalisation conditions :

$$\langle \Psi_0 | \Psi_0 \rangle = 1 \tag{12}$$

$$\langle \Psi_j(k^2) | \Psi_{j'}(k'^2) \rangle = \pi \, \delta_{jj'} \, \delta(k^2 - k'^2) \tag{13}$$

When introducing the close-coupling approximation to describe the final state in photoionisation :

$$\Psi_j = A \sum_i \sum_{M_i m_i} C^{S_i \ 1/2 \ S_j}_{\mu_i \ \mu \ \mu_j} C^{L_i l_i L_j}_{M_i m_i M_j} \, \Phi_i(\bar{x}_p) \, \varphi_{ij}(x_p) \tag{14}$$

where $\Phi_i(\bar{x}_p) = \Phi_i(x_1 \cdots x_{p-1} x_{p+1} \cdots x_{N+1})$

But Ψ_j must represent an outgoing spherical wave in channel j plus incoming spherical waves in all channels. So, $\varphi_{ij}(x_p)$ defined according to (3) is such as :

$$F_{ij}(r) \xrightarrow{r \to \infty} k_i^{-1/2}(e^{i\theta_i} \delta_{ij} - S^+ e^{-i\theta_i}) \tag{15}$$

S^+ is the adjoint of the scattering matrix S. When we compare this asymptotic amplitude with (4), we conclude that Ψ_j is expressed in terms of the real, standing wave solution Ψ_k of the close-coupling equations, as follows :

$$\Psi_j = i \sum_k (1 - i K)^{-1}_{kj} \Psi_k \tag{16}$$

and the partial cross section relative to the final state j :

$$_j\sigma_L = \frac{4\pi\alpha \, a_0^2}{3\omega}(I + k^2) \sum_{M_0 \mu_0 M_j \mu_j} \sum_{kk'} \langle \Psi_k | \sum_{t=1} r_t | \Psi_0 \rangle$$

$$(1 - iK)^{-1}_{jk}(1 + iK)^{-1}_{k'j} \langle \Psi_0 | \sum_{t=1}^{N+1} r_t | \Psi_{k'} \rangle \tag{17}$$

The total cross section is obtained by summing over all the channels j, and using the relation :

$$\sum_j (1 - i K)^{-1}_{jk}(1 + i K)^{-1}_{k'j} = (1 + K^2)^{-1}_{k'k} \tag{18}$$

If we separate in (17) the angular and radial parts :

$$_j\sigma_L = \frac{4\pi\alpha \, a^2}{3\omega}(I + k^2) B(L_0 S_0 L_j S_j) \sum_{kk'} g_k(k^2)(1 + iK)^{-1}_{k'j}$$

$$(1 - iK)^{-1}_{jk} g_{k'}(k^2) \tag{19}$$

B is determined from the angular momentum coupling coefficients of the initial and final states and both $g_k(k^2)$ and $g_{k'}(k^2)$ depend only on the radial part of the initial and final wave functions.

VARIOUS METHODS FOR SOLVING THE C.C.EQUATIONS

Many techniques (iterative or not) have been developed to solve the finite set of coupled second order integro-differential equations (10). They were listed by Smith[5] and largely described in a review paper by Burke and Seaton[6]. Here, we will only speak about those methods which have been introduced in some computer codes now available to the Physics community and allowing for the computation of cross sections relative to both electron scattering and photoionisation processes. Either an exact or an approximate solution is looked for, but in all cases the continuum functions F_{ij} are subject to the orthogonalisation conditions :

$$\left\langle F_{ij}(r) \mid P_{nl_\alpha}(r) \right\rangle = 0 \qquad (1_\alpha = 1_i) \tag{20}$$

Because this orthogonalisation with respect to the discrete orbitals can be interpreted as preventing the projectile from being captured in any incomplete subshell included in the eigenfunction expansion, we must include in our trial function some extra (N+1) electron wave functions with the same L S as the final state j, and (1) becomes :

$$\Psi_j = A \sum_i \Phi_i(x_1 \cdots x_N ; \widehat{x_{N+1}}) \, F_{ij}(r)/r_{N+1} \tag{21}$$

$$+ \sum_\mu c_\mu \, \Phi_\mu(L\,S ; x_1 \cdots x_{N+1})$$

where μ ranges over the bound configurations which can be formed by adding an electron to one of the incomplete subshells included in Φ_i. Physically speaking, we can regard the (N+1) electron functions Φ_μ of (21) as taking into account short range correlation effects.

A) Reduction to Differential Equations

The most ancient close-coupling computer code, written by Conneely et al.[7] (ATOMNP) solves exactly the C.C. equations to the accurary of the computer. Indeed, it uses a non iterative method which reduces the system (10) to a larger system of differential equations. It only deals with configurations (np)q for n=2 and 3, and uses in the photoionisation calculations close-coupled wave functions for the final continuum orbitals and Hartree-Fock Clementi[8] functions for the bound orbitals. The wave function of the final state is of type (21) with only one extra bound configuration in so far as the program allows to introduce in Φ_i only the ground state terms of the residual ion, without configuration interaction. The details of both the equations and the numerical techniques have been emphasized in several works 3,6,9,10.

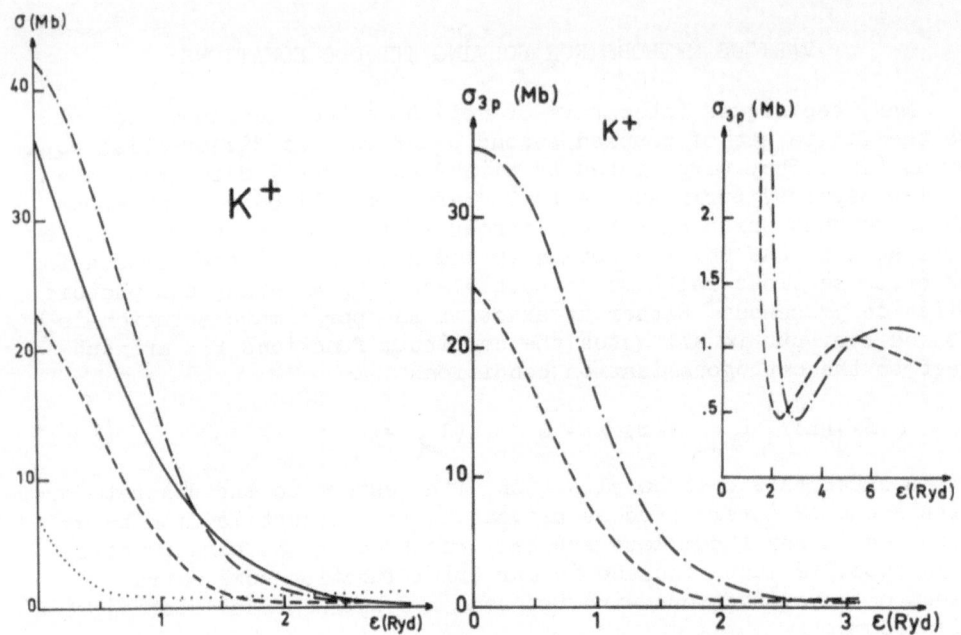

Fig 1 (left) and 2 (right) : $\overline{\sigma_{3p}}$ of K^+ by Combet Farnoux and
                              Lamoureux [16] ; ($\varepsilon = k^2$)
—.—.—dipole-length _ _ _ _dipole-velocity curves in the close-
coupling approximation (Conneely et al.[7] program in fig. 1 and R
matrix theory[22] in fig. 2)
Fig.1 : ....   Herman and Skillman[31] potential of K^+ $3p^6$
       ———— V_{3d} potential (Hartree-Fock) of K^+ $3p^5 3d$

        The first calculations by Henry and Lipsky[1] concerning Ne were
extended with ATOMNP to N by Ormonde and Conneely[11] and to neutral
atoms with incomplete outer 3p subshells by Conneely et al.[12]. For
ions, in addition to the works of Henry[13] concerning ions of C,N,O,
Ne, of Chapman and Henry[14,15] for ions of S, Al, Si, Ar, we shall
mention the recent results of Combet Farnoux and Lamoureux[16] concer-
ning $\overline{\sigma_{3p}}$ of K^+, Ca^{2+} and Sc^{3+} (Ar isoelectronic series). They are
presented in fig.1 and 3 which emphasize their comparison with other
results obtained by us with 2 different central potential models.
In our application of this close-coupling code to K^+, Ca^{2+}, Sc^{3+},
we only included the ground state configuration $3p^5$ of K^{2+}, Ca^{3+}
and Sc^{4+} ; so, there were only 2 coupled final degenerate channels
($^2P^o$ εs) and ($^2P^o$ εd) corresponding to the final state $^1P^o$.

        In fig.1 the results obtained with the V_{3d} potential are loca-
ted just between the 2 close-coupling curves, and such an agreement is still confirmed for Ca^{2+} and Sc^{3+} in fig. 3 which clearly

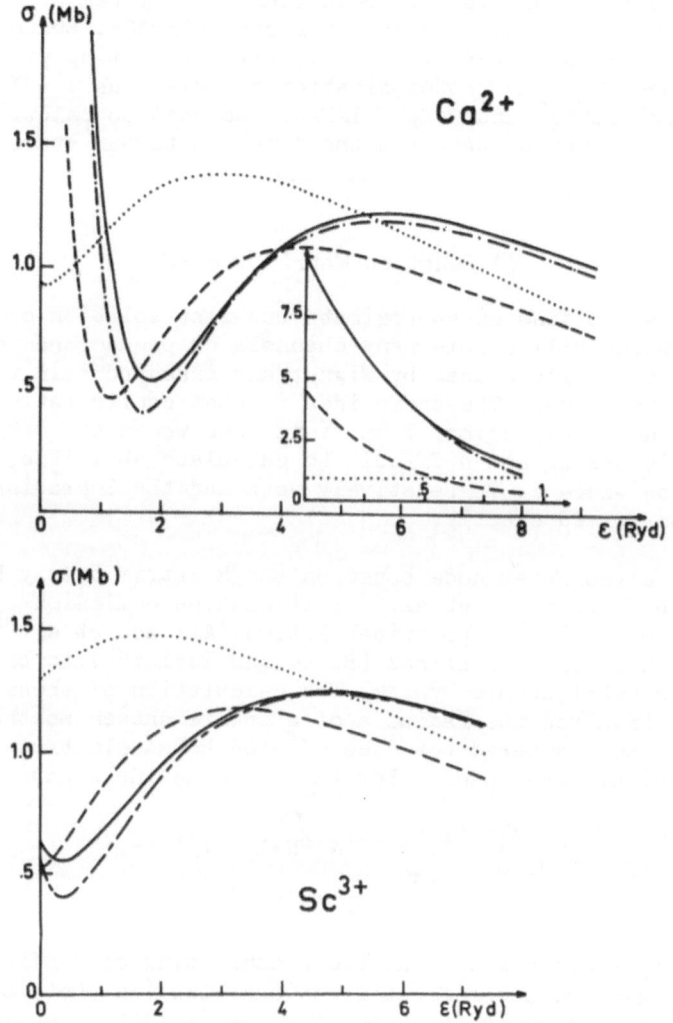

Fig.3 : σ_{3p} of Ca^{2+} and Sc^{3+} by Combet Farnoux and Lamoureux[16]
(same caption as fig.1)

shows the presence of a minimum in the continuous spectrum, while
only the second maximum is pointed out by the results obtained with
the potential of the $3p^6$ configuration. In all cases, the shift of
the various extrema towards the threshold when the ionization degree increases is pointed out.

B) Reduction to Linear Algebraic Equations

Another method to solve the C.C. equations has been developed
by Seaton[17] who considers the functions as tabulated, and by using

finite-difference formulas, reduces them to a system of algebraic equations. The corresponding computer code (IMPACT, Seaton and Wilson[18]) has been tested extensively with scattering problems, and has been used for some photoionization calculations : $- \sigma_{2p}$ of Ne including Ne$^+$ $2s^2 2p^5$ and $2s 2p^6$ (IMPACT was used to calculate both the initial Ne ground state and the final continuum state by Luke[19]); $- \sigma_{2s}$ of Be and σ_{3s} of Mg by Dubau and Wells[20].

C) Reaction Matrix Theory

It allows to find an approximate but fast solution of the C.C. equations which will enable many channels or pseudo-channels to be handled ; it was introduced by Wigner and Eisenbud[21] in the theory of nuclear reactions. The basic idea is that configuration space can be divided in 2 regions : the inner one where the manybody interaction is strong and difficult to calculate ab initio, and the outer region where it is relatively weak and the Schrödinger equation can be solved exactly.

A general computer code based on the R matrix theory has been published by Berrington et al.[22] ; it enables collision cross sections (Burke et al.[23]), polarisabilities (Allison et al.[24]) and photoionisation cross sections (Burke and Taylor[25]) to be calculated for a general atomic system. The calculation of these various quantities involves the radius a of a sphere chosen so that exchange between the scattered (or ejected) and bound electrons is only important within the sphere. For $r \leqslant a$, we define a wave function Ψ_k:

$$\Psi_k = A \sum_{ij} c_{ijk} \overline{\Phi}_i (x_1 x_2 \cdots x_N ; \widehat{x_{N+1}}) u_{ij}(r_{N+1})$$
$$+ \sum_j d_{jk} \Phi_j (x_1 \cdots x_{N+1}) \tag{22}$$

where the $\overline{\Phi}_i$ are channel functions consisting of configuration interaction wave functions for the residual ion (in photoionization) coupled with spin-angle functions for the (N+1)the electron to give an eigenstate of L and S ; the Φ_j describe (N+1) electron bound configurations (eigenstates of the same L and S). The u_{ij} are zero-order orbitals, eigenfunctions of a potential V(r) which represents the atomic charge. They are orthogonalized to the atomic orbitals and satisfy some boundary conditions :

$$u_{ij}(0) = 0 \quad \text{and} \quad \frac{a}{u_{ij}(a)} \left. \frac{du_{ij}}{dr} \right|_{r=a} = b \tag{23}$$

where b is an arbitrary constant. The u_{ij} together with the atomic orbitals form a complete set in the range $0 \leqslant r \leqslant a$.

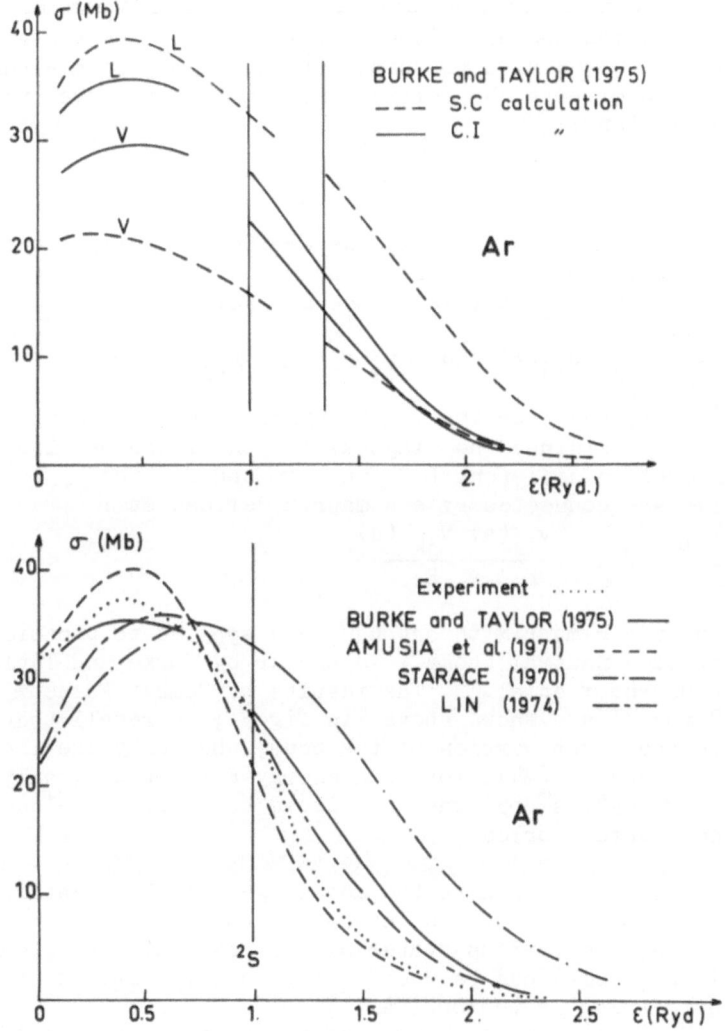

Fig.4 and 5 : σ_{3p} of Ar with the R matrix theory (Burke and Taylor[25])
In the SC. calculation both ionic states ($3p^5\ ^2P$ and $3s3p^6\ ^2S$) are
represented by single configuration while the C.I. calculation in-
troduces additional $\overline{4s}$, $\overline{4p}$, $\overline{3d}$ pseudo-orbitals to represent each
ionic state as a linear combination of configurations. L and V in-
dicate the dipole-length and velocity formulations.
In fig.5 the dipole-length C.I. results are compared with experiment
(Samson[27]) and with other theoretical results (Amusia et al.[28] using
R.P.A.E., Starace[29] who introduces only intrachannel interactions
and Lin[30] with a simplified version of the R.P.A.E. method).

The diagonalisation of the (N+1) electron Hamiltonian in the internal region gives the coefficients c_{ijk} and d_{jk} and the eigenvalues E_k. When dealing with photoionization, the Ψ_k form a basis for the expansion of both the initial and final state wave functions which can be written :

$$\Psi_o = \sum_k A_{ok} \Psi_k \quad \text{and} \quad \Psi_E = \sum_k A_{Ek} \Psi_k \quad \quad (24)$$

and the coefficients A_{ok} and A_{Ek} are determined in terms of the radial functions basis $u_{ij}(a)$ on the boundary of the internal region through the functions $v_{ik}(a)$, $y_i(a)$ and its derivative :

$$v_{ik}(r) = \sum_j c_{ijk} u_{ij}(r) \quad \text{and} \quad y_i(r) = \sum_k A_{Ek} v_{ik}(r) \quad \quad (25)$$

The E_k and $v_{ik}(a)$ occuring in the expressions giving A_{ok} and A_{Ek} are not the same since the diagonalisation of the Hamiltonian has been achieved in different L-S function spaces. Each $y_i(a)$ and its derivative are connected by a R matrix defined as :

$$R_{ij}(E) = \frac{1}{2a} \sum_k \frac{v_{ik}(a) \, v_{jk}(a)}{E_k - E} \quad \quad (26)$$

Up to now, the R matrix theory has been applied to photoionization only for some neutral atoms : Ne and Ar by Burke and Taylor[25] and Al by Le Dourneuf et al.[26]. The results by Combet Farnoux and Lamoureux [16] for K^+ are shown above in fig.2 ; we recall they were obtained with the first version of the code, when only the final state was described by (24), for each energy E ; besides, only the ground state term $^2P^o(3s^23p5)$ was introduced to describe K^{2+} and only two channels were coupled.

To conclude, we will sum up the advantages of the R matrix theory, when used in photoionization instead of the usual close-coupling procedure :   a) the various states of the residual ion (ground state + excited states) can be described by superpositions of configurations introducing pseudo-orbitals ;
          b) the wave functions of the initial and final states are chosen consistently ;
          c) because a single diagonalisation yields the R matrix at all energies, the detailed treatment of the autoionisation profiles is easier. Some of these possibilities also exist in the IMPACT code but material for a definite comparison is missing.

1- R.J.W.Henry and L.Lipsky, Phys. Rev. 153, 51 (1967)
2- M.J. Seaton, Phil. Trans. Roy. Soc. A 245, 469 (1953)
3- K. Smith, R.J.W. Henry, P.G.Burke, Phys. Rev. 147, 21 (1966)
4- A. Burgess and M.J. Seaton, Mon.Not.R.Astr.Soc.120,121 (1960)
5- K. Smith,"The Physics of Electronic and Atomic Collisions" edited by T.R. Govers and F.J. de Heer (North Holland 1972)

6- P.G. Burke and M.J. Seaton, Meth.Comp. Phys. 10,1 (1971)

7- M.J. Conneely, L. Lipsky, K. Smith, P.G. Burke,R.J.W. Henry, Comp. Phys. Comm. 1, 306 (1970)

8- E. Clementi, IBM J. Res. Develop. 9, 2 (1965)

9- K. Smith, R.J.W. Henry and P.G. Burke, Phys. Rev. 157, 51 (1967)

10- K. Smith, "The calculation of Atomic Collision Processes" (Wiley-Interscience, New York, 1971)

11- S. Ormonde and M.J. Conneely, Phys. Rev. A4, 1432 (1971)

12- M.J. Conneely, K. Smith and L. Lipsky, J. Phys. B (Atom. and Mol. Phys.) 3, 493 (1970)

13- R.J.W. Henry,   Astrophy. Jour. 161, 1153 (1970)

14- R.D. Chapman and R.J.W. Henry,   Astrophys.Jour.168, 169 (1971)

15- R.D. Chapman and R.J.W. Henry,   Astrophys.Jour.173, 243 (1972)

16- F. Combet Farnoux and M. Lamoureux, J. Phys. B. In press

17- M.J. Seaton, J. Phys.B (Atom. Mol. Phys.) 7, 1817 (1974)

18- M.J. Seaton and P.M.H. Wilson, J.Phys.B. (Atom. Mol. Phys.) 5 , L1 (1972)

19- T.M. Luke,   J. Phys.B (Atom. Mol. Phys.) 6, 30 (1973)

20- J. Dubau and J. Wells, J. Phys.B. (Atom. Mol. Phys.) 6, L31(1973)

21- E.P. Wigner and L. Eisenbud, Phys. Rev. 72, 29 (1947)

22- K.A. Berrington, P.G. Burke, J.J. Chang, A.T. Chivers, W.D. Robb and K.T. Taylor, Comp. Phys. Comm. 8, 149 (1974)

23- P.G. Burke, A. Hibbert and W.D. Robb, J. Phys. B (Atom. and Mol. Phys.) 4, 153 (1971)

24- D.C.S. Allison, P.G. Burke and W.D. Robb, J. Phys. B. (Atom. Mol. Phys.) 5, 55 (1972)

25- P.G. Burke and K.T. Taylor, J. Phys. B. 16, 2620 (1975)

26- M. le Dourneuf, Vo Ky Lan, P.G. Burke and K.T. Taylor, J. Phys. B. 16, 2640 (1975)

27- J.A.R. Samson, J.Opt. Soc. Am. 55, 935 (1965)

28- M. Ya Amusia, N.A. Cherepkhov and L.V. Chernysheva, Sov. Phys. J.E.T.P. 33, 90 (1971)

29- A.F. Starace, Phys. Rev. 2A, 118 (1970)

30- C.D. Lin, Phys. Rev. 9A, 181 (1974)

31- F. Herman and S. Skillman, "Atomic Structure Calculations" (1963 - Prentice-Hall Inc. Englewood Cliffs, New Jersey)

FUTURE EXPERIMENTAL PROBLEMS IN PHOTOIONIZATION

James A. R. Samson

University of Nebraska-Lincoln

Behlen Laboratory of Physics, Lincoln, Nebraska 68588

In trying to determine where we go from here on the subject of photoionization it is perhaps profitable to ask what are the goals that we seek.

One ultimate goal in photoionization is to answer the simple question "how do photons interact with atoms and molecules?" We would like to reach the point where we can tabulate and catalog precisely how the energy of a photon is consumed by all the accessible channels in the photoionization process. For example, what fraction of the incident photons of a given energy eject electrons only from the 2p shell of neon or from the 2s shell; what fraction causes double or triple ionization, etc.? This is a goal we would like to see accomplished for all atoms, molecules, and free radicals, in all states of excitation and ionization, and for all incident wavelengths. This would be a considerable task! But experimental progress along these lines is very good. For example, in Fig. 1 we see the complete picture of the interaction of radiation with neon over the spectral range from the ionization threshold to a photon energy of 2000 eV (1).

From the point of view of basic physics we should investigate the photon interaction with a representative number of atoms (for example, closed vs open shell atoms, low Z vs high Z atoms, etc) at various wavelengths in order to test the different theoretical approximations. The goal in this case is to arrive at a unified theory of the photoionization process; a theory that will take into account threshold phenomena, resonances, multiple excitation and ionization, and predict partial cross sections.

In the past 15 years considerable progress has been made in

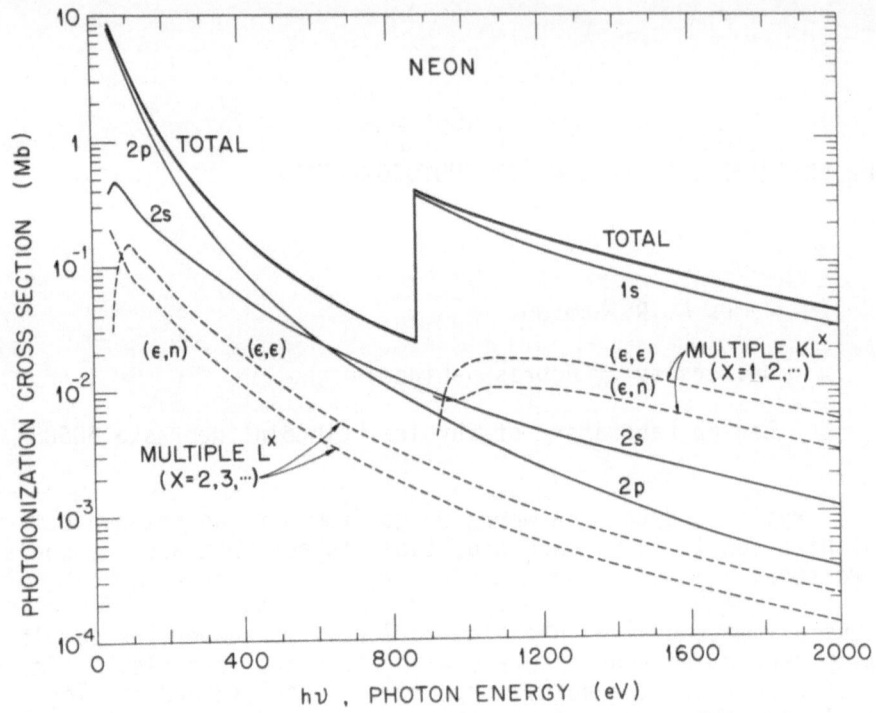

Fig. 1. Partition of total photoionization cross section of neon into its components of single ionization in 2p, 2s, and 1s sub-shells, multiple ionization, and simultaneous excitation and ionization in the various subshells (ref. 1).

meeting these goals, both experimentally and theoretically. In Table I a list is given of the various photoionization studies that should be undertaken. Many of these items already have re-ceived considerable attention whereas some have barely been examined. In all cases these studies should provide valuable information about these primary photoionization processes. A brief review will be given of some of the latest experimental data, with comparison to theory whenever possible, to illustrate the present state of our knowledge of photoionization.

Photoionization cross sections and absorption spectra have been measured for 43 of the elements over various wavelength ranges. These elements and the wavelength ranges are given in Table 2. Some of the studies consist purely of photographic absorption spec-tra with no quantitative data on cross sections. These are indicated with an asterix (2). The other measurements contain cross section data of varying degrees of completeness. The rare gases have been

TABLE 1: PHOTOIONIZATION STUDIES TO BE UNDERTAKEN

1. Total photoionization cross sections σ_{tot} for open and closed shell atoms.

2. Partial cross sections, σ_j, and branching ratios, R_j, where the branching ratio is defined as the number of ions formed in a given state j relative to the total number of ions produced, and $\sigma_j = R_j \cdot \sigma_{tot}$. Measurements should be made
    (a) within the continuum and
    (b) across resonances.

3. Multiple photoionization cross sections.

4. Angular distribution of photoelectrons from
    (a) the ionization continuum
    (b) autoionizing resonances
    (c) Auger and shake-off transitions
    (d) multiple ionization.

5. Total photoionization cross sections of excited atoms and ions (negative and positive).

6. Polarization state of the photoelectrons.

7. Multiphoton ionization.

8. Production of an atlas of absorption spectra with wavelength identification.

studied extensively, to the point where the absolute experimental photoionization cross sections are known to an accuracy of ±3 or 4% over the range from 100 Å to their ionization thresholds (3). The autoionizing structure of these gases also has been studied extensively (4). Less complete information is available for most of the other elements listed in Table 2. From the theoretical aspects tremendous strides have been made in recent years, particular with the rare gases where an accurate comparison with experimental data is possible. The various forms of the many body theory are providing data in very good agreement with experiment. In fact, the point has been reached where it is now necessary to achieve as high an accuracy as possible in cross section measurements in order to distinguish between the correctness of the various approximations. In the case of He most of the recent sophisticated calculations (5) agree with the experimental data. For the heavier rare gases the spread in the various theoretical calculations

TABLE 2: ATOMS FOR WHICH ABSORPTION SPECTRA AND/OR CROSS
SECTIONS ARE AVAILABLE. THE NUMBERS IN PARENTHESES REPRESENT
THE WAVELENGTH RANGE IN ANGSTROMS AND THE ASTERIX INDICATES
ONLY PHOTOGRAPHIC ABSORPTION SPECTRA AVAILABLE

Closed Shells

s^2, s^2p^6

  He (44-524)
  Ne (1.5-600)
  Ar (1.5-800)
  Kr (1.5-890)
  Xe (1.5-1030)

s^2s^2, p^6s^2

* Be (500-2000)
  Mg (500-2000)
  Ca (10-120), (350-2028)
* Sr (40-95), (1640-2200)
  Ba (10-150), (350-1000),
     (1700-2450)

$d^{10}s^2$

  Zn (247-1320)
  Cd (400-1450)
  Hg (20-120), (172-1188)

$ds^2, f^nd^os^2$

* La (320-1000)
* Ce (10-120)
* Pr (320-1000)
* Nd (320-1000)
* Sm (320-1000)
* Eu (320-1000)
* Dy (320-1000)
* Ho (320-1000)
* Er (320-1000)
* Tm (320-1000)
* Yb (320-1000)

ds^2

* U (320-1000)

d^5s^2

* Mn (100-300)

Open Shells

s^2p

  Al (1758-2100)
  Ga (1480-2100)
  In (350-1000), (1600-2200)
  Tl (10-120), (1400-2100)

s^2p^2

  Si (1250-1550)
  Pb (10-120), (350-2500)

s^2p^3

  N (400-900)

s^2p^4

  O (400-1000)
  S (900-1830)

s^2p^5

* F (600-1500)
* Cl (320-1000)
* Br (600-1500)
* I (600-1500)

s^2s, p^6s

  Li (170-210), (500-2500)
  Na (150-2500)
  K (10-120), (600-2860)
  Rb (10-120), (1000-3500)
  Cs (10-3500)

$d^{10}s$

* Cu (350-1400)
* Ag (350-1400)

d^5s

* Cr (100-300)

is larger. Figure 2 shows the cross section of Ar as obtained
with the R-Matrix (6) and RPAE (7) approximations in comparison
with experimental data. Figure 3 compares similar theories with
the experimental cross sections for the ejection of a 3s - subshell electron from Ar. The theoretical cross sections for ejection of an s - subshell electron are very sensitive to the inclusion
of correlation effects between electrons within the same subshell
and between neighboring shells. Thus, the theoretical curves
marked σ^0 in Figure 3, which do not take into account electron
correlation effects, do not agree with the experimental results
(8,9). A more complete theoretical analysis by Amusia (10) of the
cross section for ejection of a 5s electron from Xe is shown in
Figure 4. Again σ^0 indicates omission of correlation effects,
whereas the dashed curved is the result of considering the influence
of the inner 4d shell on the 5s electrons. When the influence
of both the 4d and the 5p shell is considered (solid curve) there
is excellent agreement with experiment (8) from threshold to 3 Ry.
The recent experiments by West et al. (11), shown in Figure 5,
indicate that the relatively large rise in the photoionization

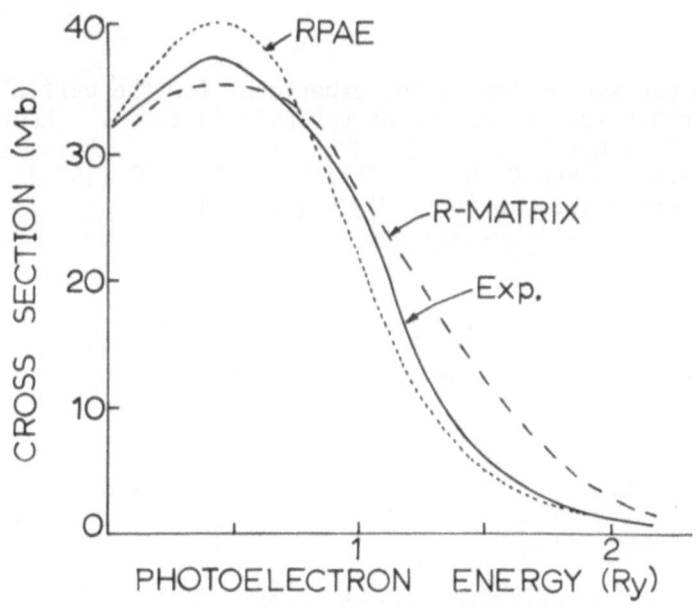

Fig. 2. Comparison of theory and experiment for the total photoionization cross section of argon. RPAE (ref. 7); R-Matrix (ref.
6); and Exp. (present work).

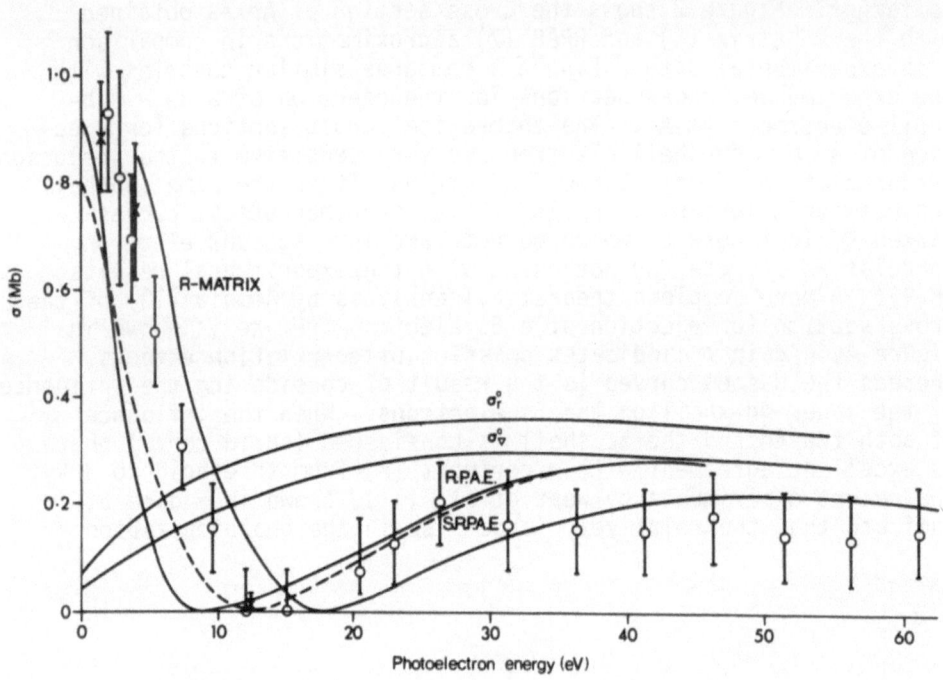

Fig.3. Comparison of theory and experiment for the partial photoionization cross section of the 3s subshell in argon. (Reproduced
from ref. 9.) R-Matrix (ref. 6); RPAE Amusia et al. Phys. Letters
40A, 361 (1972); SRPAE C. D. Lin Phys. Rev. A9, 181 (1974). Experimental data points, o (ref. 9); x (ref. 8).

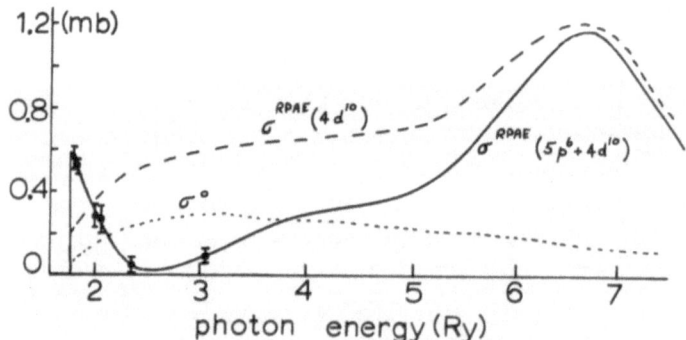

Fig. 4. Partial photoionization cross section of the 5s subshell
in xenon (ref. 10). The dashed curve represents the RPAE calculation taking into account correlations between the 5s and 4d
shells, while the solid curve represents correlations between
the 5s and 4d+5s shells. The dotted curve does not take
correlations into account. Experimental data I (ref. 8).

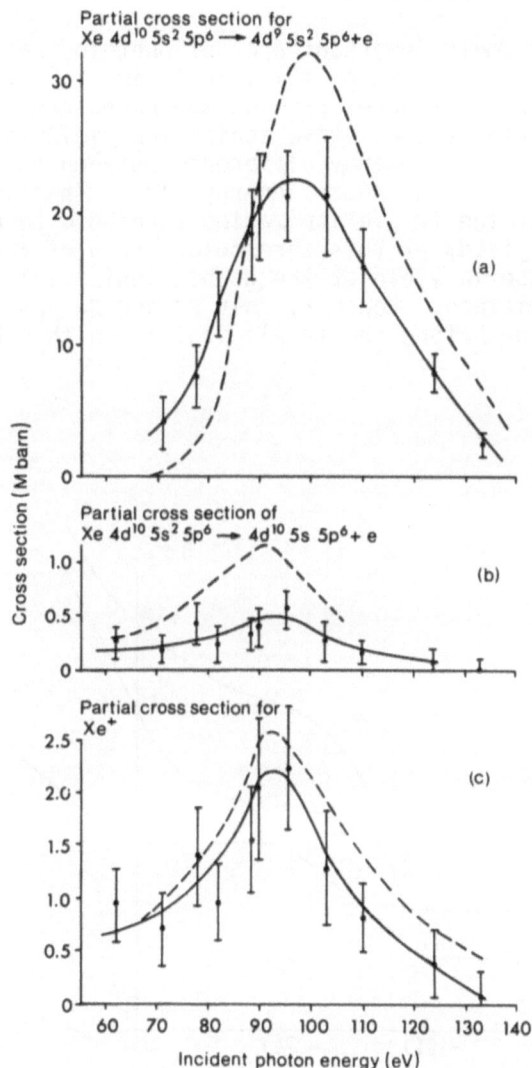

Fig. 5. Partial photoionization cross section of Xe for (a) the 4d shell, (b) the 5s shell, and (c) the 5p shell (ref. 11). The dashed curves represent the data of Amusia et al. (ref. 7 and 10).

cross section of the 5s electrons in the vicinity of d-shell ionization (between 6 and 7 Ry) may be too large. Nevertheless,
there is clear indication that correlation effects are very important and must be considered.

Another extremely important area of photoionization that shows
correlation effects is that of multiple ionization by a single
photon. Again the rare gases provide the necessary examples.
Figure 6 illustrates the relative ionization yields of Ar, Kr, and
Xe as a function of the energy difference between the triple ionization threshold and the photon energy (12). That is, the curves
are normalized at the triple ionization threshold to emphasize the
behavior of the yields at this threshold. Each of the gases shows
an abrupt increase in yield at the double ionization threshold that
increases to a plateau. However, only Kr and Xe show an increase in
this plateau value before the triple ionization threshold. This

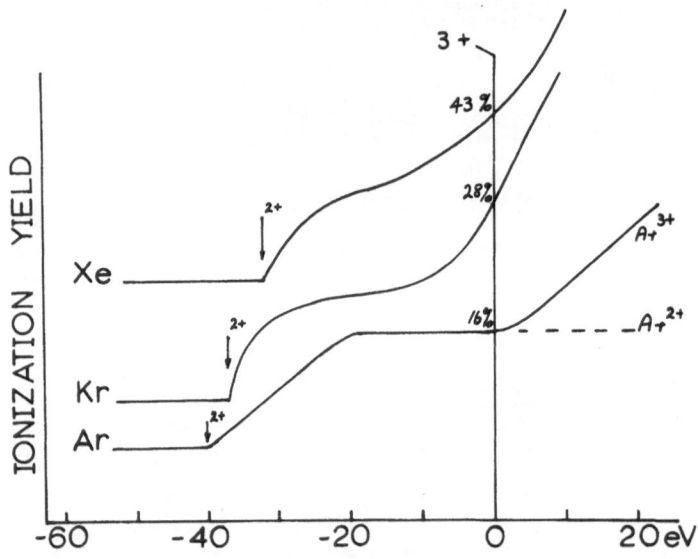

Fig. 6. Multiple ionization of Ar, Kr, and Xe. The ionization
yield of each gas is 100% until the double ionization threshold
is reached (indicated by the vertical arrow and +2). The amount
of double ionization is indicated on each curve at the triple
ionization threshold. The abscissa represents the energy difference between the photon energy and the triple ionization threshold.

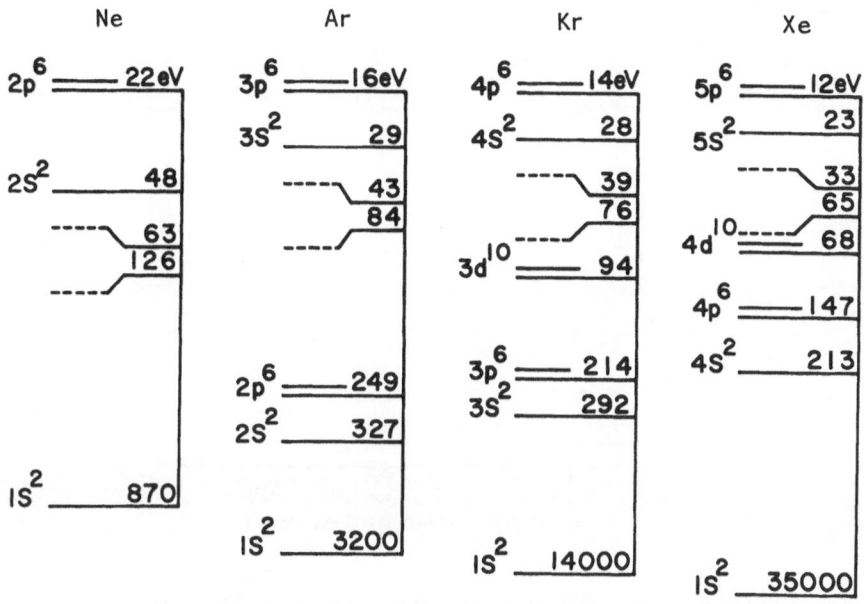

Fig. 7. Energy level diagram for Ne, Ar, Kr, and Xe. The dashed lines indicate the double and triple ionization thresholds.

occurs about 25 eV before their respective d-shell thresholds. The percentages shown on the yield curves where they cross the triple ionization threshold represent the number of double charged ions with respect to the total number of ions formed. It is likely that this increase in the double ionization yield of Kr and Xe is caused by correlation effects. Figure 7 illustrates the energy levels of the rare gases Ne, Ar, Kr, and Xe. The dotted levels indicate the thresholds for double and triple ionization. Except for the d-shells in Kr and Xe there are no shells close to the triple ionization threshold and therefore we might expect a plateau in the yield curves as is found in Ne and Ar. However, the entire process of double ionization is likely to involve correlation effects. The theoretical (13) and experimental curves (12, 14) for doubly charged Ne are shown in Figure 8. The theoretical results show good agreement with experiment after taking into account core rearrangement, ground state correlations within the 2p-shell and between the 2s and 2p-shells, virtual auger transitions and inelastic internal collisions. The recent experimental results of Schmidt et al. (15) are in good agreement with the experimental data shown in Figure 8.

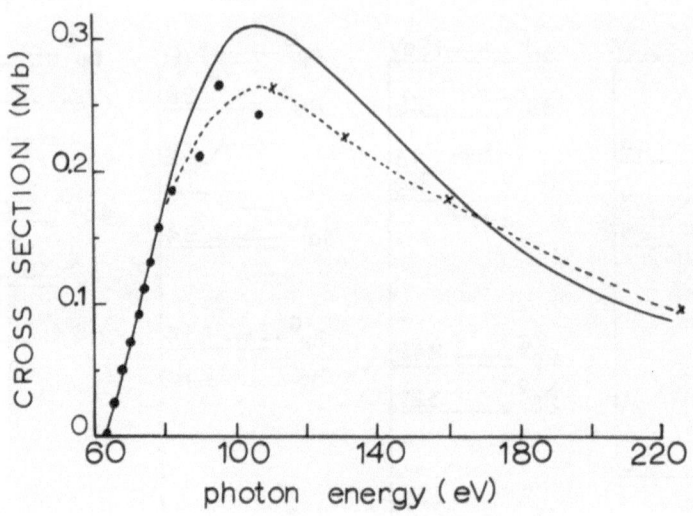

Fig. 8. Cross section for double ionization in Ne. The solid curve represents the theoretical results of T. N. Chang (ref. 13). Experimental data: o (ref. 8); x (ref. 14).

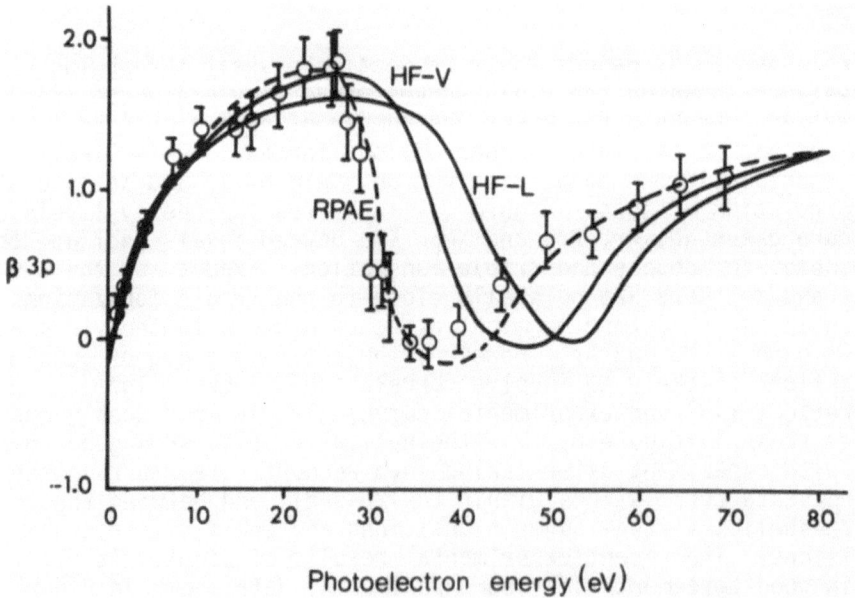

Fig. 9. Angular distribution of the 3p electrons in argon (ref. 16).

As a final example of the progress in photoionization Figure 9 shows the variation of the angular assymetry parameter β, which defines the angular distribution of photoelectrons, plotted as a function of the photoelectron energy for the 3p electrons in argon (16). This is the largest energy range over which β has been measured to date for any atom. As can be observed in the Figure it is necessary to study β beyond 25 eV from threshold in order to see a discrepancy between different theoretical approaches. There is no doubt of the superiority of the RPAE method in this case. No experimental β-values have been obtained yet in the interesting regions of resonances and multiple ionization, nor even for continuum ionization of open shell atoms.

In this brief review only a few selected experimental and theoretical results have been presented to illustrate the progress in photoionization studies and their usefulness in understanding correlation effects. Much experimental work needs to be done as indicated in Table 1. It is hoped that results on these various measurements will soon be forthcoming.

I would like to thank Drs. F. Wuilleumier, J. B. West, and J. P. Connerade for supplying some photographs and tabular material used in this article. This research was supported by the National Aeronautics and Space Administration under Grant NGR 28-004-021 and by the Atmospheric Sciences Section of the National Science Foundation.

REFERENCES

1. F. Wuilleumier and M. O. Krause, Phys. Rev. A, <u>10</u>, 242 (1974).

2. We would like to thank Professor Garton and Dr. Connerade of Imperial College, London for supplying a list of the atoms they have studied.

3. J. A. R. Samson, in "Advances in Atomic and Molecular Physics" Eds. Bates and Estermann (Academic Press, NY, 1966) Vol. 2, p. 177 and J. A. R. Samson and Haddad, unpublished data.

4. R. P. Madden and K. Codling, Phys. Rev. Letters <u>10</u> 516 (1963); K. Codling and R. P. Madden, Phys. Rev. A <u>4</u>, 2261 (1971); D. L. Ederer, Phys. Rev. A, <u>4</u>, 2263 (1971); and references therein.

5. A. L. Stewart and T. G. Webb, Proc. Phys. Soc., <u>82</u>, 532 (1963); P. G. Burke and D. D. McVicar, Proc. Phys. Soc. <u>86</u>, 989 (1965); K. L. Bell and A. E. Kingston, Proc. Phys. Soc. <u>90</u>, 31 (1967); T. N. Rescigno, C. W. McCurdy, Jr., and V. McKoy, Phys. Rev. A, <u>9</u>, 2409 (1974).

6. P. G. Burke and K. T. Taylor, J. Phys. B16, 2620 (1975).

7. Ya. Amusia, N. A. Cherepkov, and L. V. Chernysheva, Soviet Physics J.E.T.P. 33, 90 (1971).

8. J. A. R. Samson and J. L. Gardner, Phys. Rev. Letters 33, 671 (1974).

9. J. B. West, R. G. Houlgate, K. Codling, and G. Marr to be published.

10. Ya. Amusia, in "Proceedings of the IV International Conference on Vacuum UV Radiation Physics" Eds. E. E. Koch, R. Haensel, and C. Kunz (Pergamon-Vieweg, West Germany, 1974) p. 205.

11. J. B. West, P. R. Woodruff, K. Codling, and R. G. Houlgate, J. Phys. B. (to be published).

12. J. A. R. Samson and G. N. Haddad, Phys. Rev. Letters 33, 875 (1974).

13. Tu Nan Chang and R. T. Poe, Phys. Rev. A 12, 1432 (1975).

14. T. Carlson, Phys. Rev. 156, 142 (1967).

15. V. Schmidt, N. Sandner, H. Kuntzemüller, P. Dhez, F. Wuilleumier, and E. Kallne, Phys. Rev. A (to be published), April 1976.

16. J. B. West, private communication.

CONCLUDING REMARKS[*]

Mitio Inokuti

Argonne National Laboratory

9700 S. Cass Avenue, Argonne, Ill., 60439, U. S. A.

Our Director, Dr. Wuilleumier, told me to say something at the closing of the Institute. Perhaps he expects me to sum up all that we have discussed here, but such a task goes beyond my ability. I merely wish to enumerate some of the unsolved problems in our general topic of discussion in an attempt to help you maintain the strong enthusiasm which I have observed in the past two weeks and which I cherish most as a driving force for our progress.

TO EXPERIMENTALISTS

I simply expound <u>the underlying unity</u> of the diverse subjects of the lectures we heard, as stressed by Professor Fano's introduction [1]. Let us recall his Eq. (1) showing the many-branched reaction diagrams, which include arbitrary species A or A^-. I believe it is instructive to construct some specific examples of those reaction diagrams and to consider pertinent experiments at the same time.

Figure 1 represents the simplest example, i.e., the system of two electrons in the field of a proton. The upper half of Figure 1 is the reaction diagram. The central object $(H^-)^{**}$ means an excited complex containing two energetic electrons. The outer portion shows alternative reaction channels, which may be an exit or an entrance,

---

[*] Work performed under the auspices of the U. S. Energy Research and Development Administration.

and which are arranged and numbered with channel labels, roughly in
the order of increasing threshold energies as one goes clockwise around
the center. The single asterisk in H* represents any of discrete excited
states. The wavy line attached to channel 1, i.e., H⁻ + hν, is meant
to indicate the invariably weak (dipole) coupling with the photon hν.
In contrast, each of the straight lines attached to the other channels
is meant to remind us of much stronger couplings, which usually require
a nonperturbative description, unless reaction partners merge or emerge
with exceedingly high kinetic energies.

Figure 1

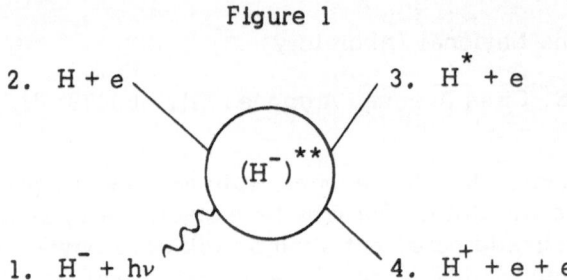

2. $H + e$ 3. $H^* + e$

$(H^-)^{**}$

1. $H^- + h\nu$ 4. $H^+ + e + e$

OUT

		1	2	3	4
	1		×		
I	2		×	×	×
N	3				×
	4				

(1,2)   S. J. Smith and D. S. Burch, Phys. Rev. <u>116</u>, 1125 (1959).

(2,2)   J. F. Williams, J. Phys. B <u>8</u>, 2191 (1975).

(2,3)   J. F. Williams and B. A. Willis, J. Phys. B <u>8</u>, 1641 (1975);
        W. R. Ott, W. E. Kaupilla, and W. L. Fite, Phys. Rev. A <u>1</u>,
        1089 (1970).

(2,4)   W. L. Fite and R. T. Brackmann, Phys. Rev. <u>112</u>, 1141 (1958);
        E. W. Rothe et al., Phys. Rev. <u>125</u>, 582 (1962).

(3,4)   A. J. Dixon, A. von Engel, and M. F. A. Harrison, Proc. Roy.
        Soc. <u>A 343</u>, 333 (1975).

In the lower half of Figure 1 we have a matrix whose rows and columns represent incoming and outgoing channels, respectively. For example, the element at row 2 and column 3 represents electron-impact excitation of a hydrogen-atom discrete level, and the element at row 1 and column 2 represents photodetachment of the hydrogen negative ion leading to the ground-state hydrogen atom. This matrix is nothing but the S matrix, a basic concept familiar to theoreticians [1] since Heisenberg's introduction [2]. (Actually, each element of my matrix in Figure 1 is in turn a matrix whose every element may be labeled by quantum numbers such as energy, angular momenta, and parity. But this detailed consideration is unimportant to my present discussion.)

In the simplified S matrix, you find a number of crosses. Each of the crosses indicates the presence of some significant experimental data pertaining to the reaction corresponding to a matrix element. By the qualifier "significant," I mean "reasonably extensive and reliable, as far as I know"—a highly personal definition. For each crossed matrix element in Figures 1—3 I have put notes concerning data sources. The brief entry "many experiments" indicates that a large number of data sources are easily found in standard references such as the treatise by Massey, Burhop, and Gilbody [3]. In other entries, only illustrative references are given; an exhaustive documentation would be beyond the scope of this discourse. (I am afraid I may be missing some important experiments. Fully realizing the limit of my knowledge in this respect, I consulted with Professor K. T. Dolder, Dr. M. J. van der Wiel, and other colleagues for the general accuracy of my understanding. I greatly thank them for their assistance, but they should not be considered responsible for any error or omission.) Despite all the qualifications, I stress the very few fractions of the matrix elements that have crosses; only five out of the sixteen possible elements have been looked at experimentally, in this simplest example. Experiments on double photodetachment of the hydrogen negative ion seem to be an important goal in the future.

Figure 2 depicts the next simplest example, i.e., two electrons in the field of an α-particle. All the notations such as the asterisk, the wavy line, the straight line, and the ordering of channel numbers, mean the same as before. The only major difference is the new channel He^* $(+ h\nu)$, because the helium atom has infinitely many discrete excited states, whereas the hydrogen negative ion has none. (I put the photon symbol "$+ h\nu$" in parentheses because a photon may not be necessarily present in channel 2.)

Figure 2

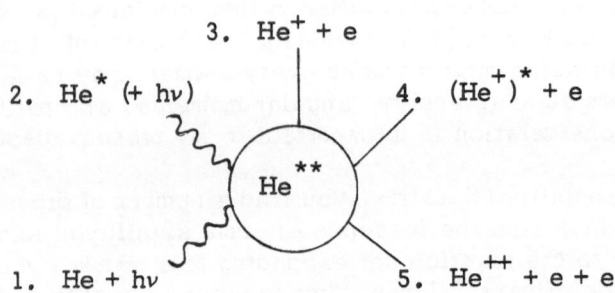

3. $He^+ + e$

2. $He^* (+ h\nu)$ 4. $(He^+)^* + e$

He^{**}

1. $He + h\nu$ 5. $He^{++} + e + e$

OUT

		1	2	3	4	5
	1	×	×	×	×	×
	2			×		
I N	3				×	×
	4					
	5					

(1,1)   many experiments on dispersion and Rayleigh scattering.
(1,2)   many experiments.
(1,3)   many experiments.
(1,4)   M. O. Krause and F. Wuilleumier, J. Phys. B 5, L143 (1972).
(1,5)   T. A. Carlson, Phys. Rev. 156, 142 (1967); V. Schmidt,
        presented at the present Institute.
(2,3)   R. F. Stebbings, F. B. Dunning, F. K. Tittel, and R. D. Rundel,
        Phys. Rev. Letters 30, 815 (1973).
(3,4)   D. H. Dance, M. F. A. Harrison, and A. C. H. Smith, Proc.
        Roy. Soc. A290, 74 (1966); K. T. Dolder and B. Peart,
        J. Phys. B 6, 2415 (1973).
(3,5)   B. Peart, D. S. Walton, and K. T. Dolder, J. Phys. B 2,
        1347 (1969).

The S matrix (in the lower portion) of Figure 2 differs from the corresponding S matrix of Figure 1 in one important respect. Here, all elements in row 1 are crossed; in other words, every one of the possible outgoing channels of photoabsorption already has been the subject of some significant experiment. Thus we confirm here the dominant role of photoabsorption experiments in our whole proceeding. At the same time, we identify here the great potential role of studies on electron collision with ions and excited species.

My last example concerns two electrons in the field of two protons. As seen in Figure 3, the variety of reaction channels increases remarkably. Indeed, the ten explicitly named channels represent here only a coarse classification of all conceivable phenomena; notice here that I have ignored enumeration of vibrational and rotational degrees of freedom.

Figure 3

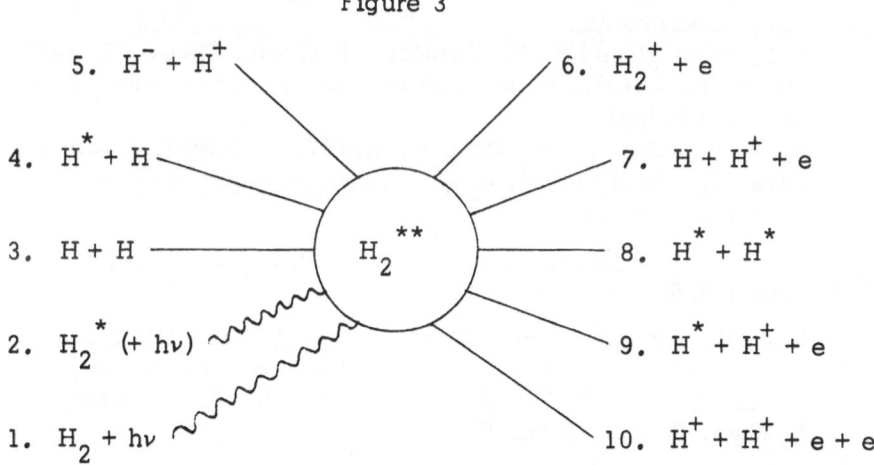

5. $H^- + H^+$
6. $H_2^+ + e$
4. $H^* + H$
7. $H + H^+ + e$
3. $H + H$ — H_2^{**} — 8. $H^* + H^*$
2. $H_2^* (+ h\nu)$
9. $H^* + H^+ + e$
1. $H_2 + h\nu$
10. $H^+ + H^+ + e + e$

Figure 3 (contd.)

OUT

IN	1	2	3	4	5	6	7	8	9	10
1	×	×		×	×	×	×		×	×
2										
3										
4										
5			×	×						
6			×	×			×		×	×
7										
8										
9										
10										

(1,1)   many experiments on dispersion and Rayleigh scattering.

(1,2)   many experiments.

(1,4)   J. E. Mentall and E. P. Gentieu, J. Chem. Phys. $\underline{52}$, 5641 (1970); P. Borrell, P. M. Guyon, and M. Glass-Maujean, to be published.

(1,5)   W. A. Chupka, P. M. Dehmer, and W. T. Jivery, J. Chem. Phys. $\underline{63}$, 3929 (1975).

(1,6)   many experiments.

(1,7)
and *   J. L. Gardner and J. A. R. Samson, Phys. Rev. A $\underline{12}$, 1404 (1975).
(1,9)

(1,10) J. A. R. Samson, Chem. Phys. Letters $\underline{12}$, 625 (1972).

(5,3)
and *   J. Moseley, W. Aberth, and J. R. Peterson, Phys. Rev. Letters $\underline{24}$, 435 (1970); T. D. Gaily and M. F. A. Harrison,
(5,4)   J. Phys. B. $\underline{3}$, L25 (1970).

(6,3)
and *   B. Peart and K. T. Dolder, J. Phys. B $\underline{7}$, 236 (1974); J. W. McGowan et al., unpublished.
(6,4)

(6,7)
and *   B. Peart and K. T. Dolder, J. Phys. B $\underline{4}$, 1496 (1971); $\underline{5}$, 860 (1972); $\underline{5}$, 1554 (1972).
(6,9)

(6,10) B. Peart and K. T. Dolder, J. Phys. B $\underline{6}$, 2409 (1973)

*Alternative exit channels are incompletely resolved or unresolved.

The S matrix of Figure 3 indicates again the foremost position of photoionization experiments, shown by the crosses in row 1. At the same time, the crosses in rows 5 and 6 (mostly due to Professor Dolder and co-workers) are notable. But the huge blank space in the matrix awaits exploration in years to come. In this respect, I am delighted to learn that Professor M. Barat and co-workers plan to conduct many experiments on the H + H entrance channel.

In conclusion, the merit of the S-matrix diagrams I have shown is twofold. First, these diagrams point out intimate connections among seemingly different experiments, e.g., photoabsorption, electron impact, and even atom-atom (or ion-atom) collisions. Second, we automatically see in the diagrams the deficiencies of our current knowledge, or alternatively, some goals of future experiments. I encourage interested students to draw similar diagrams for other systems.

TO THEORETICIANS

The following part of my remarks is even more personal than the foregoing part; all I can offer here is an excerpt from notes I jotted down for my own thinking and possible work in the future.

Let me start by telling how I see various theories of electron correlations we have heard about. What struck me most in the past two weeks is an evolution of viewpoints. We start with the one-electron approximation, which, despite its simplicity, enables one to understand the basic features of many phenomena such as photoionization. The first step beyond the one-electron potential approximation is the shake-up and shake-off theory, which treats certain many-electron phenomena in a minimal way, yet often remarkably successfully. The second step seems to be the treatment of correlation effects in terms of perturbation theories. Despite formidable-looking diagrams and other technicalities, this treatment is conceptually simple and straightforward; moreover, it has proven effective, provided correlation effects are sufficiently weak to permit a truncation of perturbation series at a reasonably low order. The third step is the random-phase approximation and is in many ways closely related to the many-body perturbation theory—to the extent that I am not sure which of the two steps I ought to call second or third. Anyway, I regard these steps as rather conservative extensions of the standard Hartree-Fock theory. The fourth step is a class of formulations such as the close-coupling theory and the R-matrix theory. A key point here is the nonperturbative treatment; in other words, correlation effects need not be weak. Obviously, these formulations permit applications to a greater variety of physical circumstances, but technical problems apparently limit the scope as well as the numerical accuracy of practical results at present. The last step is to depart sharply from the traditional one-electron orbital as a starting point of formulation. An example of this ambitious approach is the recent work of Professor Fano, Dr. Lin, and others, who study properties of the wavefunctions in the hyperspherical (or Fock) coordinates. This class of approach has brought forth some of the key features of correlation effects, but remains at the level of prototype studies.

This is what I mean by the evolution of viewpoints. Alternatively, I may describe the same idea by a continuous search for good basis sets of functions; then the same idea becomes vividly familiar to anyone who has studied quantum chemistry, i.e., the study of wavefunctions of atoms and molecules in their bound states. To stress this view, I can quote Professor Barat's lectures on ion-atom collisions at low energies: his detailed discussion on the alternative basis functions (i.e., adiabatic vs. diabatic) is exactly a case in point. In general, an answer to the question, "Which basis set is the best to use?" depends upon the problem under consideration, and more importantly, upon a compromise between the efficiency of solution and its accuracy.

Now I wish to discuss some specific problem areas I have recognized.

a. Foundation of the shake-up and shake-off theory

We have heard from Professor Åberg, Dr. Carlson, and others how remarkably good results were obtained in so many examples. Nevertheless, I am puzzled by some basic questions of the following kind. Suppose the shake theory gives a reasonably good result for a specific problem at hand. How should we compute a correction to that result? How should we obtain more detailed results, e.g., angular distributions of shaken-off electrons? As far as I can see, the shake theory has not been presented as an initial stage of successively improving approximations. (For contrast, I remind you of the first Born approximation, which invariably can be regarded as a first step of systematic successive approximations.) Another question (which was originally raised by Professor Fano) is how to incorporate the energy-conservation relation into the shake theory. In the standard formula of the shake-off theory, I see no explicit provision that would dictate the energy of the electron that gets shaken-off. To put the query in more picturesque terms, I can alternatively ask, "At which region of the configuration space does the shaken-off electron decide how much

energy to carry with it?" Admittedly, this question is phrased in
perhaps too classical terms, but I hope my idea behind this question
is fully meaningful in quantum mechanics.

b. One-step vs. many-step formulation of the theory of Auger effects

The usually great disparity of time scales between inner-shell
excitation and decays of a resulting inner-shell vacancy leads us to
theorize this class of phenomena in two or more steps. In other words,
we usually say that the initial excitation is practically instantaneous
compared to the succeeding decay process. This sharp distinction
between the initial excitation and the decay process seems to be highly
artificial to me. Professor W. Mehlhorn pointed out a few examples in
which that sharp distinction makes little sense. To interpret his
examples in full detail, we theoreticians should develop a new formalism
that will encompass both the initial excitation and the decay of inner-
shell holes, and show under what conditions the many-step formulation
is a reasonable approximation.

In this connection, we must pay attention to the recent work of
Read and others on the so-called post-collision interaction effects
(discussed here by Professor Dolder). In a class of phenomena including
electron collisions, the decay of an unstable (or resonance) state proceeds earlier than the departure of an originally incident particle, which
(after the excitation of that resonance state) has lost virtually all of its
kinetic energy and therefore is going away extremely slowly. Under this
admittedly special circumstance, one must expect to see some consequences of the interactions between the decay process and the motion
of the scattered particle. The development of a fully detailed quantum-
mechanical theory of these interactions is likely to take years of most
imaginative effort.

c. Treatment of continuum electron wavefunctions in the field of a molecule or in a solid

Most of our detailed theoretical discussions in the past two
weeks concerned electrons in the field of an atom, which is essentially
spherical. As soon as we consider continuum states of electrons in a
nonspherical potential, we face a truly serious deficiency of our theory,
even in the most primitive scheme of the one-electron orbital approximation.
The nonsphericity of the molecular potential is an essentially new element
of any sound theory. Work in this direction is beginning [4]. Yet, the

incorporation of many other effects such as coupling of numerous channels (including those due to vibrational degrees of molecules) will require many years to come.

On behalf of all the participants, I wish to express our gratitude to Dr. and Mrs. Wuilleumier, who arranged the Institute excellently and made our sojourn here so pleasant and memorable.

References

1. U. Fano, Introduction to the School Program, the present volume.

2. W. Heisenberg, Z. Physik 120, 513; 673 (1943).

3. H. S. W. Massey, E. H. S. Burhop, and H. B. Gilbody, Electronic and Ionic Impact Phenomena, Second Edition in Five Volumes, (Oxford University Press, London, 1969—1974).

4. D. Dill and J. L. Dehmer, J. Chem. Phys. 61, 692 (1974). For alternative approaches, see references 11—24 therein.

AUTHOR INDEX

Numbers underlined indicate the page on which the complete reference is listed.

A

Aarons L.J., 347, $\underline{353}$
Aberg T., 49, 51, $\overline{53}$, 54, 55, 56, 58, $\underline{59}$, 153, 157, 162, 163, 188, $\overline{197}$, 274, $\overline{281}$, $\overline{283}$, $\underline{284}$, $\underline{285}$, 287, 289, 294, $\overline{299}$, $\overline{300}$, 301, 302, 305, 306, $\underline{307}$, 319, 320, 321, 323, $\overline{329}$, 308
Aberth W., $\underline{436}$
Abrahams E., 364, $\underline{365}$
Accad Y., 173, $\underline{183}$
Adamczyk B., $\overline{381}$, 384, $\underline{386}$
Adelman S.A., $\underline{9}$
Afrosimov V.V., 245, $\underline{253}$, 325, $\underline{330}$
Aksela H., $\overline{280}$, 284
Aksela S., 280, $\overline{284}$
Alan C.J., 345, 350, $\underline{353}$
Allison D.C.S., 414, $\overline{417}$
Altick P.L., 126, 131
Amaldi U., 169, $\underline{182}$

Amusia M.Ya, 50, 53, $\underline{58}$, 62, 66, 67, 74, 75, $76, \overline{81}$, 82, 95, 105, 108, $\underline{109}$, $\overline{123}$, $\overline{124}$, 128, $131, \overline{132}, \overline{158}$, 159, 163, $\overline{171}$, $\overline{183}$, 192, 193, $\overline{194}$, 198, $\overline{383}$, 386, 415, 417, $\overline{423}$, 424, $\overline{425}$, $\underline{430}$
Andrä J., 326, 330
Andersen N., $2\overline{49}$, 250, $\underline{252}$
Andresen B., 241, 252
Andreyev E.P., $19\overline{0}$, 191, 195, 196, $\underline{197}$
Andriessen J., 376, 378
Armbruster P., 248, $\overline{252}$
Armstead R.L., 177, $\overline{184}$
Armstrong L., 64, 81, $\overline{371}$, 374, $\underline{377}$, 378
Arnau C., 90, $\overline{108}$
Asaad W.N., $3\overline{13}$, 315, 328, $\underline{329}$
Ashley C., 362, $\underline{365}$

441

SUBJECT INDEX

Each element studied of the Mendeleev Table is listed alphabetically under its chemical symbol.